The Haynes
Suspension, Steering and Driveline Manual

by Jeff Killingsworth, Eric Godfrey and John H Haynes

Member of the Guild of Motoring Writers

The Haynes Automotive Repair Manual
for maintaining, troubleshooting and repairing
suspension, steering and driveline systems

ABCDE
FGHIJ
KLMNO
PQRS

Haynes Publishing Group
Sparkford Nr Yeovil
Somerset BA22 7JJ England

Haynes North America, Inc
861 Lawrence Drive
Newbury Park
California 91320 USA

Acknowledgements

We are grateful for the help and cooperation of the Chrysler Corporation and Toyota Motor Corporation for assistance with technical information and certain illustrations. Thanks to Len Tucci of Ground Force Inc. for supplying the photos used in the manual and on the back cover. We also wish to thank Rick Sadler (Prothane), Fred Lisle (Lisle Corp.), Garry Moore (MAC Tools), Mike Cross (Cross Ent.) and Wayne Harris (Cal Tire) for their help and cooperation. Contributing to this project was Bob Henderson.

© **Haynes North America, Inc. 1998**

With permission from J.H. Haynes & Co. Ltd.

A book in the Haynes Automotive Repair Manual Series

Printed in the U.S.A.

ISBN 1 56392 293 2

Library of Congress Catalog Card Number 98-84464

While every attempt is made to ensure that the information in this manual is correct, no liability can be accepted by the authors or publishers for loss, damage or injury caused by any errors in, or omissions from, the information given.

98-352

Contents

Chapter 1
Introduction 1-1

Chapter 2
Tools and equipment 2-1

Chapter 3
Troubleshooting 3-1

Chapter 4
Routine maintenance 4-1

Chapter 5
Driveline systems 5-1

Chapter 6
Suspension systems 6-1

Chapter 7
Steering systems 7-1

Chapter 8
Modifications 8-1

Wheel alignment specifications WA-1

Glossary GL-1

Index IND-1

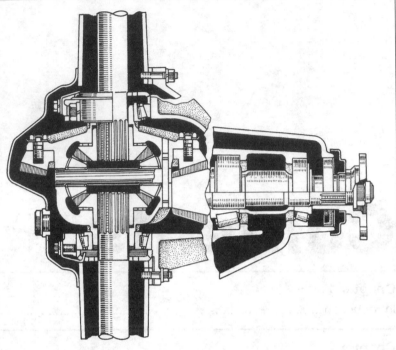

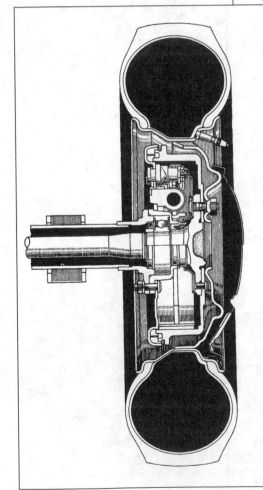

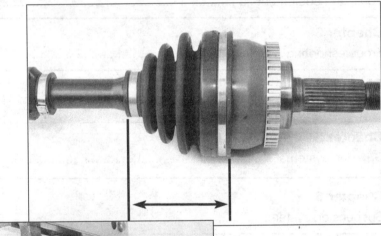

1 Introduction

Years ago, the common derogatory phrase "Your car drives like a truck!" was sometimes applied to a harsh-riding automobile. Not to cast aspersions on trucks, but this phrase was usually synonymous with "something is wrong with your car - you really should fix it!"

Modern cars - and trucks, for that matter - ride and drive better than even the most expensive machines of yesteryear. Thanks to technology and advances in automotive design, today's vehicles are better insulated from the road yet handle, accelerate and decelerate better than those of the past. However, as time passes, new vehicles grow old and things begin to wear out. This wear happens so gradually, though, that most drivers don't notice the decline in their car's performance until something becomes very noisy or actually breaks.

A comprehensive routine maintenance program and periodic inspections of your vehicle's steering, suspension and driveline components will ensure that your car or truck remains in top condition. Potential trouble areas will be noticed, allowing you to make necessary repairs before they leave you stranded on the side of the road.

That's where this manual comes in. It will help you diagnose problems with your suspension, steering and driveline systems and repair them, saving you a great deal of money. Or, if you feel that the work required exceeds your abilities, you'll at least be able to deal intelligently with the shop that does perform the repairs. Either way, you'll have the satisfaction knowing the job is done right.

Included in this manual are Chapters on tools and equipment, troubleshooting, maintenance, driveline repairs (including clutch, driveshaft, driveaxles, hubs, bearings and differentials), suspension repairs, steering repairs and finally, a Chapter describing modifications that can be carried out to enhance your vehicle's handling characteristics. At the end of the manual there's also a glossary of terms used throughout the book. In short, everything you need to know to successfully maintain and repair your suspension, steering and driveline systems to preserve the performance of your vehicle and ensure your safety and the safety of others you share the road with, as well.

Haynes suspension, steering and driveline manual

How to use this manual

Its purpose

The purpose of this manual is to help you get the best value from your vehicle. It can do so in several ways. It can help you decide what work must be done, even if you choose to have it done by a dealer service department or a repair shop; it provides information and procedures for routine maintenance and servicing; and it offers diagnostic and repair procedures to follow when trouble occurs.

We hope you use the manual to tackle the work yourself. For many simpler jobs, doing it yourself may be quicker than arranging an appointment to get the vehicle into a shop and making the trips to leave it and pick it up. More importantly, a lot of money can be saved by avoiding the expense the shop must pass on to you to cover its labor and overhead costs. An added benefit is the sense of satisfaction and accomplishment that you feel after doing the job yourself.

Using the manual

The manual is divided into Chapters. Each Chapter is subdivided into Sections, some of which consist of consecutively numbered paragraphs (usually referred to as "Steps," since they're normally part of a procedure). If the material is basically informative in nature, rather than a step-by-step procedure, the paragraphs aren't numbered.

The term **(see illustration)** is used in the text to indicate that a photo or drawing has been included to make the information easier to understand (the old cliché "a picture is worth a thousand words" is especially true when it comes to how - to procedures). Also every attempt is made to position illustrations on the same page as the corresponding text to minimize confusion. Some procedures are largely made up of illustrations and captions, with little or no accompanying text. The two types of illustrations used (photographs and line drawings) are referenced by a number proceeding the caption. Illustration numbers denote Chapter and numerical sequence within the Chapter (i.e. 3.4 means Chapter 3, illustration number 4 in order).

The terms **"Note,"** **"Caution,"** and **"Warning"** are used throughout the book with a specific purpose in mind - to attract the reader's attention. A "Note" simply provides information required to properly complete a procedure or information which will make the procedure easier to understand. A "Caution" outlines a special procedure or special steps which must be taken when completing the procedure where the Caution is found. Failure to pay attention to a Caution can result in damage to the component being repaired or the tools being used. A "Warning" is included where personal injury can result if the instructions aren't followed exactly as described.

References to the left or right side of the vehicle assume you are sitting in the driver's seat, facing forward.

Even though extreme care has been taken during the preparation of this manual, neither the publisher nor the authors can accept responsibility for any errors in, or omissions from, the information given

Buying parts

Replacement parts are available from many sources, which generally fall into one of three categories - authorized dealer parts departments, independent retail auto parts stores and used parts from wrecking yards. Our advice concerning these parts is as follows:

Retail auto parts stores: Good auto parts stores will stock frequently needed components which wear out relatively fast, such as clutch components, exhaust systems, steering racks, shocks, suspension parts, struts, etc. These stores often supply new or reconditioned parts on an exchange basis, which can save a considerable amount of money. Discount auto parts stores are often very good places to buy materials and parts needed for general vehicle maintenance such as oil, grease, filters, spark plugs, belts, touch-up paint, bulbs, etc. They also usually sell tools and general accessories, have convenient hours, charge lower prices and can often be found not far from home.

Authorized dealer parts department: This is the best source for parts which are unique to the vehicle and not generally available elsewhere (such as major engine parts, transmission parts, differential gears, trim pieces, etc.).

Wrecking yards: are a good source for major parts that would otherwise only be available through a dealer service department (where the price would likely be high). Transmission cases, steering columns, steering gearbox, axle housings, etc. are commonly available for reasonable prices. Although, you must be very careful when selecting used parts. Running changes are often made during the model year and a newly designed component from the same vehicle type may not be compatible with your vehicle. To insure the used part will be an exact match, select a used transmission or rear axle assembly for your parts source with the same identification code as the one you're rebuilding. Then as a final precaution, visually compare the replacement part with the damaged component to make sure they are identical. The parts people at wrecking yards have parts interchange books they can use to quickly identify parts from other models and years that are the same as the ones on your vehicle.

Warranty information: If the vehicle is still covered under warranty, be sure that any replacement parts purchased - regardless of the source - do not invalidate the warranty!

To be sure of obtaining the correct parts, have engine, chassis and differential tag numbers available and, if possible, take the old parts along for positive identification.

Maintenance techniques

There are a number of techniques involved in maintenance and repair that will be referred to throughout this manual. Application of these techniques will enable the home mechanic to be more efficient, better organized and capable of performing the various tasks properly, which will ensure that the repair job is thorough and complete.

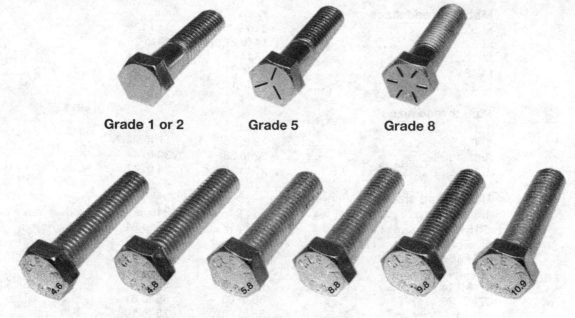

Grade 1 or 2   Grade 5   Grade 8

Bolt strength marking (standard/SAE/USS; bottom - metric)

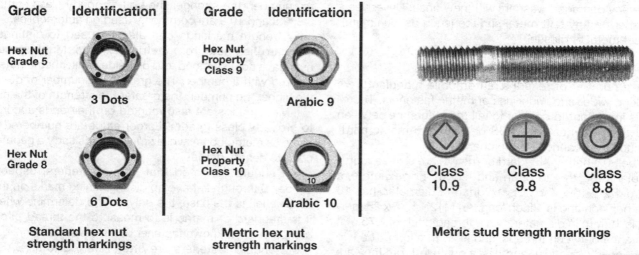

Grade	Identification
Hex Nut Grade 5	3 Dots
Hex Nut Grade 8	6 Dots

Standard hex nut strength markings

Grade	Identification
Hex Nut Property Class 9	Arabic 9
Hex Nut Property Class 10	Arabic 10

Metric hex nut strength markings

Class 10.9   Class 9.8   Class 8.8

Metric stud strength markings

Bolt strength marking (standard/SAE/USS; bottom - metric)

00-1 HAYNES

Fasteners

Fasteners are nuts, bolts, studs and screws used to hold two or more parts together. There are a few things to keep in mind when working with fasteners. Almost all of them use a locking device of some type, either a lockwasher, locknut, locking tab or thread adhesive. All threaded fasteners should be clean and straight, with undamaged threads and undamaged corners on the hex head where the wrench fits. Develop the habit of replacing all damaged nuts and bolts with new ones. Special locknuts with nylon or fiber inserts can only be used once. If they are removed, they lose their locking ability and must be replaced with new ones.

Rusted nuts and bolts should be treated with a pene-trating fluid to ease removal and prevent breakage. Some mechanics use turpentine in a spout-type oil can, which works quite well. After applying the rust penetrant, let it work for a few minutes before trying to loosen the nut or bolt. Badly rusted fasteners may have to be chiseled or sawed off or removed with a special nut breaker, available at tool stores.

If a bolt or stud breaks off in an assembly, it can be drilled and removed with a special tool commonly available for this purpose. Most automotive machine shops can perform this task, as well as other repair procedures, such as the repair of threaded holes that have been stripped out.

Flat washers and lockwashers, when removed from an assembly, should always be replaced exactly as removed.

Metric thread sizes	Ft-lbs	Nm
M-6	6 to 9	9 to 12
M-8	14 to 21	19 to 28
M-10	28 to 40	38 to 54
M-12	50 to 71	68 to 96
M-14	80 to 140	109 to 154

Pipe thread sizes		
1/8	5 to 8	7 to 10
1/4	12 to 18	17 to 24
3/8	22 to 33	30 to 44
1/2	25 to 35	34 to 47

U.S. thread sizes		
1/4 - 20	6 to 9	9 to 12
5/16 - 18	12 to 18	17 to 24
5/16 - 24	14 to 20	19 to 27
3/8 - 16	22 to 32	30 to 43
3/8 - 24	27 to 38	37 to 51
7/16 - 14	40 to 55	55 to 74
7/16 - 20	40 to 60	55 to 81
1/2 - 13	55 to 80	75 to 108

Replace any damaged washers with new ones. Never use a lockwasher on any soft metal surface (such as aluminum), thin sheet metal or plastic.

Fastener sizes

For a number of reasons, automobile manufacturers are making wider and wider use of metric fasteners. Therefore, it is important to be able to tell the difference between standard (sometimes called US or SAE) and metric hardware, since they cannot be interchanged.

All bolts, whether standard or metric, are sized according to diameter, thread pitch and length. For example, a standard 1/2 - 13 x 1 bolt is 1/2 inch in diameter, has 13 threads per inch and is 1 inch long. An M12 - 1.75 x 25 metric bolt is 12 mm in diameter, has a thread pitch of 1.75 mm (the distance between threads) and is 25 mm long. The two bolts are nearly identical, and easily confused, but they are not interchangeable (see illustration).

In addition to the differences in diameter, thread pitch and length, metric and standard bolts can also be distinguished by examining the bolt heads. To begin with, the distance across the flats on a standard bolt head is measured in inches, while the same dimension on a metric bolt is sized in millimeters (the same is true for nuts). As a result, a standard wrench should not be used on a metric bolt and a metric wrench should not be used on a standard bolt. Also, most standard bolts have slashes radiating out from the center of the head to denote the grade or strength of the bolt, which is an indication of the amount of torque that can be applied to it. The greater the number of slashes, the greater the strength of the bolt. Grades 0 through 5 are commonly used on automobiles. Metric bolts have a property class (grade) number, rather than a slash, molded into their heads to indicate bolt strength. In this case, the higher the number, the stronger the bolt. Property class numbers 8.8, 9.8 and 10.9 are commonly used on automobiles.

Strength markings can also be used to distinguish standard hex nuts from metric hex nuts. Many standard nuts have dots stamped into one side, while metric nuts are marked with a number. The greater the number of dots, or the higher the number, the greater the strength of the nut.

Metric studs are also marked on their ends according to property class (grade). Larger studs are numbered (the same as metric bolts), while smaller studs carry a geometric code to denote grade.

It should be noted that many fasteners, especially Grades 0 through 2, have no distinguishing marks on them. When such is the case, the only way to determine whether it is standard or metric is to measure the thread pitch or compare it to a known fastener of the same size.

Standard fasteners are often referred to as SAE, as opposed to metric. However, it should be noted that SAE technically refers to a non-metric fine thread fastener only. Coarse thread non-metric fasteners are referred to as USS sizes.

Since fasteners of the same size (both standard and metric) may have different strength ratings, be sure to reinstall any bolts, studs or nuts removed from your vehicle in their original locations. Also, when replacing a fastener with a new one, make sure that the new one has a strength rating equal to or greater than the original.

Tightening sequences and procedures

Warning: *These are general torque specifications for conventional fasteners. Some fasteners used (most notably the disc brake caliper bolts or guide pins) are specifically designed for their purposes and should not fall into the categories listed below.*

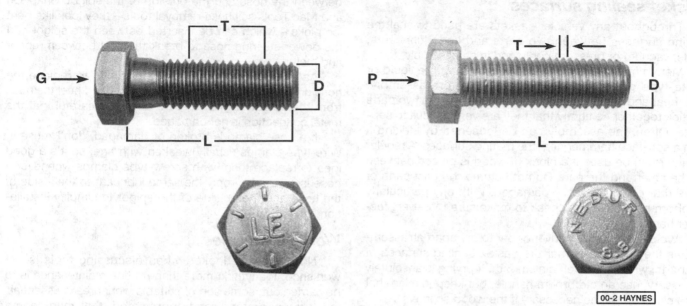

Standard (SAE and USS) bolt dimensions/grade marks

G   Grade marks (bolt strength)
L   Length (in inches)
T   Thread pitch (number of threads per inch)
D   Nominal diameter (in inches)

Metric bolt dimensions/grade marks

P   Property class (bolt strength)
L   Length (in millimeters)
T   Thread pitch (distance between threads in millimeters)
D   Diameter

Most threaded fasteners should be tightened to a specific torque value (torque is the twisting force applied to a threaded component such as a nut or bolt). Overtightening the fastener can weaken it and cause it to break, while undertightening can cause it to eventually come loose. Bolts, nuts, screws and studs, have specific torque values. A general torque value chart is presented here as a guide. These torque values are for dry (unlubricated) fasteners threaded into steel or cast iron (not aluminum). The size and grade of a fastener determine the amount of torque that can safely be applied to it. The figures listed here are approximate for Grade 2 and Grade 3 fasteners. Higher grades fasteners can tolerate higher torque values, provided both the bolt and nut have the same grade **(see illustration)**.

Fasteners laid out in a pattern, such as cylinder head bolts, oil pan bolts, differential cover bolts, etc., must be loosened or tightened in sequence to avoid warping the component. This sequence will normally be shown in the appropriate Chapter. If a specific pattern is not given, the following procedures can be used to prevent warping.

Initially, the bolts or nuts should be assembled finger-tight only. Next, they should be tightened one full turn each, in a criss-cross or diagonal pattern. After each one has been tightened one full turn, return to the first one and tighten them all one-half turn, following the same pattern. Finally, tighten each of them one-quarter turn at a time until each fastener has been tightened to the proper torque. To loosen and remove the fasteners, the procedure would be reversed.

Component disassembly

Component disassembly should be done with care and purpose to help ensure that the parts go back together properly. Always keep track of the sequence in which parts are removed. Make note of special characteristics or marks on parts that can be installed more than one way, such as a grooved thrust washer on a shaft. It is a good idea to lay the disassembled parts out on a clean surface in the order that they were removed. It may also be helpful to make sketches or take instant photos of components before removal.

When removing fasteners from a component, keep track of their locations. Sometimes threading a bolt back in a part, or putting the washers and nut back on a stud, can prevent mix-ups later. If nuts and bolts cannot be returned to their original locations, they should be kept in a compartmented box or a series of small boxes. A cupcake or muffin tin is ideal for this purpose, since each cavity can hold the bolts and nuts from a particular area (i.e. oil pan bolts, valve cover bolts, engine mount bolts, etc.). A pan of this type is especially helpful when working on assemblies with very small parts, such as the carburetor, alternator, valve train or interior dash and trim pieces. The cavities can be marked with paint or tape to identify the contents.

Whenever wiring looms, harnesses or connectors are separated, it is a good idea to identify the two halves with numbered pieces of masking tape so they can be easily reconnected.

Gasket sealing surfaces

Throughout any vehicle, gaskets are used to seal the mating surfaces between two parts and keep lubricants, fluids, vacuum or pressure contained in an assembly.

Many times these gaskets are coated with a liquid or paste-type gasket sealing compound before assembly. Age, heat and pressure can sometimes cause the two parts to stick together so tightly that they are very difficult to separate. Often, the assembly can be loosened by striking it with a soft-face hammer near the mating surfaces. A regular hammer can be used if a block of wood is placed between the hammer and the part. Do not hammer on cast parts or parts that could be easily damaged. With any particularly stubborn part, always recheck to make sure that every fastener has been removed.

Avoid using a screwdriver or bar to pry apart an assembly, as they can easily mar the gasket sealing surfaces of the parts, which must remain smooth. If prying is absolutely necessary, use an old broom handle, but keep in mind that extra clean up will be necessary if the wood splinters.

After the parts are separated, the old gasket must be carefully scraped off and the gasket surfaces cleaned. Stubborn gasket material can be soaked with rust penetrant or treated with a special chemical to soften it so it can be easily scraped off. A scraper can be fashioned from a piece of copper tubing by flattening and sharpening one end. Copper is recommended because it is usually softer than the surfaces to be scraped, which reduces the chance of gouging the part. Some gaskets can be removed with a wire brush, but regardless of the method used, the mating surfaces must be left clean and smooth. If for some reason the gasket surface is gouged, then a gasket sealer thick enough to fill scratches will have to be used during reassembly of the components. For most applications, a non-drying (or semi-drying) gasket sealer should be used.

Hose removal tips

Warning: *If the vehicle is equipped with air conditioning, do not disconnect any of the A/C hoses without first having the system depressurized by a dealer service department or a service station.*

Hose removal precautions closely parallel gasket removal precautions. Avoid scratching or gouging the surface that the hose mates against or the connection may leak. This is especially true for power steering reservoirs. Because of various chemical reactions, the rubber in hoses can bond itself to the metal spigot that the hose fits over. To remove a hose, first loosen the hose clamps that secure it to the spigot. Then, with slip-joint pliers, grab the hose at the clamp and rotate it around the spigot. Work it back and forth until it is completely free, then pull it off. Silicone or other lubricants will ease removal if they can be applied between the hose and the outside of the spigot. Snap–On and Mac Tools sell hose removal tools – they look like bent ice picks – which can be inserted between the spigot and the power steering hose to break the seal between rubber and metal.

As a last resort – or if you're planning to replace the hose anyway – slit the rubber with a knife and peel the hose from the spigot. If this must be done, be careful that the metal connection is not damaged.

If a hose clamp is broken or damaged, don't reuse it. Wire-type clamps usually weaken with age, so it's a good idea to replace them with screw-type clamps whenever a hose is removed. Apply the same lubricant to the inside of the hose and the outside of the spigot to simplify installation.

Working facilities

Not to be overlooked when discussing tools is the workshop. If anything more than routine maintenance is to be carried out, some sort of suitable work area is essential.

It is understood, and appreciated, that many home mechanics do not have a good workshop or garage available, and end up removing an engine or doing major repairs outside. It is recommended, however, that the overhaul or repair be completed under the cover of a roof.

A clean, flat workbench or table of comfortable working height is an absolute necessity. The workbench should be equipped with a vise that has a jaw opening of at least four inches. For more information see Chapter 2.

As mentioned previously, some clean, dry storage space is also required for tools, as well as the lubricants, fluids, cleaning solvents, etc. which soon become necessary.

Sometimes waste oil and fluids, drained from the engine or cooling system during normal maintenance or repairs, present a disposal problem. To avoid pouring them on the ground or into a sewage system, pour the used fluids into large containers, seal them with caps and take them to an authorized disposal site or recycling center. Plastic jugs, such as old antifreeze containers, are ideal for this purpose.

Always keep a supply of old newspapers and clean rags available. Old towels are excellent for mopping up spills. Many mechanics use rolls of paper towels for most work because they are readily available and disposable. To help keep the area under the vehicle clean, a large cardboard box can be cut open and flattened to protect the garage or shop floor.

Whenever working over a painted surface, such as when leaning over a fender to service something under the hood, always cover it with an old blanket or bedspread to protect the finish. Vinyl covered pads, made especially for this purpose, are available at auto parts stores.

Jacking and towing

Jacking

Warning: *The jack supplied with the vehicle should only be used for raising the vehicle when changing a tire or placing jackstands under the frame. Never work under the vehicle or start the engine while the jack is being used as the only means of support.*

The vehicle must be on a level surface with the wheels blocked and the transmission in Park. Apply the parking brake if the front of the vehicle must be raised. Make sure no one is in the vehicle as it's being raised with the jack.

Remove the jack and lug nut wrench and spare tire.

To replace the tire, use the tapered end of the lug wrench to pry loose the wheel cover. **Note:** *If the vehicle is equipped with aluminum wheels, it may be necessary to pry out the special lug nut covers. Also, aluminum wheels normally have anti-theft lug nuts (one per wheel) which require using a special "key" between the lug wrench and lug nut. The key is usually in the glove compartment.* Loosen the lug nuts one-half turn, but leave them in place until the tire is raised off the ground.

Position the jack under the side of the vehicle at the indicated jacking points. There's a front and rear jacking point on each side of the vehicle **(see illustrations)**.

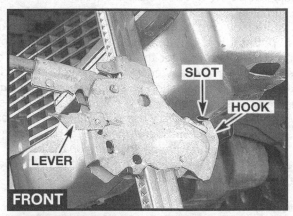

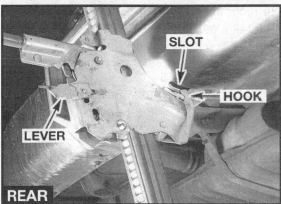

On earlier models, a bumper jack is used to raise the vehicle - it either fits into a slot in the bumper, as shown, or hooks onto the lower edge of the bumper at a specific location (see the vehicles owner's manual for the location)

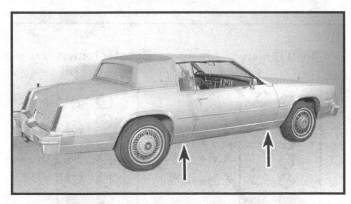

The head of the scissors jack should engage the notch in the rocker panel flange - there's one at the front and one at the rear on each side of the vehicle

Later model jacking details

Turn the jack handle clockwise (the lug wrench also serves as the jack handle) until the tire clears the ground. Remove the lug nuts and pull the tire off. Clean the mating surfaces of the hub and wheel, then install the spare. Replace the lug nuts with the beveled edges facing in and tighten them snugly. Don't attempt to tighten them completely until the vehicle is lowered or it could slip off the jack.

Turn the jack handle counterclockwise to lower the vehicle. Remove the jack and tighten the lug nuts in a criss-cross pattern. If possible, tighten the nuts with a torque wrench (see Chapter 1 for the torque figures). If you don't have access to a torque wrench, have the nuts checked by a service station or repair shop as soon as possible. **Caution:** *If the vehicle is equipped with a compact spare, the spare tire is intended for temporary use only. Have the tire repaired and reinstall it on the vehicle at the earliest opportunity and don't exceed 50 mph with the spare tire on the car.*

Install the wheel cover, then stow the tire, jack and wrench and unblock the wheels.

Towing

We recommend rear-wheel drive vehicles be towed from the rear, with the rear wheels off the ground. If it's absolutely necessary, these vehicles can be towed from the front with the front wheels off the ground, provided that speeds don't exceed 35 mph and the distance is less than 50 miles; the transmission can be damaged if these mileage/speed limitations are exceeded.

On front-wheel drive vehicles with an automatic transaxle, do not tow the vehicle with all four wheels on the ground - transaxle damage may occur if you do. Use a towing dolly to keep the front wheels off the road. Make sure the parking brake is released and the ignition switch is in the ACC position. Safety is a major consideration when towing and all applicable state and local laws must be obeyed.

Equipment specifically designed for towing should be used. It must be attached to the main structural members of the vehicle, not the bumpers or brackets.

Safety is a major consideration when towing and all applicable state and local laws must be obeyed. A safety chain must be used at all times.

The parking brake must be released and the transmission must be in Neutral. The steering must be unlocked (ignition switch in the Off position). Remember that power steering and power brakes won't work with the engine off.

Booster battery (jump) starting

Observe these precautions when using a booster battery to start a vehicle:

a) *Before connecting the booster battery, make sure the ignition switch is in the Off position.*
b) *Turn off the lights, heater and other electrical loads.*
c) *Your eyes should be shielded. Safety goggles are a good idea.*
d) *Make sure the booster battery is the same voltage as the dead one in the vehicle.*
e) *The two vehicles MUST NOT TOUCH each other!*
f) *Make sure the transaxle is in Neutral (manual) or Park (automatic).*
g) *If the booster battery is not a maintenance-free type, remove the vent caps and lay a cloth over the vent holes.*

Connect the red jumper cable to the positive (+) terminals of each battery **(see illustration)**.

Connect one end of the black jumper cable to the negative (-) terminal of the booster battery. The other end of this cable should be connected to a good ground on the vehicle to be started, such as a bolt or bracket on the body.

Start the engine using the booster battery, then, with the engine running at idle speed, disconnect the jumper cables in the reverse order of connection.

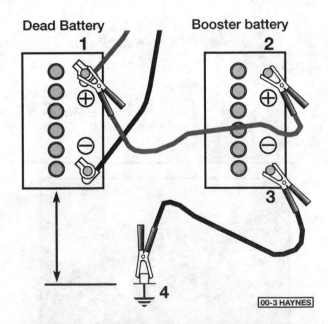

Make the booster battery cable connections in the numerical order shown (note that the negative cable of the booster battery is NOT attached to the negative terminal of the dead battery)

Cleaners, lubricants and sealants

A number of automotive chemicals and lubricants are available for use during vehicle maintenance and repair. They include a wide variety of products ranging from cleaning solvents and degreasers to lubricants and protective sprays for rubber, plastic and vinyl.

Cleaners

Carburetor cleaner and choke cleaner is a strong solvent for gum, varnish and carbon. Most carburetor cleaners leave a dry-type lubricant film which will not harden or gum up. Because of this film it is not recommended for use on electrical or brake components.

Brake system cleaner is used to remove grease and brake fluid from the brake system, where clean surfaces are absolutely necessary. It leaves no residue and often eliminates brake squeal caused by contaminants.

Electrical cleaner removes oxidation, corrosion and carbon deposits from electrical contacts, restoring full current flow. It can also be used to clean spark plugs, carburetor jets, voltage regulators and other parts where an oil-free surface is desired.

Demoisturants remove water and moisture from electrical components such as alternators, voltage regulators, electrical connectors and fuse blocks. They are non-conductive, non-corrosive and non-flammable.

Degreasers are heavy-duty solvents used to remove grease from the outside of the engine and from chassis components. They can be sprayed or brushed on and, depending on the type, are rinsed off either with water or solvent.

Lubricants

Motor oil is the lubricant formulated for use in engines. It normally contains a wide variety of additives to prevent corrosion and reduce foaming and wear. Motor oil comes in various weights (viscosity ratings) from 0 to 50. The recommended weight of the oil depends on the season, temperature and the demands on the engine. Light oil is used in cold climates and under light load conditions. Heavy oil is used in hot climates and where high loads are encountered. Multi-viscosity oils are designed to have characteristics of both light and heavy oils and are available in a number of weights from 5W-20 to 20W-50.

Gear oil is designed to be used in differentials, manual transmissions and other areas where high-temperature lubrication is required.

Chassis and wheel bearing grease is a heavy grease used where increased loads and friction are encountered, such as for wheel bearings, balljoints, tie-rod ends and universal joints.

High-temperature wheel bearing grease is designed to withstand the extreme temperatures encountered by wheel bearings in disc brake equipped vehicles. It usually contains molybdenum disulfide (moly), which is a dry-type lubricant.

White grease is a heavy grease for metal-to-metal applications where water is a problem. White grease stays soft under both low and high temperatures (usually from -100 to +190-degrees F), and will not wash off or dilute in the presence of water.

Assembly lube is a special extreme pressure lubricant, usually containing moly, used to lubricate high-load parts (such as main and rod bearings and cam lobes) for initial start-up of a new engine. The assembly lube lubricates the parts without being squeezed out or washed away until the engine oiling system begins to function.

Silicone lubricants are used to protect rubber, plastic, vinyl and nylon parts.

Graphite lubricants are used where oils cannot be used due to contamination problems, such as in locks. The dry graphite will lubricate metal parts while remaining uncontaminated by dirt, water, oil or acids. It is electrically conductive and will not foul electrical contacts in locks such as the ignition switch.

Moly penetrants loosen and lubricate frozen, rusted and corroded fasteners and prevent future rusting or freezing.

Heat-sink grease is a special electrically non-conductive grease that is used for mounting electronic ignition modules where it is essential that heat is transferred away from the module.

Sealants

RTV sealant is one of the most widely used gasket compounds. Made from silicone, RTV is air curing, it seals, bonds, waterproofs, fills surface irregularities, remains flexible, doesn't shrink, is relatively easy to remove, and is used as a supplementary sealer with almost all low and medium temperature gaskets.

Anaerobic sealant is much like RTV in that it can be used either to seal gaskets or to form gaskets by itself. It remains flexible, is solvent resistant and fills surface imperfections. The difference between an anaerobic sealant and an RTV-type sealant is in the curing. RTV cures when exposed to air, while an anaerobic sealant cures only in the absence of air. This means that an anaerobic sealant cures only after the assembly of parts, sealing them together.

Thread and pipe sealant is used for sealing hydraulic and pneumatic fittings and vacuum lines. It is usually made from a Teflon compound, and comes in a spray, a paint-on liquid and as a wrap-around tape.

Automotive chemicals

Anti-seize compound prevents seizing, galling, cold welding, rust and corrosion in fasteners. High-temperature ant-seize, usually made with copper and graphite lubricants, is used for exhaust system and exhaust manifold bolts.

Anaerobic locking compounds are used to keep fasteners from vibrating or working loose and cure only after installation, in the absence of air. Medium strength locking compound is used for small nuts, bolts and screws that may be removed later. High-strength locking compound is for large nuts, bolts and studs which aren't removed on a regular basis.

Oil additives range from viscosity index improvers to chemical treatments that claim to reduce internal engine friction. It should be noted that most oil manufacturers caution against using additives with their oils.

Gas additives perform several functions, depending on their chemical makeup. They usually contain solvents that help dissolve gum and varnish that build up on carburetor, fuel injection and intake parts. They also serve to break down carbon deposits that form on the inside surfaces of the combustion chambers. Some additives contain upper cylinder lubricants for valves and piston rings, and others contain chemicals to remove condensation from the gas tank.

Miscellaneous

Brake fluid is specially formulated hydraulic fluid that can withstand the heat and pressure encountered in brake systems. Care must be taken so this fluid does not come in contact with painted surfaces or plastics. An opened container should always be resealed to prevent contamination by water or dirt.

Weatherstrip adhesive is used to bond weatherstripping around doors, windows and trunk lids. It is sometimes used to attach trim pieces.

Undercoating is a petroleum-based, tar-like substance that is designed to protect metal surfaces on the underside of the vehicle from corrosion. It also acts as a sound-deadening agent by insulating the bottom of the vehicle.

Safety first

Regardless of how enthusiastic you may be about getting on with the job at hand, take the time to ensure that your safety is not jeopardized. A moment's lack of attention can result in an accident, as can failure to observe certain simple safety precautions. The possibility of an accident will always exist, and the following points should not be considered a comprehensive list of all dangers. Rather, they are intended to make you aware of the risks and to encourage a safety conscious approach to all work you carry out on your vehicle.

Essential DOs and DON'Ts

DON'T rely on a jack when working under the vehicle. Always use approved jackstands to support the weight of the vehicle and place them under the recommended lift or support points.

DON'T attempt to loosen extremely tight fasteners (i.e. wheel lug nuts) while the vehicle is on a jack - it may fall.

DON'T start the engine without first making sure that the transmission is in Neutral (or Park where applicable) and the parking brake is set.

DON'T remove the radiator cap from a hot cooling system - let it cool or cover it with a cloth and release the pressure gradually.

DON'T attempt to drain the engine oil until you are sure it has cooled to the point that it will not burn you.

DON'T touch any part of the engine or exhaust system until it has cooled sufficiently to avoid burns.

DON'T siphon toxic liquids such as gasoline, antifreeze and brake fluid by mouth, or allow them to remain on your skin.

DON'T inhale brake lining dust - it is potentially hazardous (see *Asbestos* below).

DON'T allow spilled oil or grease to remain on the floor - wipe it up before someone slips on it.

DON'T use loose fitting wrenches or other tools which may slip and cause injury.

DON'T push on wrenches when loosening or tightening nuts or bolts. Always try to pull the wrench toward you. If the situation calls for pushing the wrench away, push with an open hand to avoid scraped knuckles if the wrench should slip.

DON'T attempt to lift a heavy component alone - get someone to help you.

DON'T rush or take unsafe shortcuts to finish a job.

DON'T allow children or animals in or around the vehicle while you are working on it.

DO wear eye protection when using power tools such as a drill, sander, bench grinder, etc. and when working under a vehicle.

DO keep loose clothing and long hair well out of the way of moving parts.

DO make sure that any hoist used has a safe working load rating adequate for the job.

DO get someone to check on you periodically when working alone on a vehicle.

DO carry out work in a logical sequence and make sure that everything is correctly assembled and tightened.

DO keep chemicals and fluids tightly capped and out of the reach of children and pets.

DO remember that your vehicle's safety affects that of yourself and others. If in doubt on any point, get professional advice.

Asbestos

Certain friction, insulating, sealing, and other products - such as brake linings, brake bands, clutch linings, torque converters, gaskets, etc. - contain asbestos. Extreme care must be taken to avoid inhalation of dust from such products, since it is hazardous to health. If in doubt, assume that they do contain asbestos.

Fire

Remember at all times that gasoline is highly flammable. Never smoke or have any kind of open flame around when working on a vehicle. But the risk does not end there. A spark caused by an electrical short circuit, by two metal surfaces contacting each other, or even by static electricity built up in your body under certain conditions, can ignite gasoline vapors, which in a confined space are highly explosive. Do not, under any circumstances, use gasoline for cleaning parts. Use an approved safety solvent.

Always disconnect the battery ground (-) cable at the battery before working on any part of the fuel system or electrical system. Never risk spilling fuel on a hot engine or exhaust component. It is strongly recommended that a fire extinguisher suitable for use on fuel and electrical fires be kept handy in the garage or workshop at all times. Never try to extinguish a fuel or electrical fire with water.

Fumes

Certain fumes are highly toxic and can quickly cause unconsciousness and even death if inhaled to any extent. Gasoline vapor falls into this category, as do the vapors from some cleaning solvents. Any draining or pouring of such volatile fluids should be done in a well ventilated area.

When using cleaning fluids and solvents, read the instructions on the container carefully. Never use materials from unmarked containers.

Never run the engine in an enclosed space, such as a garage. Exhaust fumes contain carbon monoxide, which is extremely poisonous. If you need to run the engine, always do so in the open air, or at least have the rear of the vehicle outside the work area.

If you are fortunate enough to have the use of an inspection pit, never drain or pour gasoline and never run the engine while the vehicle is over the pit. The fumes, being heavier than air, will concentrate in the pit with possibly lethal results.

The battery

Never create a spark or allow a bare light bulb near a battery. They normally give off a certain amount of hydrogen gas, which is highly explosive.

Always disconnect the battery ground (-) cable at the battery before working on the fuel or electrical systems.

If possible, loosen the filler caps or cover when charging the battery from an external source (this does not apply to sealed or maintenance-free batteries). Do not charge at an excessive rate or the battery may burst.

Take care when adding water to a non maintenance-free battery and when carrying a battery. The electrolyte, even when diluted, is very corrosive and should not be allowed to contact clothing or skin.

Always wear eye protection when cleaning the battery to prevent the caustic deposits from entering your eyes.

Household current

When using an electric power tool, inspection light, etc., which operates on household current, always make sure that the tool is correctly connected to its plug and that, where necessary, it is properly grounded. Do not use such items in damp conditions and, again, do not create a spark or apply excessive heat in the vicinity of fuel or fuel vapor.

Environmental safety

At the time this manual was being written, several state and federal regulations governing the storage and disposal of oil and other lubricants, gasoline, solvents and antifreeze were pending (contact the appropriate government agency or your local auto parts store for the latest information). Be absolutely certain that all materials are properly stored, handled and disposed of. Never pour used or leftover oil, solvents or antifreeze down the drain or dump them on the ground. Also, don't allow volatile liquids to evaporate – keep them in sealed containers. Air conditioning refrigerant should never be expelled into the atmosphere. Have a properly equipped shop discharge and recharge the system for you.

2 Tools and equipment

A place to work

Establish a place to work. It doesn't have to be particularly large, but it should be clean, safe, well-lit, organized and adequately equipped for the job. True, without a good workshop or garage, you can still service and repair vehicles, even if you have to work outside. But major repairs should be carried out in a sheltered area with a roof. Some of the procedures in this book require an environment totally free of dirt, which could cause contamination and subsequent failure if it finds its way into a delicate component or system.

The workshop

The size, shape and location of a shop building is usually dictated by circumstance rather than personal choice. Every do-it-yourselfer dreams of having a spacious, clean, well-lit building specially designed and equipped for working on everything from small engines on lawn and garden equipment to cars and other vehicles. In reality, however, most of us must content ourselves with a garage, basement or shed in the backyard.

Spend some time considering the potential - and drawbacks - of your current facility. Even a well-established workshop can benefit from intelligent design. Lack of space is the most common problem, but you can significantly increase usable space by carefully planning the locations of work and storage areas. One strategy is to look at how others do it. Ask local repair shop owners if you can see their shops. Note how they've arranged their work areas, storage and lighting, then try to scale down their solutions to fit your own shop space, finances and needs.

General workshop requirements

A solid concrete floor is the best surface for a shop area. The floor should be even, smooth and dry. A coat of paint or sealant formulated for concrete surfaces will make oil spills and dirt easier to remove and help cut down on dust - always a problem with concrete.

Paint the walls and ceiling white for maximum reflection. Use gloss or semi-gloss enamel. It's washable and reflective. If your shop has windows, situate workbenches to take advantage of them. Skylights are even better. You can't have too much natural light. Artificial light is also good, but you'll need a lot of it to equal ordinary daylight.

Make sure the building is adequately ventilated. This is critical during the winter months, to prevent condensation problems. It's also a vital safety consideration where solvents, gasoline and other volatile liquids are being used. You should be able to open one or more windows for ventilation. In addition, opening vents in the walls are desirable.

Electricity and lights

Electricity is essential in a shop. It's relatively easy to install if the workshop is part of the house, but it can be difficult and expensive to install if it isn't. Safety should be your primary consideration when dealing with electricity; unless you have a very good working knowledge of electrical installations, have an electrician do any work required to provide power and lights in the shop.

Consider the total electrical requirements of the shop, making allowances for possible later additions of lights and equipment. Don't substitute extension cords for legal and safe permanent wiring. If the wiring isn't adequate, or is substandard, have it upgraded.

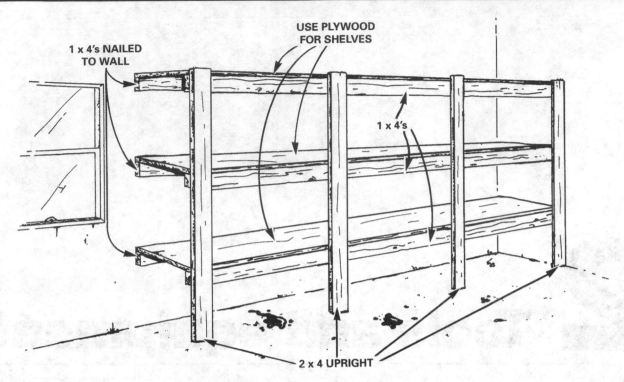

2.1 Homemade wood shelves are relatively inexpensive to build and you can design them to fit the available space, but all that wood can be a fire hazard

Give careful consideration to lights for the workshop. A pair of 150-watt incandescent bulbs, or two 48-inch long, 40-watt fluorescent tubes, suspended approximately 48-inches above the workbench, are the minimum you can get by with. As a general rule, fluorescent lights are probably the best choice. Their light is bright, even, shadow-free and fairly economical, although some people don't care for the bluish tinge they cast on everything. The usual compromise is a good mix of fluorescent and incandescent fixtures.

The position of the lights is important. Don't place a fixture directly above the work area. It will cause shadows, even with fluorescent lights. Attach the light(s) slightly to the rear - or to each side - of the workbench or garage to provide shadow-free lighting. A portable "trouble-light" is very helpful for use when overhead lights are inadequate. If gasoline, solvents or other flammable liquids are present - not an unusual situation in a shop - use special fittings to minimize the risk of fire. And don't use fluorescent lights above machine tools (like a drill press). The flicker produced by alternating current is especially pronounced with this type of light and can make a rotating chuck appear stationary at certain speeds - a very dangerous situation.

Storage and shelves

Set up an organized storage area to avoid losing parts. You'll need storage space for hardware, lubricants and other chemicals, rags, tools and equipment.

If space and finances allow, install metal shelves along the walls. Arrange the shelves so they're widely spaced near the bottom to take large or heavy items. Metal shelf units are pricey, but they make the best use of available space. And the shelf height is adjustable on most units.

Wood shelves **(see illustration)** are sometimes a cheaper storage solution. But they must be built - not just assembled. They must be much heftier than metal shelves to carry the same weight, the shelves can't be adjusted vertically and you can't just disassemble them and take them with you if you move. Wood also absorbs oil and other liquids and is obviously a much greater fire hazard.

Store small parts in plastic drawers or bins mounted on metal racks attached to the wall. They're available from most hardware, home and lumber stores. Bins come in various sizes and usually have slots for labels.

All kinds of containers are useful in a shop. Glass jars are handy for storing fasteners, but they're easily broken. Cardboard boxes are adequate for temporary use, but if they become damp, the bottoms eventually weaken and fall apart if you store oily or heavy parts in them. Plastic containers come in a variety of sizes and colors for easy identification. Egg cartons are excellent organizers for tiny parts. Large ice cream tubs are suitable for keeping small parts together. Get the type with a snap cover. Old metal cake pans, bread pans and muffin tins also make good storage containers for small parts.

Workbenches

A workbench is essential - it provides a place to lay out parts and tools during repair procedures, and it's a lot more comfortable than working on a floor or the driveway. The workbench should be as large and sturdy as space and

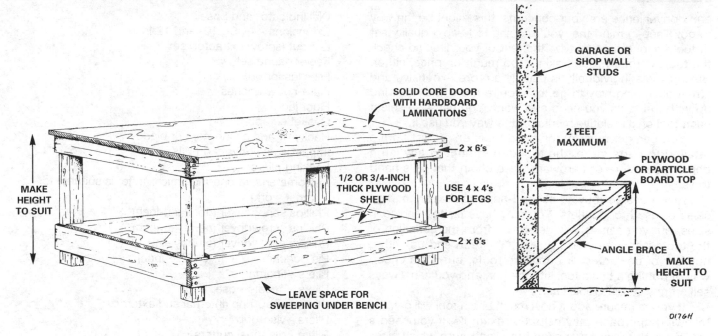

2.2 You can build a sturdy, inexpensive workbench with 4 x 4s, 2 x 6s and a solid core door with hardboard laminations - or build a bench using the wall as an integral member as shown

finances allow. If cost is no object, buy industrial steel benches. They're more expensive than home-built benches, but they're very strong, they're easy to assemble, and - if you move - they can be disassembled quickly and you can take them with you. They're also available in various lengths, so you can buy the exact size to fill the space along a wall.

If steel benches aren't in the budget, fabricate a bench frame from slotted angle-iron or Douglas fir (use 2 x 6's rather than 2 x 4's) **(see illustration)**. Cut the pieces of the frame to the required size and bolt them together with carriage bolts. A 30 or 36 by 80-inch, solid-core door with hardboard surfaces makes a good bench top. And you can flip it over when one side is worn out.

An even cheaper - and quicker - solution? Assemble a bench by attaching the bench top frame pieces to the wall with angled braces and use the wall studs as part of the framework.

Regardless of the type of frame you decide to use for the workbench, be sure to position the bench top at a comfortable working height and make sure everything is level. Shelves installed below the bench will make it more rigid and provide useful storage space.

Storage and care of tools

Good tools are expensive, so treat them well. After you're through with your tools, wipe off any dirt, grease or metal chips and put them away. Don't leave tools lying around in the work area. General purpose hand tools - screwdrivers, pliers, wrenches and sockets - can be hung on a wall panel or stored in a tool box. Store precision measuring instruments, gauges, meters, etc. in a tool box to protect them from dust, dirt, metal chips and humidity.

Tools and equipment

For some home mechanics, the idea of using the correct tool is completely foreign. They'll cheerfully tackle the most complex procedures with only a set of cheap open-end wrenches of the wrong type, a single screwdriver with a worn tip, a large hammer and an adjustable wrench. Though they often get away with it, this cavalier approach is stupid and dangerous. It can result in relatively minor annoyances like stripped fasteners, or cause catastrophic consequences. It can also result in serious injury.

A complete assortment of good tools is a given for anyone who plans to work on cars. If you don't already have most of the tools listed below, the initial investment may seem high, but compared to the spiraling costs of routine maintenance and repairs, it's a deal. Besides, you can use a lot of the tools around the house for other types of mechanical repairs. While some of the tools we'll describe aren't necessary to complete most repair operations, they are representative of the kinds of tools you would expect to find in a well-equipped shop.

Buying tools

There are two ways to buy tools. The easiest and quickest way is to simply buy an entire set. Tool sets are often priced substantially below the cost of the same individually priced tools - and sometimes they even come with a tool box. When purchasing such sets, you often wind up with some tools you don't need or want. But if low price and convenience are your concerns, this might be the way to go. Keep in mind that you're going to keep a quality set of tools a long time (maybe the rest of your life), so check the tools carefully; don't skimp too much on price, either. Buying tools individually is usually a more expensive and time-consuming way to go, but you're more likely to wind up with the tools you need and want. You can also select each tool on its relative merits for the way you use it.

You can get most of the hand tools on our list from a retail auto parts store, the tool department of any large department store or hardware store chain that sells hand tools.

Also consider buying second-hand tools from garage sales or used tool outlets. You may have limited choice in sizes, but you can usually determine from the condition of the tools if they're worth buying. You can end up with a number of unwanted or duplicate tools, but it's a cheap way of putting a basic tool kit together, and you can always sell off any surplus tools later.

If you're unsure about how much use a tool will get, the following approach may help. For example, if you need a set of combination wrenches but aren't sure which sizes you'll end up using most, buy a cheap or medium-priced set (make sure the jaws fit the fastener sizes marked on them). After some use over a period of time, carefully examine each tool in the set to assess its condition. If all the tools fit well and are undamaged, don't bother buying a better set. If one or two are worn, replace them with high-quality items - this way you'll end up with top-quality tools where they're needed most and the cheaper ones are sufficient for occasional use. On rare occasions you may conclude the whole set is poor quality. If so, buy a better set, if necessary, and remember never to buy that brand again.

In summary, try to avoid cheap tools, especially when you're purchasing high-use items like screwdrivers, wrenches and sockets. Cheap tools don't last long. Their initial cost plus the additional expense of replacing them will exceed the initial cost of better-quality tools.

Hand tools

Note: *The information that follows is for Standard fastener sizes. On some late-model vehicles, you'll need Metric wrenches, sockets and Allen wrenches. Generally, manufacturers began integrating metric fasteners into their vehicles around 1975.*

A list of general-purpose hand tools you should have in your shop

Adjustable wrench - 10-inch
Allen wrench set (1/8 to 3/8-inch or 4 mm to 10 mm)
Ball peen hammer - 12 oz (any steel hammer will do)
Brass hammer
Brushes (various sizes, for cleaning small parts)
Combination (slip-joint) pliers - 6-inch
Center punch
Cold chisels - 1/4 and 1/2-inch
Combination wrench set (1/4 to 1-inch or 7 mm to 19 mm)

Dial indicator and base
Extensions - 1-, 6-, 10- and 12-inch
E-Z out (screw extractor) set
Feeler gauge set
Files (assorted)
Flare-nut wrenches
Floor jack
Gasket scraper
Hacksaw and assortment of blades
Impact screwdriver and bits
Locking pliers
Micrometer(s) (a one-inch micrometer is suitable for most work)
Phillips screwdriver (no. 2 x 6-inch)
Phillips screwdriver (no. 3 x 8-inch)
Phillips screwdriver (stubby - no. 2)
Pin punches (1/16, 1/8, 3/16-inch)
Pliers - lineman's
Pliers - needle-nose
Pliers - snap-ring (internal and external)
Pliers - vise-grip
Pliers - diagonal cutters
Ratchet (reversible)
Scraper (made from flattened copper tubing)
Scribe
Socket set (6-point sockets are preferred, but some fasteners require the use of 12-point sockets)
Soft-face hammer (plastic/rubber)
Spark plug socket (with rubber insert)
Spark plug gap adjusting tool
Standard screwdriver (1/4-inch x 6-inch)
Standard screwdriver (5/16-inch x 6-inch)
Standard screwdriver (3/8-inch x 10-inch)
Standard screwdriver (5/16-inch - stubby)
Steel ruler - 6-inch
Tap and die set
Thread gauge
Torque wrench (same size drive as sockets)
Torx socket(s)
Universal joint
Vacuum gauge/pump (hand-held)
Wire brush (large)
Wire cutter pliers

What to look for when buying hand tools and general purpose tools

Wrenches and sockets

Wrenches vary widely in quality. One indication of their cost is their quality: The more they cost, the better they are. Buy the best wrenches you can afford. You'll use them a lot.

Start with a set containing wrenches from 1/4 to 1-inch in size. The size, stamped on the wrench **(see illustration)**, indicates the distance across the nut or bolt head, or the distance between the wrench jaws - not the diameter of the

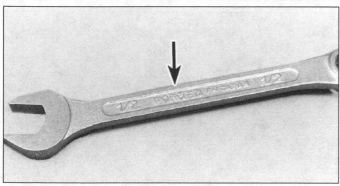

2.3 One quick way to determine whether you're looking at a quality wrench is to read the information printed on the handle - if it says "chrome vanadium" or "forged", it's made out of the right material

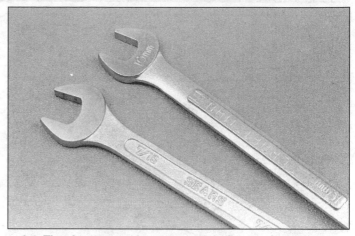

2.4 The size stamped on a wrench indicates the distance across the nut or bolt head (or the distance between the wrench jaws) in inches, not the diameter of the threads on the fastener

threads on the fastener - in inches. For example, a 1/4-inch bolt usually has a 7/16-inch hex head - the size of the wrench required to loosen or tighten it. However, the relationship between thread diameter and hex size doesn't always hold true. In some instances, an unusually small hex may be used to discourage over-tightening or because space around the fastener head is limited. Conversely, some fasteners have a disproportionately large hex-head.

Wrenches are similar in appearance, so their quality level can be difficult to judge just by looking at them. There are bargains to be had, just as there are overpriced tools with well-known brand names. On the other hand, you may buy what looks like a reasonable value set of wrenches only to find they fit badly or are made from poor-quality steel.

With a little experience, it's possible to judge the quality of a tool by looking at it. Often, you may have come across the brand name before and have a good idea of the quality. Close examination of the tool can often reveal some hints as to its quality. Prestige tools are usually polished and chrome-plated over their entire surface, with the working faces ground to size. The polished finish is largely cosmetic, but it does make them easy to keep clean. Ground jaws normally indicate the tool will fit well on fasteners.

A side-by-side comparison of a high-quality wrench with a cheap equivalent is an eye opener. The better tool will be made from a good-quality material, often a forged chrome-vanadium/steel alloy **(see illustration)**. This, together with careful design, allows the tool to be kept as small and compact as possible. If, by comparison, the cheap tool is thicker and heavier, especially around the jaws, it's usually because the extra material is needed to compensate for its lower quality. If the tool fits properly, this isn't necessarily bad - it is, after all, cheaper - but in situations where it's necessary to work in a confined area, the cheaper tool may be too bulky to fit.

Open-end wrenches

Because of its versatility, the open-end wrench is the most common type of wrench. It has a jaw on either end, connected by a flat handle section. The jaws either vary by a size, or overlap sizes between consecutive wrenches in a

set. This allows one wrench to be used to hold a bolt head while a similar-size nut is removed. A typical fractional size wrench set might have the following jaw sizes: 1/4 x 5/16, 3/8 x 7/16, 1/2 x 9/16, 9/16 x 5/8 and so on.

Typically, the jaw end is set at an angle to the handle, a feature which makes them very useful in confined spaces; by turning the nut or bolt as far as the obstruction allows, then turning the wrench over so the jaw faces in the other direction, it's possible to move the fastener a fraction of a turn at a time **(see illustration)**. The handle length is generally determined by the size of the jaw and is calculated to allow a nut or bolt to be tightened sufficiently by hand with minimal risk of breakage or thread damage (though this doesn't apply to soft materials like brass or aluminum).

Common open-end wrenches are usually sold in sets and it's rarely worth buying them individually unless it's to replace a lost or broken tool from a set. Single tools invariably cost more, so check the sizes you're most likely to need regularly and buy the best set of wrenches you can

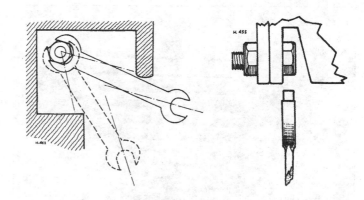

2.5 Open-end wrenches can do several things other wrenches can't - for example, they can be used on bolt heads with limited clearance (above) and they can be used in tight spots where there's little room to turn a wrench by flipping the offset jaw over every few degrees of rotation

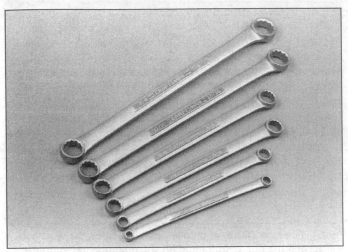

2.6 Box-end wrenches have a ring-shaped "box" at each end - when space permits, they offer the best combination of "grip" and strength

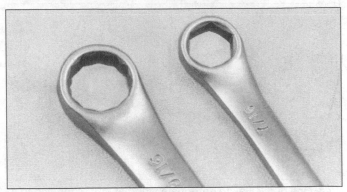

2.7 Box-end wrenches are available in 12 (left) and 6-point (right) openings; even though the 12-point design offers twice as many wrench positions, buy the 6-point first - it's less likely to strip off the corners of a nut or bolt head

afford in that range of sizes. If money is limited, remember that you'll use open-end wrenches more than any other type - it's a good idea to buy a good set and cut corners elsewhere.

Box-end wrenches

Box-end wrenches **(see illustration)** have ring-shaped ends with a 6-point (hex) or 12-point (double hex) opening **(see illustration)**. This allows the tool to fit on the fastener hex at 15 (12-point) or 30-degree (6-point) intervals. Normally, each tool has two ends of different sizes, allowing an overlapping range of sizes in a set, as described for open-end wrenches.

Although available as flat tools, the handle is usually offset at each end to allow it to clear obstructions near the fastener, which is normally an advantage. In addition to normal length wrenches, it's also possible to buy long handle types to allow more leverage (very useful when trying to loosen rusted or seized nuts). It is, however, easy to shear

off fasteners if not careful, and sometimes the extra length impairs access.

As with open-end wrenches, box-ends are available in varying quality, again often indicated by finish and the amount of metal around the ring ends. While the same criteria should be applied when selecting a set of box-end wrenches, if your budget is limited, go for better-quality open-end wrenches and a slightly cheaper set of box-ends.

Combination wrenches

These wrenches **(see illustration)** combine a box-end and open-end of the same size in one tool and offer many of the advantages of both. Like the others, they're widely available in sets and as such are probably a better choice than box-ends only. They're generally compact, short-handled tools and are well suited for tight spaces where access is limited.

Adjustable wrenches

Adjustable wrenches **(see illustration)** come in several sizes. Each size can handle a range of fastener sizes. Adjustable wrenches aren't as effective as one-size tools

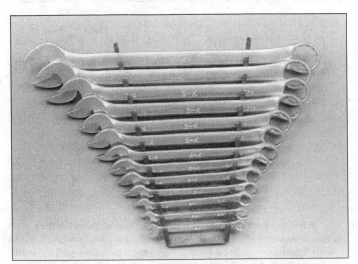

2.8 Buy a set of combination wrenches from 1/4 to 1-inch or 7 mm to 19 mm

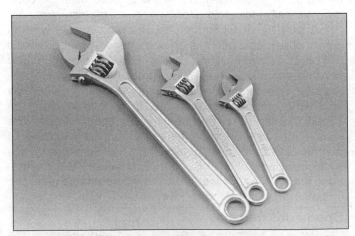

2.9 Adjustable wrenches can handle a range of fastener sizes - they're not as good as single-size wrenches but they're handy for loosening and tightening those odd-sized fasteners for which you haven't yet bought the correct wrench

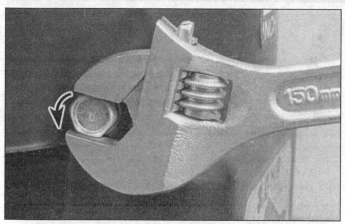

2.10 When you use an adjustable wrench, make sure the movable jaw points in the direction the wrench is being turned (arrow) so the wrench doesn't distort and slip off the fastener head

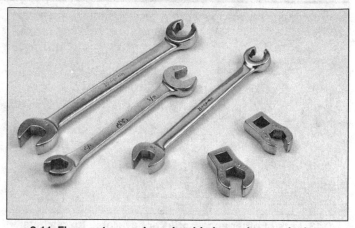

2.11 Flare nut wrenches should always be used when loosening or tightening line fittings (tube nuts) - the two odd-shaped items on the right are flare-nut *crows feet*, which are useful for loosening hard-to-reach fittings when used with a long extension

and it's easy to damage fasteners with them. However, they can be an invaluable addition to any tool kit - if they're used with discretion. **Note:** *If you attach the wrench to the fastener with the movable jaw pointing in the direction of wrench rotation* **(see illustration)**, *an adjustable wrench will be less likely to slip and damage the fastener head.*

The most common adjustable wrench is the open-end type with a set of parallel jaws that can be set to fit the head of a fastener. Most are controlled by a threaded spindle, though there are various cam and spring-loaded versions available. Don't buy large tools of this type; you'll rarely be able to find enough clearance to use them.

Flare nut wrenches

These wrenches, sometimes called line wrenches, are used for loosening and tightening hydraulic line fittings (tube nuts). Construction is similar to a six-point box end wrench, but a portion of one of the flats is cut out to allow the wrench to pass over a line or hose **(see illustration)**. This design offers much more surface area of the wrench to be in contact with the flats on the fitting, which will prevent

the fittings from being rounded off because the load is distributed over as much area as possible. They are also thicker than most open or box end wrenches, which adds to the surface area in contact with the tube nut. **Caution:** *When loosening a tube nut on a fitting connected to a flexible hose, always use a backup wrench to hold the larger, female fitting stationary. This will prevent the hydraulic line from twisting.*

Ratchet and socket sets

Ratcheting socket wrenches **(see illustration)** are highly versatile. Besides the sockets themselves, many other interchangeable accessories - extensions, U-drives, step-down adapters, screwdriver bits, Allen bits, crow's feet, etc. - are available. Buy six-point sockets - they're less likely to slip and strip the corners off bolts and nuts. **Note:** *Some driveline fasteners (driveaxle and driveshaft mounting bolts on some vehicles, for example) require the use of 12-point sockets.* Don't buy sockets with extra-thick walls - they might be stronger but they can be hard to use on recessed fasteners or fasteners in tight quarters.

A 3/8-inch drive set is adequate for most work. Sometimes a 1/2-inch drive set is required when working with large fasteners. Although the larger drive is bulky and more expensive, it has the capacity of accepting a very wide range of large sockets. Later, you may want to consider a 1/4-inch drive for little stuff like ignition and carburetor work.

Interchangeable sockets consist of a forged-steel alloy cylinder with a hex or double-hex formed inside one end. The other end is formed into the square drive recess that engages over the corresponding square end of various socket drive tools.

Sockets are available in 1/4, 3/8, 1/2 and 3/4-inch drive sizes. A 3/8-inch drive set is most useful, although 1/4-inch drive sockets and accessories may occasionally be needed.

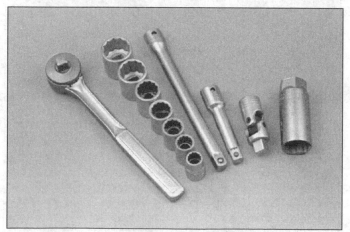

2.12 A typical ratchet and socket set includes a ratchet, a set of sockets, a long and a short extension, a universal joint and a spark plug socket

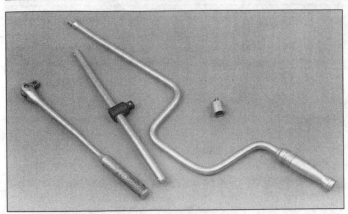

2.13 Lots of other accessories are available for ratchets; From left to right, a breaker bar, a sliding T-handle, a speed handle and a 3/8-to-1/4-inch adapter

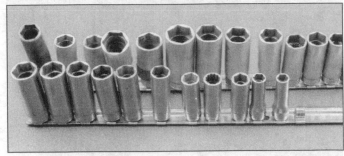

2.14 Deep sockets enable you to loosen or tighten an elongated fastener, or to get at a nut with a long bolt protruding from it

2.15 Standard and Phillips bits, Allen-head and Torx drivers will expand the versatility of your ratchet and extensions even further

The most economical way to buy sockets is in a set. As always, quality will govern the cost of the tools. Once again, the "buy the best" approach is usually advised when selecting sockets. While this is a good idea, since the end result is a set of quality tools that should last a lifetime, the cost is so high it's difficult to justify the expense for home use.

As far as accessories go, you'll need a ratchet, at least one extension (buy a three or six-inch size), a spark plug socket and maybe a T-handle or breaker bar. Other desirable, though less essential items, are a speeder handle, a U-joint, extensions of various other lengths and adapters from one drive size to another **(see illustration)**. Some of the sets you find may combine drive sizes; they're well worth having if you find the right set at a good price, but avoid being dazzled by the number of pieces.

Above all, be sure to completely ignore any label that reads "86-piece Socket Set," which refers to the number of pieces, not to the number of sockets (sometimes even the metal box and plastic insert are counted in the total!).

Apart from well-known and respected brand names, you'll have to take a chance on the quality of the set you buy. If you know someone who has a set that has held up well, try to find the same brand, if possible. Take a pocketful of nuts and bolts with you and check the fit in some of the sockets. Check the operation of the ratchet. Good ones operate smoothly and crisply in small steps; cheap ones are coarse and stiff - a good basis for guessing the quality of the rest of the pieces.

One of the best things about a socket set is the built-in facility for expansion. Once you have a basic set, you can purchase extra sockets when necessary and replace worn or damaged tools. There are special deep sockets for reaching recessed fasteners or to allow the socket to fit over a projecting bolt or stud **(see illustration)**. You can also buy screwdriver, Allen and Torx bits to fit various drive tools (they can be very handy in some applications) **(see illustration)**. Most socket sets include a special deep socket for 14 millimeter spark plugs. They have rubber inserts to protect the spark plug porcelain insulator and hold the plug in the socket to avoid burned fingers.

Torque wrenches

Torque wrenches **(see illustration)** are essential for tightening critical fasteners pitman arms, pinion nut differential bearing caps, etc. A fastener that's not tight enough

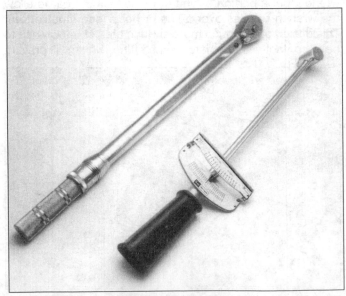

2.16 Torque wrenches (click-type on left, beam-type on right) are the only way to accurately tighten critical fasteners like pinion nuts and carrier bearing cap bolts, etc.

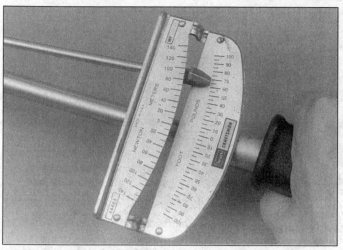

2.17 The deflecting beam-type torque wrench is inexpensive and simple to use - just tighten the fastener until the pointer points to the specified torque setting

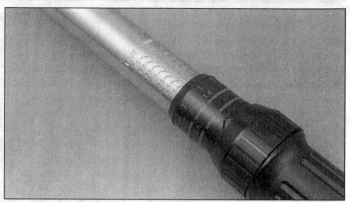

2.18 "Click" type torque wrenches can be set to "give" at a pre-set torque, which makes them very accurate and easy to use

may eventually escape from its hole. An overtightened fastener, on the other hand, could break under stress.

There are several different types of torque wrenches on the market. The most common are; the "beam" type, which indicates torque loads by deflecting a flexible shaft and the "click" type **(see illustrations)**, which emits an audible click when the torque resistance reaches the specified resistance. Another type is the "dial" type **(see illustration)**; torque is indicated by a needle on a dial, similar to a dial indicator. Dial types are very accurate down to the inch-pound range.

Torque wrenches are available in a variety of drive sizes, including 1/4, 3/8 and 1/2 inch. Torque ranges vary for particular applications. For most work, 0 to 100 ft-lbs should be adequate. You will need a 3/8 "dial" type to set the pinion bearing preload on most differential overhauls. Keep in mind that "click" types are usually more accurate (and more expensive). **Note:** Don't use a "click" type to set pinion bearing preloads.

Impact drivers

The impact driver **(see illustration)** belongs with the screwdrivers, but it's mentioned here since it can also be used with sockets (impact drivers normally are 3/8-inch square drive). As explained later, an impact driver works by converting a hammer blow on the end of its handle into a sharp twisting movement. While this is a great way to jar a seized fastener loose, the loads imposed on the socket are excessive. Use sockets only with discretion and expect to have to replace damaged ones on occasion.

Using wrenches and sockets

Although you may think the proper use of tools is self-evident, it's worth some thought. After all, when did you last see instructions for use supplied with a set of wrenches?

Which wrench?

Before you start working, figure out the best tool for the job; in this instance the best wrench for a hex-head fastener. Sit down with a few nuts and bolts and look at how various tools fit the bolt heads.

A golden rule is to choose a tool that contacts the largest area of the hex-head. This distributes the load as evenly as possible and lessens the risk of damage. The

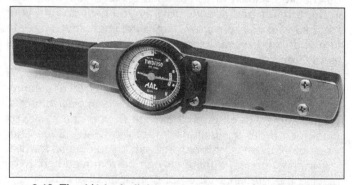

2.19 The 1/4-inch dial-type torque wrench is the most accurate for the inch-pound settings - just tighten the fastener until the pointer points to the specified torque setting

2.20 The impact driver converts a sharp blow into a twisting motion - this is a handy addition to your socket arsenal for those fasteners that won't let go - you can use it with any bit that fits a 38-inch drive ratchet

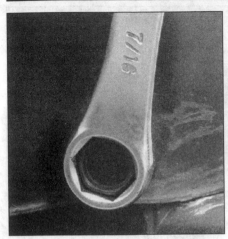

2.21 Try to use a six-point box wrench (or socket) whenever possible - its shape matches that of the fastener, which means maximum grip and minimum slip

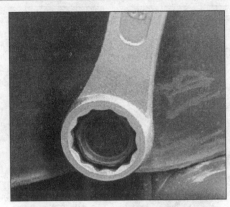

2.22 Sometimes a six-point tool just doesn't offer you any grip when you get the wrench at the angle it needs to be in to loosen or tighten a fastener - when this happens, pull out the 12-point sockets or wrenches - but remember; they're much more likely to strip the corners off a fastener

2.23 Open-end wrenches contact only two sides of the fastener and the jaws tend to open up when you put some muscle on the wrench handle - that's why they should only be used as a last resort

shape most closely resembling the bolt head or nut is another hex, so a 6-point socket or box-end wrench is usually the best choice **(see illustration)**. Many sockets and box-end wrenches have double hex (12-point) openings. If you slip a 12-point box-end wrench over a nut, look at how and where the two are in contact. The corners of the nut engage in every other point of the wrench. When the wrench is turned, pressure is applied evenly on each of the six corners **(see illustration)**. This is fine unless the fastener head was previously rounded off. If so, the corners will be damaged and the wrench will slip. If you encounter a damaged bolt head or nut, always use a 6-point wrench or socket if possible. If you don't have one of the right size, choose a wrench that fits securely and proceed with care.

If you slip an open-end wrench over a hex-head fastener, you'll see the tool is in contact on two faces only **(see illustration)**. This is acceptable provided the tool and fastener are both in good condition. The need for a snug fit between the wrench and nut or bolt explains the recommendation to buy good-quality open-end wrenches. If the wrench jaws, the bolt head or both are damaged, the wrench will probably slip, rounding off and distorting the head. In some applications, an open-end wrench is the only possible choice due to limited access, but always check the fit of the wrench on the fastener before attempting to loosen it; if it's hard to get at with a wrench, think how hard it will be to remove after the head is damaged.

The last choice is an adjustable wrench or self-locking pliers/wrench (Vise-Grips). Use these tools only when all else has failed. In some cases, a self-locking wrench may be able to grip a damaged head that no wrench could deal with, but be careful not to make matters worse by damaging it further.

Bearing in mind the remarks about the correct choice of tool in the first place, there are several things worth noting about the actual use of the tool. First, make sure the wrench head is clean and undamaged. If the fastener is rusted or

coated with paint, the wrench won't fit correctly. Clean off the head and, if it's rusted, apply some penetrating oil. Leave it to soak in for a while before attempting removal.

It may seem obvious, but take a close look at the fastener to be removed before using a wrench. On many mass-produced machines, one end of a fastener may be fixed or captive, which speeds up initial assembly and usually makes removal easier. If a nut is installed on a stud or a bolt threads into a captive nut or tapped hole, you may have only one fastener to deal with. If, on the other hand, you have a separate nut and bolt, you must hold the bolt head while the nut is removed. In some areas this can be difficult, in this type of situation you may need an assistant to hold the bolt head with a wrench while you remove the nut from the other side. If this isn't possible, you'll have to try to position a box-end wrench so it wedges against some other component to prevent it from turning.

Be on the lookout for left-hand threads. They aren't common, but are sometimes used on the ends of rotating shafts to make sure the nut doesn't come loose during operation. If you can see the shaft end, the thread type can be checked visually. If you're unsure, place your thumbnail in the threads and see which way you have to turn your hand so your nail "unscrews" from the shaft. If you have to turn your hand counterclockwise, it's a conventional right-hand thread.

Beware of the upside-down fastener syndrome. If you're loosening a fastener from the under side of a something, it's easy to get confused about which way to turn it. What seems like counterclockwise to you can easily be clockwise (from the fastener's point of view). Even after years of experience, this can still catch you once in a while.

In most cases, a fastener can be removed simply by placing the wrench on the nut or bolt head and turning it. Occasionally, though, the condition or location of the fastener may make things more difficult. Make sure the wrench is square on the head. You may need to reposition the tool

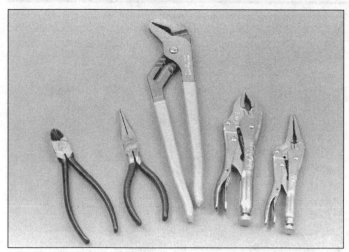

2.24 A typical assortment of the types of pliers you need to have in your box - from the left; diagonal cutters (dikes), needle-nose pliers, Channel-lock pliers, Vise-grip pliers, needle-nose Vise-grip pliers

or try another type to obtain a snug fit. Make sure the part you're working on is secure and can't move when you turn the wrench. If necessary, get someone to help steady it for you. Position yourself so you can get maximum leverage on the wrench.

If possible, locate the wrench so you can pull the end towards you. If you have to push on the tool, remember that it may slip, or the fastener may move suddenly. For this reason, don't curl your fingers around the handle or you may crush or bruise them when the fastener moves; keep your hand flat, pushing on the wrench with the heel of your thumb. If the tool digs into your hand, place a rag between it and your hand or wear a heavy glove.

If the fastener doesn't move with normal hand pressure, stop and try to figure out why before the fastener or wrench is damaged or you hurt yourself. Stuck fasteners may require penetrating oil, heat or an impact driver or air tool.

Using sockets to remove hex-head fasteners is less likely to result in damage than if a wrench is used. Make sure the socket fits snugly over the fastener head, then attach an extension, if needed, and the ratchet or breaker bar. Theoretically, a ratchet shouldn't be used for loosening a fastener or for final tightening because the ratchet mechanism may be overloaded and could slip. In some instances, the location of the fastener may mean you have no choice but to use a ratchet, in which case you'll have to be extra careful.

Never use extensions where they aren't needed. Whether or not an extension is used, always support the drive end of the breaker bar with one hand while turning it with the other. Once the fastener is loose, the ratchet can be used to speed up removal.

Pliers

Some tool manufacturers make 25 or 30 different types of pliers. You only need a fraction of this selection **(see**

illustration). Get a good pair of slip-joint pliers for general use. A pair of needle-nose models is handy for reaching into hard-to-get-at places. A set of diagonal wire cutters (dikes) is essential for electrical work and pulling out cotter pins. Vise-Grips are adjustable, locking pliers that grip a fastener firmly - and won't let go - when locked into place. Parallel-jaw, adjustable pliers have angled jaws that remain parallel at any degree of opening. They're also referred to as Channel-lock (the original manufacturer) pliers, arc-joint pliers and water pump pliers. Whatever you call them, they're terrific for gripping things with a lot of force.

Slip-joint pliers have two open positions; a figure eight-shaped, elongated slot in one handle slips back-and-forth on a pivot pin on the other handle to change them. Good-quality pliers have jaws made of tempered steel and there's usually a wire-cutter at the base of the jaws. The primary uses of slip-joint pliers are for holding objects, bending and cutting throttle wires and crimping and bending metal parts, not loosening nuts and bolts.

Arc-joint or "Channel-lock" pliers have parallel jaws you can open to various widths by engaging different tongues and grooves, or channels, near the pivot pin. Since the tool expands to fit many size objects, it has countless uses for vehicle and equipment maintenance. Channel-lock pliers come in various sizes. The medium size is adequate for general work; small and large sizes are nice to have as your budget permits. You'll use all three sizes frequently.

Vise-Grips (a brand name) come in various sizes; the medium size with curved jaws is best for all-around work. However, buy a large and small one if possible, since they're often used in pairs. Although this tool falls somewhere between an adjustable wrench, a pair of pliers and a portable vise, it can be invaluable for loosening and tightening fasteners - it's the only pliers that should be used for this purpose.

The jaw opening is set by turning a knurled knob at the end of one handle. The jaws are placed over the head of the fastener and the handles are squeezed together, locking the tool onto the fastener **(see illustration)**. The design of

2.25 To adjust the jaws on a pair of vise-grips, grasp the part you want to hold with the jaws, tighten them down by turning the knurled knob on the end of one handle and snap the handles together - if you tightened the knob all the way down, you'll probably have to open it up (back it off) a little before you can close the handles

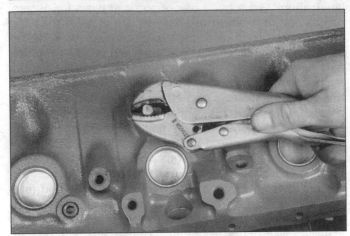

2.26 If you're persistent and careful, most fasteners can be removed with vise-grips

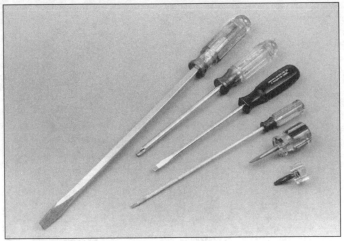

2.27 Screwdrivers come in a myriad of lengths, sizes and styles

the tool allows extreme pressure to be applied at the jaws and a variety of jaw designs enable the tool to grip firmly even on damaged heads **(see illustration)**. Vise-Grips are great for removing fasteners that have been rounded off by badly-fitting wrenches.

As the name suggests, needle-nose pliers have long, thin jaws designed for reaching into holes and other restricted areas. Most needle-nose, or long-nose, pliers also have wire cutters at the base of the jaws.

Look for these qualities when buying pliers: Smooth operating handles and jaws, jaws that match up and grip evenly when the handles are closed, a nice finish and the word "forged" somewhere on the tool.

Screwdrivers

Screwdrivers **(see illustration)** come in a wide variety of sizes and price ranges. Good advice is to avoid inexpensive and inferior brands. Even if they look exactly like more expensive brands, the metal tips and shafts are made with inferior alloys and aren't properly heat treated. They usually bend the first time you apply some serious torque.

A screwdriver consists of a steel blade or shank with a drive tip formed at one end. The most common tips are standard (also called straight slot and flat-blade) and Phillips. The other end has a handle attached to it. Traditionally, handles were made from wood and secured to the shank, which had raised tangs to prevent it from turning in the handle. Most screwdrivers now come with plastic handles, which are generally more durable than wood.

The design and size of handles and blades vary considerably. Some handles are specially shaped to fit the human hand and provide a better grip. The shank may be either round or square and some have a hex-shaped bolster under the handle to accept a wrench to provide more leverage when trying to turn a stubborn screw. The shank diameter, tip size and overall length vary too.

If access is restricted, a number of special screwdrivers are designed to fit into confined spaces. The "stubby" screwdriver has a specially shortened handle and blade. There are also offset screwdrivers and special screwdriver

bits that attach to a ratchet or extension.

The important thing to remember when buying screwdrivers is that they really do come in sizes designed to fit different size fasteners. The slot in any screw has definite dimensions - length, width and depth. Like a bolt head or a nut, the screw slot must be driven by a tool that uses all of the available bearing surface and doesn't slip. Don't use a big wide blade on a small screw and don't try to turn a large screw slot with a tiny, narrow blade. The same principles apply to Allen heads, Phillips heads, Torx heads, etc. Don't even think of using a slotted screwdriver on one of these heads! And don't use your screwdrivers as levers, chisels or punches! This kind of abuse turns them into very bad screwdrivers. It's also dangerous.

Standard screwdrivers

These are used to remove and install conventional slotted screws and are available in a wide range of sizes denoting the width of the tip and the length of the shank (for example: a 3/8- x 10-inch screwdriver is 3/8-inch wide at the tip and the shank is 10-inches long). You should have a variety of screwdrivers so screws of various sizes can be dealt with without damaging them. The blade end must be

Misuse of a screwdriver – the blade shown is both too narrow and too thin and will probably slip or break off

The left-hand example shows a snug-fitting tip. The right-hand drawing shows a damaged tip which will twist out of the slot when pressure is applied

2.28 Standard screwdrivers - wrong size (left), correct fit in screw slot (center) and worn tip (right)

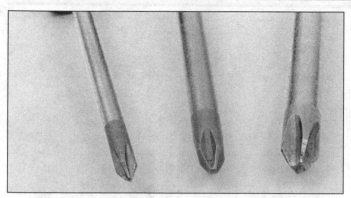

2.29 The tip size on a Phillips screwdriver is indicated by a number from 1 to 4, with 1 the smallest (left - No. 1: center - Nos. 2; eight - No. 3)

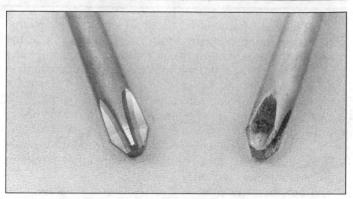

2.30 New (left) and worn (right) Phillips screwdriver tips

the same width and thickness as the screw slot to work properly, without slipping. When selecting standard screwdrivers, choose good-quality tools, preferably with chrome moly, forged steel shanks. The tip of the shank should be ground to a parallel, flat profile (hollow ground) and not to a taper or wedge shape, which will tend to twist out of the slot when pressure is applied **(see illustration)**.

All screwdrivers wear in use, but standard types can be reground to shape a number of times. When reshaping a tip, start by grinding the very end flat at right angles to the shank. Make sure the tip fits snugly in the slot of a screw of the appropriate size and keep the sides of the tip parallel. Remove only a small amount of metal at a time to avoid overheating the tip and destroying the temper of the steel.

Phillips screwdrivers

Phillips screws are sometimes installed during initial assembly with air tools and are next to impossible to remove later without ruining the heads, particularly if the wrong size screwdriver is used. And don't use other types of cross-head screwdrivers (Torx, Posi-drive, etc.) on Phillips screws - they won't work.

The only way to ensure the screwdrivers you buy will fit properly, is to take a couple of screws with you to make sure the fit between the screwdriver and fastener is snug. If the fit is good, you should be able to angle the blade down almost vertically without the screw slipping off the tip. Use only screwdrivers that fit exactly - anything else is guaranteed to chew out the screw head instantly.

The idea behind all cross-head screw designs is to make the screw and screwdriver blade self-aligning. Provided you aim the blade at the center of the screw head, it'll engage correctly, unlike conventional slotted screws, which need careful alignment. This makes the screws suitable for machine installation on an assembly line (which explains why they're sometimes so tight and difficult to remove). The drawback with these screws is the driving tangs on the screwdriver tip are very small and must fit very precisely in the screw head. If this isn't the case, the huge loads imposed on small flats of the screw slot simply tear the metal away, at which point the screw ceases to be removable by normal methods. The problem is made worse by the normally soft material chosen for screws.

To deal with these screws on a regular basis, you'll need high-quality screwdrivers with various size tips so you'll be sure to have the right one when you need it. Phillips screwdrivers are sized by the tip number and length of the shank (for example: a number 2 x 6-inch Phillips screwdriver has a number 2 tip - to fit screws of only that size recess - and the shank is 6-inches long). Tip sizes 1, 2 and 3 should be adequate for most repair work **(see illustration)**. If the tips get worn or damaged, buy new screwdrivers so the tools don't destroy the screws they're used on **(see illustration)**.

Here's a tip that may come in handy when using Phillips screwdrivers - if the screw is extremely tight and the tip tends to back out of the recess rather than turn the screw, apply a small amount of valve lapping compound to the screwdriver tip so it will grip the screw better.

Hammers

Resorting to a hammer should always be the last resort. When nothing else will do the job, a medium-size ball peen hammer, a heavy rubber mallet and a heavy soft-brass hammer **(see illustration)** are often the only way to loosen or install a part.

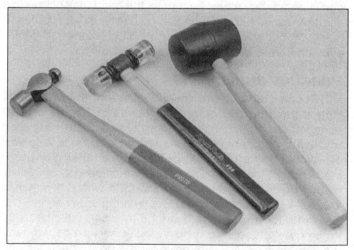

2.31 A ball-peen hammer, soft-face hammer and rubber mallet (left-to-right) will be needed for various tasks (any steel hammer can be used in place of the ball peen hammer)

A ball-peen hammer has a head with a conventional cylindrical face at one end and a rounded ball end at the other and is a general-purpose tool found in almost any type of shop. It has a shorter neck than a claw hammer and the face is tempered for striking punches and chisels. A fairly large hammer is preferable to a small one. Although it's possible to find small ones, you won't need them very often and it's much easier to control the blows from a heavier head. As a general rule, a single 12 or 16-ounce hammer will work for most jobs, though occasionally larger or smaller ones may be useful.

A soft-face hammer is used where a steel hammer could cause damage to the component or other tools being used. A steel hammer head might crack an aluminum part, but a rubber or plastic hammer can be used with more confidence. Soft-face hammers are available with interchangeable heads (usually one made of rubber and another made of relatively hard plastic). When the heads are worn out, new ones can be installed. If finances are really limited, you can get by without a soft-face hammer by placing a small hardwood block between the component and a steel hammer head to prevent damage.

Hammers should be used with common sense; the head should strike the desired object squarely and with the right amount of force. For many jobs, little effort is needed - simply allow the weight of the head to do the work, using the length of the swing to control the amount of force applied. With practice, a hammer can be used with surprising finesse, but it'll take a while to achieve. Initial mistakes include striking the object at an angle, in which case the hammer head may glance off to one side, or hitting the edge of the object. Either one can result in damage to the part or to your thumb, if it gets in the way, so be careful. Hold the hammer handle near the end, not near the head, and grip it firmly but not too tightly.

Check the condition of your hammers on a regular basis. The danger of a loose head coming off is self-evident, but check the head for chips and cracks too. If damage is noted, buy a new hammer - the head may chip in use and the resulting fragments can be extremely dangerous. It goes without saying that eye protection is essential whenever a hammer is used.

Punches and chisels

Punches and chisels (see illustration) are used along with a hammer for various purposes in the shop. Drift punches are often simply a length of round steel bar used to drive a component or fastener out of a bore. A typical use would be for removing or installing a bearing or bushing. A drift of the same diameter as the bearing outer race is placed against the bearing and tapped with a hammer to knock it in or out of the bore. Most manufacturers offer special drifts for the various bearings in a particular housing. While they're useful to a busy dealer service department, they are prohibitively expensive for the do-it-yourselfer who may only need to use them once. In such cases, it's better to improvise. For bearing removal and installation, it's usually possible to use a socket of the appropriate diameter to

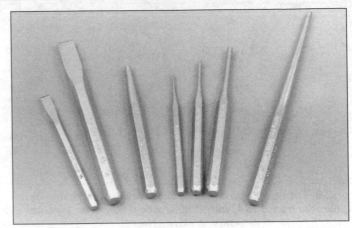

2.32 Cold chisels, center-punches, pin punches and line-up punches (left-to-right) will be needed sooner or later for many jobs

tap the bearing in or out; an unorthodox use for a socket, but it works.

Smaller diameter drift punches can be purchased or fabricated from steel bar stock. In some cases, you'll need to drive out items like caliper retaining pins. Here, it's essential to avoid damaging the pin or caliper, so the drift must be made from a soft material. Brass or copper is the usual choice for such jobs; the drift may be damaged in use, but the pin and surrounding components will be protected.

Punches are available in various shapes and sizes and a set of assorted types will be very useful. One of the most basic is the center punch, a small cylindrical punch with the end ground to a point. It'll be needed whenever a hole is drilled. The center of the hole is located first and the punch is used to make a small indentation at the intended point. The indentation acts as a guide for the drill bit so the hole ends up in the right place. Without a punch mark the drill bit will wander and you'll find it impossible to drill with any real accuracy. You can also buy automatic center punches. They're spring loaded and are pressed against the surface to be marked, without the need to use a hammer.

Pin punches are intended for removing items like roll pins (semi-hard, hollow pins that fit tightly in their holes). Pin punches have other uses, however. You may occasionally have to remove rivets or bolts by cutting off the heads and driving out the shanks with a pin punch. They're also very handy for aligning holes in components while bolts or screws are inserted.

Of the various sizes and types of metal-cutting chisels available, a simple cold chisel is essential in any mechanic's workshop. One about 6-inches long with a 1/2-inch wide blade should be adequate. The cutting edge is ground to about 80-degrees (see illustration), while the rest of the tip is ground to a shallower angle away from the edge. The primary use of the cold chisel is rough metal cutting - this can be anything from sheet metal work to cutting off the heads of seized or rusted bolts or splitting nuts. A cold chisel is also useful for turning out screws or bolts with messed-up heads.

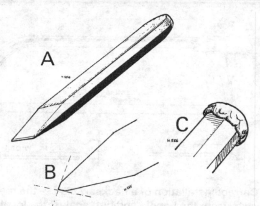

2.33 A typical general purpose cold chisel (A) - note the angle of the cutting edge (B), which should be checked and resharpened on a regular basis; the mushroomed head (C) is dangerous and should be filed to restore it to its original shape

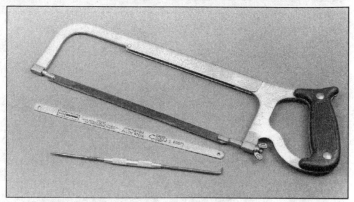

2.34 Hacksaws are handy for little cutting jobs like sheet metal and rusted fasteners

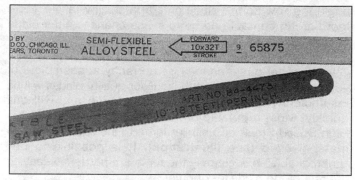

2.35 Hacksaw blades are marked with the number of teeth per inch (TPI) - use a relatively coarse blade for aluminum and thicker items such as bolts or bar stock; use a finer blade for materials like thin sheet steel

All of the tools described in this Section should be good quality items. They're not particularly expensive, so it's not really worth trying to save money on them. More significantly, there's a risk that with cheap tools, fragments may break off in use - a potentially dangerous situation.

Even with good-quality tools, the heads and working ends will inevitably get worn or damaged, so it's a good idea to maintain all such tools on a regular basis. Using a file or bench grinder, remove all burrs and mushroomed edges from around the head. This is an important task because the build-up of material around the head can fly off when it's struck with a hammer and is potentially dangerous. Make sure the tool retains its original profile at the working end, again, filing or grinding off all burrs. In the case of cold chisels, the cutting edge will usually have to be reground quite often because the material in the tool isn't usually much harder than materials typically being cut. Make sure the edge is reasonably sharp, but don't make the tip angle greater than it was originally; it'll just wear down faster if you do.

The techniques for using these tools vary according to the job to be done and are best learned by experience. The one common denominator is the fact they're all normally struck with a hammer. It follows that eye protection should be worn. Always make sure the working end of the tool is in contact with the part being punched or cut. If it isn't, the tool will bounce off the surface and damage may result.

Hacksaws

A hacksaw **(see illustration)** consists of a handle and frame supporting a flexible steel blade under tension. Blades are available in various lengths and most hacksaws can be adjusted to accommodate the different sizes. The most common blade length is 10-inches.

Most hacksaw frames are adequate. There's little difference between brands. Pick one that's rigid and allows easy blade changing and repositioning.

The type of blade to use, indicated by the number of teeth per inch, (TPI) **(see illustration)**, is determined by the material being cut. The rule of thumb is to make sure at least three teeth are in contact with the metal being cut at any one time **(see illustration)**. In practice, this means a fine blade for cutting thin sheet materials, while a coarser blade can be used for faster cutting through thicker items

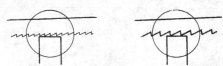

When cutting thin materials, check that at least three teeth are in contact with the workpiece at any time. Too coarse a blade will result in a poor cut and may break the blade. If you do not have the correct blade, cut at a shallow angle to the material

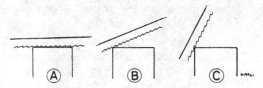

The correct cutting angle is important. If it is too shallow (A) the blade will wander. The angle shown at (B) is correct when starting the cut, and may be reduced slightly once under way. In (C) the angle is too steep and the blade will be inclined to jump out of the cut

2.36 Correct procedure for use of a hacksaw

2.37 Good quality hacksaw blades are marked like this

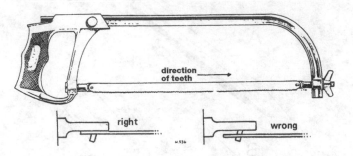

2.38 Correct installation of a hacksaw blade - the teeth must point away from the handle and butt against the locating lugs

such as bolts or bar stock. When cutting thin materials, angle the saw so the blade cuts at a shallow angle. More teeth are in contact and there's less chance of the blade binding and breaking, or teeth breaking.

When you buy blades, choose a reputable brand. Cheap, generic blades may be perfectly acceptable, but you can't tell by looking at them. Poor quality blades will be insufficiently hardened on the teeth edge and will dull quickly. Most reputable brands will be marked "Flexible High Speed Steel" or a similar term, to indicate the type of material used **(see illustration)**. It is possible to buy "unbreakable" blades (only the teeth are hardened, leaving the rest of the blade less brittle).

Sometimes, a full-size hacksaw is too big to allow access to a frozen nut or bolt. On most saws, you can overcome this problem by turning the blade 90-degrees. Occasionally you may have to position the saw around an obstacle and then install the blade on the other side of it. Where space is really restricted, you may have to use a handle that clamps onto a saw blade at one end. This allows access when a hacksaw frame would not work at all and has another advantage in that you can make use of broken off hacksaw blades instead of throwing them away. Note that

because only one end of the blade is supported, and it's not held under tension, it's difficult to control and less efficient when cutting.

Before using a hacksaw, make sure the blade is suitable for the material being cut and installed correctly in the frame **(see illustration)**. Whatever it is you're cutting must be securely supported so it can't move around. The saw cuts on the forward stroke, so the teeth must point away from the handle. This might seem obvious, but it's easy to install the blade backwards by mistake and ruin the teeth on the first few strokes. Make sure the blade is tensioned adequately or it'll distort and chatter in the cut and may break. Wear safety glasses and be careful not to cut yourself on the saw blade or the sharp edge of the cut.

Files

Files **(see illustration)** come in a wide variety of sizes and types for specific jobs, but all of them are used for the same basic function of removing small amounts of metal in a controlled fashion. Files are used by mechanics mainly for deburring, marking parts, removing rust, filing the heads off rivets, restoring threads and fabricating small parts.

File shapes commonly available include flat, half-round, round, square and triangular. Each shape comes in a range of sizes (lengths) and cuts ranging from rough to smooth. The file face is covered with rows of diagonal ridges which form the cutting teeth. They may be aligned in one direction

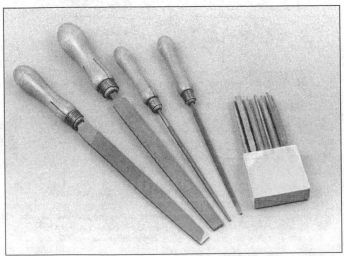

2.39 Get a good assortment of files - they're handy for deburring, marking parts, removing rust, filing the heads off rivets, restoring threads and fabricating small parts

2.40 Files are either single-cut (left) or double-cut (right) - generally speaking, use a single-cut file to produce a very smooth surface; use a double-cut file to remove large amounts of material quickly

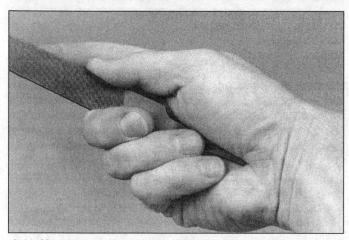

2.41 Never use a file without a handle - the tang is sharp and could puncture your hand

2.42 Adjustable handles that will work with many different size files are also available

only (single cut) or in two directions to form a diamond-shaped pattern (double-cut) **(see illustration)**. The spacing of the teeth determines the file coarseness, again, ranging from rough to smooth in five basic grades: Rough, coarse, bastard, second-cut and smooth.

You'll want to build up a set of files by purchasing tools of the required shape and cut as they're needed. A good starting point would be flat, half-round, round and triangular files (at least one each - bastard or second-cut types). In addition, you'll have to buy one or more file handles (files are usually sold without handles, which are purchased separately and pushed over the tapered tang of the file when in use) **(see illustration)**. You may need to buy more than one size handle to fit the various files in your tool box, but don't attempt to get by without them. A file tang is fairly sharp and you almost certainly will end up stabbing yourself in the palm of the hand if you use a file without a handle and it catches in the workpiece during use. Adjustable handles are also available for use with files of various sizes, eliminating the need for several handles **(see illustration)**.

Exceptions to the need for a handle are fine Swiss pattern files, which have a rounded handle instead of a tang. These small files are usually sold in sets with a number of different shapes. Originally intended for very fine work, they can be very handy for use in inaccessible areas. Swiss files are normally the best choice if piston ring ends require filing to obtain the correct end gap.

The correct procedure for using files is fairly easy to master. As with a hacksaw, the work should be clamped securely in a vise, if needed, to prevent it from moving around while being worked on. Hold the file by the handle, using your free hand at the file end to guide it and keep it flat in relation to the surface being filed. Use smooth cutting strokes and be careful not to rock the file as it passes over the surface. Also, don't slide it diagonally across the surface or the teeth will make grooves in the workpiece. Don't drag a file back across the workpiece at the end of the stroke - lift it slightly and pull it back to prevent damage to the teeth.

Files don't require maintenance in the usual sense, but

they should be kept clean and free of metal filings. Steel is a reasonably easy material to work with, but softer metals like aluminum tend to clog the file teeth very quickly, which will result in scratches in the workpiece. This can be avoided by rubbing the file face with chalk before using it. General cleaning is carried out with a file card or a fine wire brush. If kept clean, files will last a long time - when they do eventually dull, they must be replaced; there is no satisfactory way of sharpening a worn file.

Taps and dies

Taps

Tap and die sets **(see illustration)** are available in inch and metric sizes. Taps are used to cut internal threads and clean or restore damaged threads. A tap consists of a fluted shank with a drive square at one end. It's threaded along part of its length - the cutting edges are formed where the

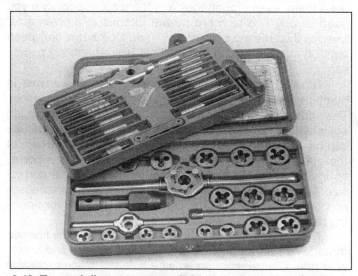

2.43 Tap and dies sets are available in inch and metric sizes - taps are used for cutting internal threads and cleaning and restoring damaged threads; dies are used for cutting, cleaning and restoring external threads

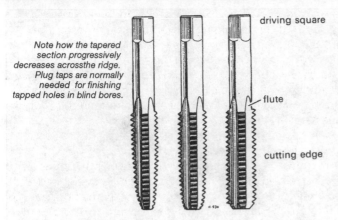

Note how the tapered section progressively decreases across the ridge. Plug taps are normally needed for finishing tapped holes in blind bores.

driving square

flute

cutting edge

2.44 Taper, plug and bottoming taps (left-to-right)

2.45 If you need to drill and tap a hole, the drill bit size to use for a given bolt (tap) size is marked on the tap

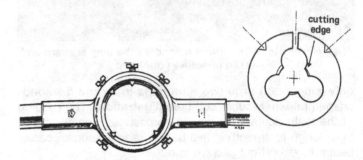

cutting edge

2.46 A die (right) is used for cutting external threads (this one is a split-type/adjustable die) and is held in a tool called a die stock (left)

flutes intersect the threads (see illustration). Taps are made from hardened steel so they will cut threads in materials softer than what they're made of.

Taps come in three different types: Taper, plug and bottoming. The only real difference is the length of the chamfer on the cutting end of the tap. Taper taps are chamfered for the first 6 or 8 threads, which makes them easy to start but prevents them from cutting threads close to the bottom of a hole. Plug taps are chamfered up about 3 to 5 threads, which makes them a good all around tap because they're relatively easy to start and will cut nearly to the bottom of a hole. Bottoming taps, as the name implies, have a very short chamfer (1-1/2 to 3 threads) and will cut as close to the bottom of a blind hole as practical. However, to do this, the threads should be started with a plug or taper tap.

Although cheap tap and die sets are available, the quality is usually very low and they can actually do more harm than good when used on threaded holes in aluminum parts. The alternative is to buy high-quality taps if and when you need them, even though they aren't cheap, especially if you need to buy two or more thread pitches in a given size. Despite this, it's the best option - you'll probably only need taps on rare occasions, so a full set isn't absolutely necessary.

Taps are normally used by hand (they can be used in machine tools, but primarily for manufacturing purposes). The square drive end of the tap is held in a tap wrench (an adjustable T-handle). For smaller sizes, a T-handled chuck can be used. The tapping process starts by drilling a hole of the correct diameter. For each tap size, there's a corresponding twist drill that will produce a hole of the correct size. Note how the tapered section progressively decreases across the ridge. Plug taps are normally needed for finishing tapped holes in blind bores.

This is important; too large a hole will leave the finished thread with the tops missing, producing a weak and unreliable grip. Conversely, too small a hole will place excessive loads on the hard and brittle shank of the tap, which can break it off in the hole. Removing a broken off tap from a hole is no fun! The correct tap drill size is normally marked on the tap itself or the container it comes in (see illustration).

Dies

Dies are used to cut, clean or restore external threads. Most dies are made from a hex-shaped or cylindrical piece of hardened steel with a threaded hole in the center. The threaded hole is overlapped by three or four cutouts, which equate to the flutes on taps and allow metal waste to escape during the threading process. Dies are held in a T-handled holder (called a die stock) (see illustration). Some dies are split at one point, allowing them to be adjusted slightly (opened and closed) for fine control of thread clearances.

Dies aren't needed as often as taps, for the simple reason it's normally easier to install a new bolt than to salvage one. However, it's often helpful to be able to extend the threads of a bolt or clean up damaged threads with a die. Hex-shaped dies are particularly useful for mechanic's work, since they can be turned with a wrench (see illustra-

2.47 Hex-shaped dies are especially handy for mechanic's work because they can be turned with a wrench

tion) and are usually less expensive than adjustable ones.

The procedure for cutting threads with a die is broadly similar to that described above for taps. When using an adjustable die, the initial cut is made with the die fully opened, the adjustment screw being used to reduce the diameter of successive cuts until the finished size is reached. As with taps, a cutting lubricant should be used, and the die must be backed off every few turns to clear swarf from the cutouts.

Pullers

You may need a general-purpose puller for some driveline steering and suspension work. Pullers can remove seized or corroded parts and push driveaxles from wheel hubs. Universal two- and three-legged pullers are widely available in numerous designs and sizes.

The typical puller consists of a central boss with two or three pivoting arms attached. The outer ends of the arms are hooked jaws which grab the part you want to pull off (see illustration). You can reverse the arms on most pullers to use the puller on internal openings when necessary. The central boss is threaded to accept a puller bolt, which does the work. You can also get hydraulic versions of these tools which are capable of more pressure, but they're expensive.

You can adapt pullers by purchasing, or fabricating, special jaws for specific jobs. If you decide to make your own jaws, keep in mind that the pulling force should be concentrated as close to the center of the component as possible to avoid damaging it.

Before you use a puller, assemble it and check it to make sure it doesn't snag on anything and the loads on the part to be removed are distributed evenly. If you're dealing with a part held on a shaft by a nut, loosen the nut but don't remove it. Leaving the nut on helps prevent distortion of the shaft end under pressure from the puller bolt and stops the part from flying off the shaft when it comes loose.

Tighten a puller gradually until the assembly is under moderate pressure, then try to jar the component loose by striking the puller bolt a few times with a hammer. If this doesn't work, tighten the bolt a little further and repeat the process. If this approach doesn't work, stop and reconsider. At some point you must make a decision whether to continue applying pressure in this manner. Sometimes, you can apply penetrating oil around the joint and leave it overnight, with the puller in place and tightened securely. By the next day, the taper has separated and the problem has resolved itself.

If nothing else works, try heating the area surrounding the troublesome part with a propane or gas welding torch (we don't, however, recommend messing around with welding equipment if you're not already experienced in its use). Apply the heat to the hub area of the component you wish to remove. Keep the flame moving to avoid uneven heating and the risk of distortion. Keep pressure applied with the puller and make sure that you're able to deal with the resulting hot component and the puller jaws if it does come free. Be very careful to keep the flame away from aluminum parts.

If all reasonable attempts to remove a part fail, don't be afraid to give up. It's cheaper to quit now than to damage components. Either buy or borrow the correct tool, or take the component (or vehicle) to a dealer service department or other repair shop to have the part removed for you.

Drawbolt extractors

The simple drawbolt extractor is easy to make up and invaluable in every workshop. There are no commercially available tools of this type; you simply make a tool to suit a particular application. You can use a drawbolt extractor to pull out stubborn piston pins and to remove bearings and bushings.

To make a drawbolt extractor, you'll need an assortment of threaded rods in various sizes (available at hardware stores), and nuts to fit them. You'll also need assorted washers, spacers and tubing. For things like piston pins, you'll usually need a longer piece of tube.

Some typical drawbolt uses are shown in the accompanying line drawings (see illustration). They also reveal

2.48 A two- or three-jaw puller will come in handy for many tasks in the shop and can also be used for pulling the hub from a front wheel drive axle

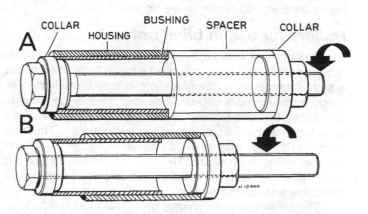

2.49 Typical drawbolt uses - in A, the nut is tightened to pull the collar and bushing into the large spacer; in B, the spacer is left out and the drawbolt is repositioned to install the new bushing

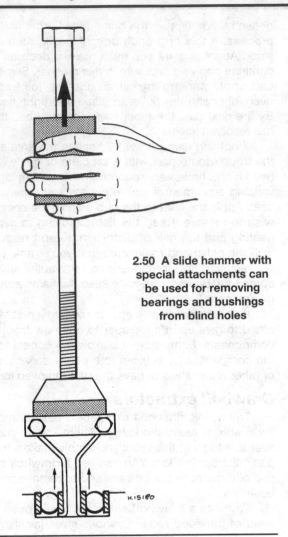

2.50 A slide hammer with special attachments can be used for removing bearings and bushings from blind holes

2.51 A bench vise is one of the most useful pieces of equipment you can have in the shop - bigger is usually better with vises, so get a vise with jaws that open at least four inches

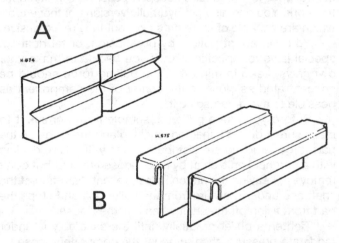

2.52 Sometimes, the parts you have to jig up in the vise are delicate, or made of soft materials - to avoid damaging them, get a pair of fiberglass or plastic "soft jaws" (A) or fabricate your own with 1/8'-inch thick aluminum sheet (B)

the order of assembly of the various pieces. The same arrangement, minus the tubular spacer section, can usually be used to install a new bushing or piston pin. Using the tool is quite simple. Just make sure you get the bush or pin square to the bore when you install it. Lubricate the part being pressed into place, where appropriate.

Pullers for use in blind bores

Bushings or bearings installed in "blind holes" often require special pullers. If you need a puller to do the job, get a slide-hammer with interchangeable tips. Slide hammers range from universal two or three-jaw puller arrangements to special bearing pullers. Bearing pullers are hardened steel tubes with a flange around the bottom edge. The tube is split at several places, which allows a wedge to expand the tool once it's in place. The tool fits inside the bearing inner race and is tightened so the flange or lip is locked under the edge of the race.

The slide-hammer consists of a steel shaft with a stop at its upper end. The shaft carries a sliding weight which slides along the shaft until it strikes the stop. This allows the tool holding the bearing to yank it out of the bore (see illustration).

Bench vise

The bench vise (see illustration) is an essential tool in a shop. Buy the best quality vise you can afford. A good vise is expensive, but the quality of its materials and workmanship are worth the extra money. Size is also important - bigger vises are usually more versatile. Make sure the jaws open at least four inches. Get a set of soft jaws to fit the vise as well - you'll need them to grip driveline or suspension parts that could be damaged by the hardened vise jaws (see illustration).

Really, the only power tool you absolutely need is an electric drill. But if you have an air compressor and electricity, there's a wide range of pneumatic and electric hand tools to make all sorts of jobs easier and faster.

2.53 Although it's not absolutely necessary, an air compressor can make many jobs easier and produce better results, especially when air powered tools are available to use with it

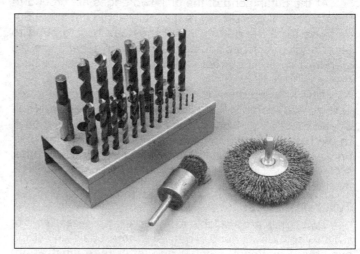

2.54 Another indispensable piece of equipment is the bench grinder (with a wire wheel mounted on one arbor) - make sure it's securely bolted down and never use it with the rests or eye shields removed

Power tools

Air compressor

An air compressor (see illustration) makes most jobs easier and faster. Drying off parts after cleaning them with solvent, blowing out passages in a block or head, running power tools - the list is endless. Once you buy a compressor, you'll wonder how you ever got along without it. Air tools really speed up tedious procedures like removing and installing wheel lug nuts.

Bench-mounted grinder

A bench grinder (see illustration) is also handy. With a wire wheel on one end and a grinding wheel on the other, it's great for cleaning up fasteners, sharpening tools and removing rust. Make sure the grinder is fastened securely to the bench or stand, always wear eye protection when operating it and never grind aluminum parts on the grinding wheel.

Electric drills

Countersinking bolt holes, enlarging oil passages, honing brake cylinder bores, removing rusted or broken off fasteners, enlarging holes and fabricating small parts - electric drills (see illustration) are great for this type of work. A 3/8-inch chuck (drill bit holder) will handle most jobs. Collect several different wire brushes to use in the drill and make sure you have a complete set of sharp metal drill bits (see illustration). Cordless drills are extremely versatile because

2.55 Electric drills can be cordless (above) or 115-volt, AC-powered (below

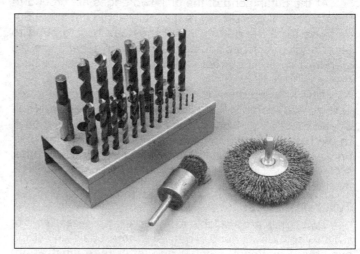

2.56 Get a set of good quality drill bits for drilling holes and wire brushes of various sizes for cleaning up metal parts - make sure the bits are designed for drilling in metal!

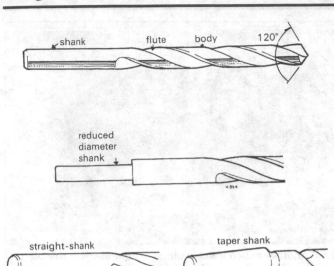

2.57 A typical drill bit (top), a reduced shank bit (center), and a tapered shank bit (bottom right)

2.58 Drill bits in the range most commonly used are available in fractional sizes (left) and number sizes (right) so almost any size hole can be drilled

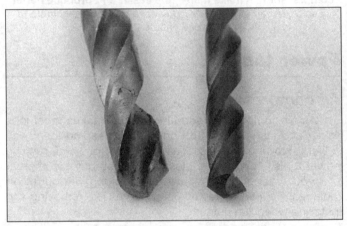

2.59 If a bit gets dull (left), discard it or resharpen it so it looks like the bit on the right

they don't force you to work near an outlet. They're also handy to have around for a variety of non-mechanical jobs.

Twist drills and drilling equipment

Drilling operations are done with twist drills, either in a hand drill or a drill press. Twist drills (or drill bits, as they're often called) consist of a round shank with spiral flutes formed into the upper two-thirds to clear the waste produced while drilling, keep the drill centered in the hole and finish the sides of the hole.

The lower portion of the shank is left plain and used to hold the drill in the chuck. In this Section, we will discuss only normal parallel shank drills (see illustration). There is another type of bit with the plain end formed into a special size taper designed to fit directly into a corresponding socket in a heavy-duty drill press. These drills are known as Morse Taper drills and are used primarily in machine shops.

At the cutting end of the drill, two edges are ground to form a conical point. They're generally angled at about 60-degrees from the drill axis, but they can be reground to other angles for specific applications. For general use the standard angle is correct - this is how the drills are supplied.

When buying drills, purchase a good-quality set (sizes 1/16 to 3/8-inch). Make sure the drills are marked "High Speed Steel" or "HSS". This indicates they're hard enough to withstand continual use in metal; many cheaper, unmarked drills are suitable only for use in wood or other soft materials. Buying a set ensures the right size bit will be available when it's needed.

Twist drill sizes

Twist drills are available in a vast array of sizes, most of which you'll never need. There are three basic drill sizing systems: Fractional, number and letter (see illustration) (we won't get involved with the fourth system, which is metric sizes).

Fractional sizes start at 1/64-inch and increase in increments of 1/64-inch. Number drills range in descending order from 80 (0.0135-inch), the smallest, to 1 (0.2280-inch), the largest. Letter sizes start with A (0.234-inch), the smallest, and go through Z (0.413-inch), the largest.

This bewildering range of sizes means it's possible to drill an accurate hole of almost any size within reason. In practice, you'll be limited by the size of chuck on your drill (normally 3/8 or 1/2-inch). In addition, very few stores stock the entire range of possible sizes, so you'll have to shop around for the nearest available size to the one you require.

Sharpening twist drills

Like any tool with a cutting edge, twist drills will eventually get dull (see illustration). How often they'll need sharpening depends to some extent on whether they're used correctly. A dull twist drill will soon make itself known. A good indication of the condition of the cutting edges is to watch the waste emerging from the hole being drilled. If the tip is in good condition, two even spirals of waste metal will be produced; if this fails to happen or the tip gets hot, it's safe to assume that sharpening is required.

With smaller size drills - under about 1/8-inch - it's easier and more economical to throw the worn drill away and

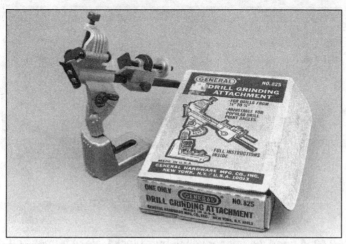

2.60 Inexpensive drill bit sharpening jigs designed to be used with a bench grinder are widely available

2.61 Before you drill a hole, use a centerpunch to make an indentation for the drill bit so it won't wander

buy another one. With larger (more expensive) sizes, sharpening is a better bet. When sharpening twist drills, the included angle of the cutting edge must be maintained at the original 120-degrees and the small chisel edge at the tip must be retained. With some practice, sharpening can be done freehand on a bench grinder, but it should be noted that it's very easy to make mistakes. For most home mechanics, a sharpening jig that mounts next to the grinding wheel should be used so the drill is clamped at the correct angle **(see illustration)**.

Drilling equipment

Tools to hold and turn drill bits range from simple, inexpensive hand-operated or electric drills to sophisticated and expensive drill presses. Ideally, all drilling should be done on a drill press with the workpiece clamped solidly in a vise. These machines are expensive and take up a lot of bench or floor space, so they're out of the question for many do-it-yourselfers.

The best tool for the home shop is an electric drill with a 3/8-inch chuck. Both cordless and AC drills (that run off household current) are available. If you're purchasing one for the first time, look for a well-known, reputable brand name and variable speed as minimum requirements. A 1/4-inch chuck, single-speed drill will work, but it's worth paying a little more for the larger, variable speed type.

All drills require a key to lock the bit in the chuck. When removing or installing a bit, make sure the cord is unplugged to avoid accidents. Initially, tighten the chuck by hand, checking to see if the bit is centered correctly. This is especially important when using small drill bits which can get caught between the jaws. Once the chuck is hand tight, use the key to tighten it securely - remember to remove the key afterwards!

Drilling and finishing holes

Preparation for drilling

If possible, make sure the part you intend to drill in is securely clamped in a vise. If it's impossible to get the work to a vise, make sure it's stable and secure. Twist drills often

dig in during drilling - this can be dangerous, particularly if the work suddenly starts spinning on the end of the drill. Make sure the work supported securely.

Start by locating the center of the hole you're drilling. Use a center punch to make an indentation for the drill bit so it won't wander. If you're drilling out a broken-off bolt, be sure to position the punch in the exact center of the bolt **(see illustration)**.

If you're drilling a large hole (above 1/4-inch), you may want to make a pilot hole first. As the name suggests, it will guide the larger drill bit and minimize drill bit wandering. Before actually drilling a hole, make sure the area immediately behind the bit is clear of anything you don't want drilled.

Drilling

When drilling steel, especially with smaller bits, no lubrication is needed. If a large bit is involved, oil can be used to ensure a clean cut and prevent overheating of the drill tip. When drilling aluminum, which tends to smear around the cutting edges and clog the drill bit flutes, use kerosene as a lubricant.

Wear safety goggles or a face shield and assume a comfortable, stable stance so you can control the pressure on the drill easily. Position the drill tip in the punch mark and make sure, if you're drilling by hand, the bit is perpendicular to the surface of the workpiece. Start drilling without applying much pressure until you're sure the hole is positioned correctly. If the hole starts off center, it can be very difficult to correct. You can try angling

the bit slightly so the hole center moves in the opposite direction, but this must be done before the flutes of the bit have entered the hole. It's at the starting point that a variable-speed drill is invaluable; the low speed allows fine adjustments to be made before it's too late. Continue drilling until the desired hole depth is reached or until the drill tip emerges at the other side of the workpiece.

Cutting speed and pressure are important - as a general rule, the larger the diameter of the drill bit, the slower the drilling speed should be. With a single-speed drill, there's little that can be done to control it, but two-speed or

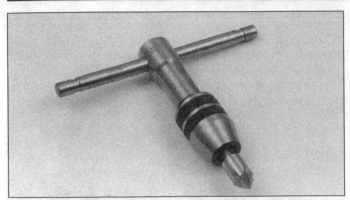

2.62 Use a large drill bit or a countersink mounted in a tap wrench to remove burrs from a hole after drilling or enlarging it

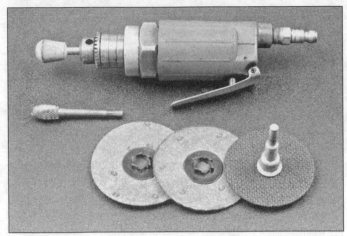

2.63 A good die grinder will deburr blocks, clean frames and do a lot of other little jobs what would be tedious if done manually

variable speed drills can be controlled. If the drilling speed is too high, the cutting edges of the bit will tend to overheat and dull. Pressure should be varied during drilling. Start with light pressure until the drill tip has located properly in the work. Gradually increase pressure so the bit cuts evenly. If the tip is sharp and the pressure correct, two distinct spirals of metal will emerge from the bit flutes. If the pressure is too light, the bit won't cut properly, while excessive pressure will overheat the tip.

Decrease pressure as the bit breaks through the workpiece. If this isn't done, the bit may jam in the hole; if you're using a hand-held drill, it could be jerked out of your hands, especially when using larger size bits.

Once a pilot hole has been made, install the larger bit in the chuck and enlarge the hole. The second bit will follow the pilot hole - there's no need to attempt to guide it (if you do, the bit may break off). It is important, however, to hold the drill at the correct angle.

After the hole has been drilled to the correct size, remove the burrs left around the edges of the hole. This can be done with a small round file, or by chamfering the opening with a larger bit or a countersink **(see illustration)**. Use a drill bit that's several sizes larger than the hole and simply twist it around each opening by hand until any rough edges are removed.

Enlarging and reshaping holes

The biggest practical size for bits used in a hand drill is about 1/2-inch. This is partly determined by the capacity of the chuck (although it's possible to buy larger drills with stepped shanks). The real limit is the difficulty of controlling large bits by hand; drills over 1/2-inch tend to be too much to handle in anything other than a drill press. If you have to make a larger hole, or if a shape other than round is involved, different techniques are required.

If a hole simply must be enlarged slightly, a round file is probably the best tool to use. If the hole must be very large, a hole saw will be needed, but they can only be used in sheet metal.

Large or irregular-shaped holes can also be made in sheet metal and other thin materials by drilling a series of small holes very close together. In this case the desired

hole size and shape must be marked with a scribe. The next step depends on the size bit to be used; the idea is to drill a series of almost touching holes just inside the outline of the large hole. Center punch each location, then drill the small holes. A cold chisel can then be used to knock out the waste material at the center of the hole, which can then be filed to size. This is a time consuming process, but it's the only practical approach for the home shop. Success is dependent on accuracy when marking the hole shape and using the center punch.

High-speed grinders

A good die grinder **(see illustration)** will clean frames, remove rust and clean parts ten times as fast as you can do any of these jobs by hand. Used in conjunction with a synthetic abrasive disc, it can be used to de-glaze brake discs.

Safety items that should be in every shop

Fire extinguishers

Buy at least one fire extinguisher **(see illustration)**

2.64 Buy at least one fire extinguisher before you open shop - make sure it's rated for flammable liquid fires and KNOW HOW TO USE IT!

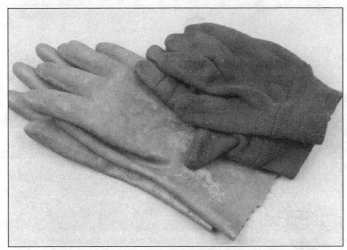

2.65 Get a pair of heavy work gloves for handling hot or sharp-edged objects and a pair of rubber gloves for washing parts with solvent or brake cleaner

before doing any maintenance or repair procedures. Make sure it's rated for flammable liquid fires. Familiarize yourself with its use as soon as you buy it - don't wait until you need it to figure out how to use it. And be sure to have it checked and recharged at regular intervals. Refer to the safety tips at the end of this Chapter for more information about the hazards of gasoline and other flammable liquids.

Gloves

If you're handling hot parts or metal parts with sharp edges, wear a pair of industrial work gloves to protect yourself from burns, cuts and splinters **(see illustration)**. Wear a pair of heavy duty rubber gloves (to protect your hands when you wash parts in solvent or brake cleaner.

Safety glasses or goggles

Never work on a bench or high-speed grinder without safety glasses **(see illustration)**. Don't take a chance on getting a metal sliver in your eye. It's also a good idea to wear safety glasses when you're washing parts.

2.67 Don't begin work on or near your brakes until you're wearing a filtering mask like this

2.66 One of the most important items you'll need in the shop is a face shield or safety goggles, especially when you're hitting metal parts with a hammer, washing parts or grinding something on the bench grinder

Filtering mask

The linings of most brake pads and shoes contain asbestos, which is extremely hazardous to your health. The dust deposited all over your brakes (and wheels) is made up of a high percentage of asbestos fibers. Be sure to always wear a filtering mask **(see illustration)** when working on or around your brakes - it'll greatly reduce the risk of inhaling asbestos fibers.

Maintenance-related tools

Grease cap pliers

When servicing the front or rear wheel bearings the dust cap must be removed. The grease cap is usually been exposed to water, heat, corrosion and many other types of elements that can make it difficult to separate the cap from the hub. The pliers work great for removing the cap; it simply clamps to the outer edge of the cap **(see illustration)** without damaging the cap or hub then pull the cap off.

Bearing packer

Using a specially designed bearing packer is the clean and easy way to work grease into bearings. Place the bear-

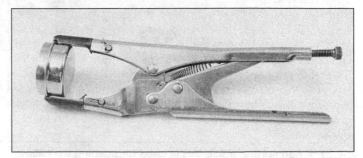

2.68 Lock the pliers to the outer edge of the cap and pull the cap off of the hub

2.69 Use a bearing packer to force clean grease between the rollers, cone and cage

ing between the two plastic cones and tighten the cones on the bearing. Wipe the grease fitting clean, use a grease gun and push the nozzle firmly into place on the bearing packer. Pump the grease into the bearing forcing the grease between the rollers, cone and cage **(see illustration)**. Be sure to use grease designed for wheel bearings.

Special brake-related tools

Torx Bits

Many owners become surprised, sometimes even frustrated, when they tackle a service or repair procedure and run into a fastener with a little six-pointed, star-shaped recess, in it's head. These are called Torx head bolts and are becoming increasingly popular with automobile manufacturers, especially for mounting brake calipers **(see illustration)**. It is very important to use the proper Torx bit on these fasteners. Never try to use an Allen wrench on a Torx fastener - you'll strip the head out and really create a problem!

Brake pad spreader

When removing calipers or replacing brake pads, the caliper piston(s) must be pushed back into the caliper to make room for the new pads. On some caliper designs (sliding calipers) this can be done before the caliper is

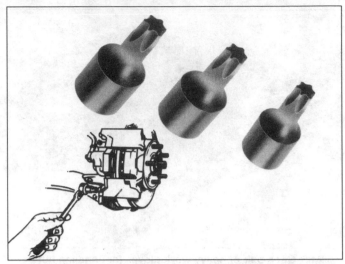

2.70 Many late model vehicles use Torx head fasteners to secure the driveaxles, driveshaft retaining bolts or brake calipers - always use the proper size bit when loosening or tightening these bolts

removed using an ordinary C-clamp. On other kinds of calipers (especially multiple-piston calipers) a brake pad spreader is very helpful in accomplishing this task **(see illustration)**. Just unbolt the caliper, slide it off the brake disc, insert this tool between the pads and turn the screw. The piston(s) will be forced to the bottom of the bore(s).

Universal disc brake caliper tool

This kit **(see illustration)** combines two kinds of caliper piston depressors. One serves the same purpose as the tool described above. The other is for use on rear disc brake calipers with threaded parking brake actuators in the

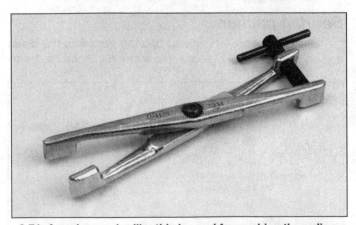

2.71 A pad spreader like this is used for pushing the caliper piston(s) back into the caliper bore(s) to make room for the new brake pads

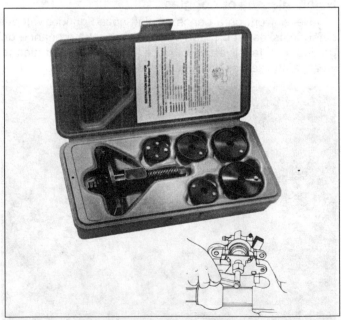

2.72 This universal brake caliper tool works on many different kinds of calipers and greatly facilitates brake pad replacement, especially on rear disc brakes

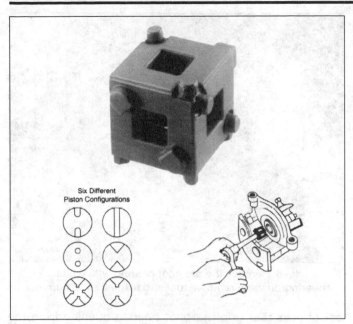

2.73 This six-in-one rear caliper piston retractor is an economical alternative to the universal disc brake caliper tool (shown in illustration 2.72) - it also simplifies the task of turning the piston on actuator-screw rear calipers

caliper pistons. On these calipers, the piston can't be pushed into their bores, they have to be rotated back in.

Rear disc brake piston tool

Here is another tool for turning the rear brake caliper pistons back into their bores **(see illustration)**. This one offers six different lug arrangements - one to fit almost any vehicle using this design rear caliper.

Hose pinchers

Hose pinchers, or clamps **(see illustration)**, are very useful tools, provided they are used correctly. Placed over a hose and tightened, they will prevent fluid from flowing through the hose. They're great for using on fluid feed

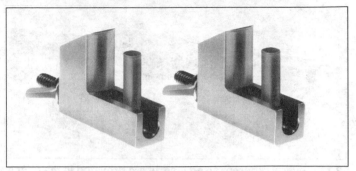

2.74 Hose pinchers are useful for clamping off the hoses to the remote brake fluid reservoir (on vehicles so equipped) when removing the master cylinder

hoses when removing the master cylinder on vehicles that employ remote reservoirs. It reduces the mess resulting from brake fluid running all over the place, and eliminates the need for completely draining the reservoir. **Warning:** *These tools should never be overtightened, and NEVER use them on high-pressure flexible brake lines, such as the ones that connect to the calipers or wheel cylinders.*

Special driveline-related tools

Clutch hydraulic fitting release tool

Many of the new vehicles built today have gone to a quick connect fitting used on many different systems such as fuel, air-conditioning and clutch hydraulics. Without these tools it is almost impossible to remove them without destroying the part you're trying to take off. The clutch release tool is used to unlock the quick-connect fitting and allow the line or slave cylinder to be easily disconnected and removed **(see illustration)**.

Clutch alignment tool

The clutch alignment tool is used to center the clutch

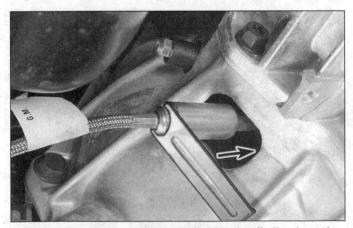

2.75 To disconnect some clutch fluid hydraulic line from the release cylinder, put a hydraulic clutch line release tool against the shoulder of the quick-connect fitting, push in on the spring-loaded fitting and pull the braided clutch line out

2.76 A clutch alignment tool set is used to properly align the clutch disc when installing the clutch assembly onto an engine

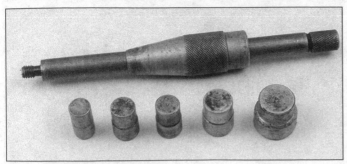

2.77 Several different sized inserts are provided in the clutch alignment tool set, which allow the alignment tool to be used on a variety of vehicles

2.78 Remove the old pilot bearing with a pilot bearing/bushing removal tool and a small slide-hammer

2.79 Remove the driveshaft or driveaxle and quickly install the plugs to prevent fluid from pouring out

disc in the pressure plate before the transmission is installed **(see illustrations)**. When used properly, the transmission/transaxle can be easily installed (the first time).

Pilot bearing/bushing removal tool

The removal tool is a small slide hammer with a reverse set of jaws that grip the pilot bearing/bushing from the inside **(see illustration)**. Once locked inside of the bearing/bushing, the force of the slide hammer pulls the bearing out of the crankshaft cavity or flywheel without damaging the crankshaft or the flywheel.

Transmission/transaxle plugs

Many jobs require the driveshaft or driveaxles to be removed. Once removed, fluid comes pouring out. To prevent this mess, simply install one of the plugs which come in a variety of sizes to fit different transmission/transaxles **(see illustration)**.

Four-wheel drive axle nut sockets

When servicing or removing the front hubs on some four wheel drive vehicles, the adjustment nut or lock nut requires a special tool to remove it. There are a variety of sockets available for most vehicles from your local auto part stores or four-wheel drive shops **(see illustrations)**.

2.80 Four wheel drive wheel bearing lock-nut tools are used to remove the front wheel bearing lock-nuts on most domestic four wheel drive vehicles)

2.81 Some late model four wheel drive trucks use ratcheting lock-nuts which require this ratcheting lock-nut tool

2.82 This four wheel drive spindle puller is necessary to remove the front spindle to access the spindle bearings, It fits all common vehicles (cut-away to show the stepped design)

2.83 Lock-rings are usually used to retain a component on a splined shaft; lock-ring pliers are used to spread the lock-ring

Four-wheel drive spindle puller

Once the front hubs are removed it is a great time to service the front spindles. The spindle remover tool has three step threads to fit most four-wheel drive spindles. Once the puller is screwed onto the spindle, connect a slide hammer to the puller and remove the spindle **(see illustration)**.

Snap-ring pliers

You will need at least two sets of snap-ring pliers. We stated earlier in this Chapter that pliers are basic hand tools, but these pliers are special. They serve one function, to remove snap-rings. Actually, one of the sets are lock-ring pliers **(see illustration)**. Large, stiff lock-rings are used to hold a component on a splined shaft, and these pliers with their outer gripping jaws, are used to expand the ring **(see illustration)**. You may also encounter either internal or external snap-rings and one set of reversible snap-ring pliers is all that's necessary for both **(see illustration)**.

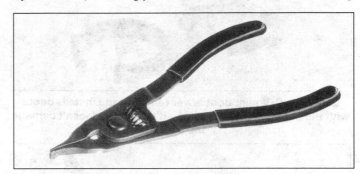

2.84 Lock-ring pliers

Hub and bearing puller set

With the advent of front-wheel drive automobiles a whole new tool category has been created. If you need to replace a damaged CV-joint boot (which is a very common repair), you may need a special tool to separate the hub from the driveaxle **(see illustration)**.

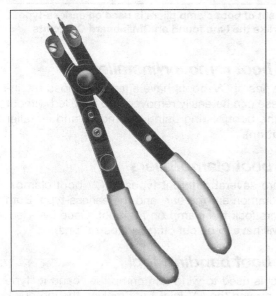

2.85 This set of snap-ring pliers is reversible and can be used on either internal or external snap-rings

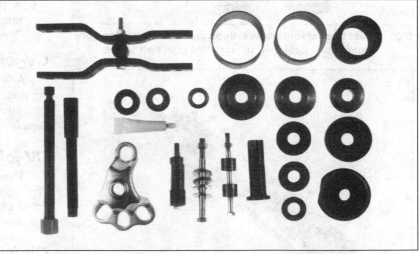

2.86 Several special tools are needed to service the driveaxle on a front wheel drive vehicle - this Lisle Corp. tool set is used to pull the hub and bearing assembly from the splined driveaxle and removes the hub from the bearing and the bearing from the steering knuckle on most domestic made vehicles - the tool is also used for reassembly of the components

2.87 This C-V joint boot driver removes and installs boots with the stamped retainer rings - it's fast and doesn't damage the boots

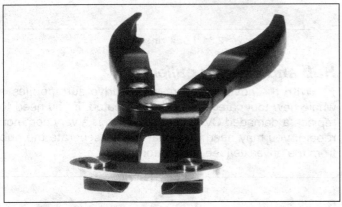

2.90 Offset boot clamp pliers crimps ear-type clamps and are a special set designed to work in a confined space

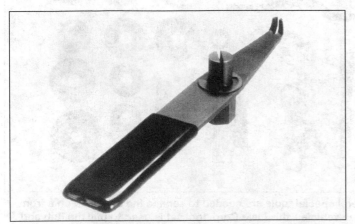

2.91 A CV-joint banding tool is needed for use on all "Band-it" type boot bands - it uses a socket or wrench to wind and crimp the new band on the boot

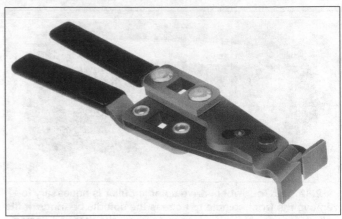

2.88 The manufacturers have used several different types of C-V joint boot clamps - this tool can be used on all ear-type clamps and is designed for use with a torque wrench which is required for stainless steel clamps

2.89 This set of boot clamp pliers is used on earless-type clamps like the type found on GM inboard C-V joints

CV-joint boot remover/installer

A few types of CV-boots have a large stamped retaining ring. These can be easily removed and installed without damaging the boot or ring using a remover and installer (see illustration).

CV-joint boot clamp/pliers

There are several different types of CV-boot clamps. The most common are the ear- and the earless-type. Both types of pliers lock the clamp on the boot. Once installed, the clamp will have to be cut off (see illustrations).

CV-joint boot banding tool

This tool is used to wind and crimp the "band-it "type boot clamp around the CV-joint boot making the job quick and easy (see illustration).

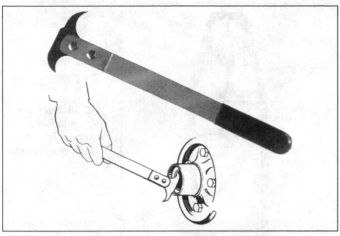

2.92 A seal removal tool like this will facilitate the removal of stubborn seals

2.93 Typical axleshaft bearing removal tool

Seal removal tool

This universal tool is great for removing many types of seals **(see illustration)**. Hook the edge of the tool into the seal and pry the seal out.

Axleshaft bearing removal tool

Rear axle bearing noise or failure is a common problem among many of today cars. Bearing replacement is made simple when using the axleshaft bearing tool. Screw the tool onto the end of a slide hammer, and insert it through the center of the bearing **(see illustrations)**. Rotate the end of the tool to catch each side of the bearing then pull the bearing out.

2.94 To remove the axle bearing, insert a bearing removal tool attached to a slide hammer through the center, pull the tool up against the back side and use the slide hammer to drive the bearing from the axle housing

Full-floating hub socket

When servicing or removing the brake drums on full-floating axles, the adjustment nut or lock nut requires a special tool to remove it. There are a variety of sockets available for most vehicles **(see illustration)**. The socket has a hub that slides into the axle housing allowing the socket tangs to fully engage into the locknut or adjustment nut. This prevents slipping while loosening or tightening the nut.

Bearing separator (splitter)

The bearing separator, or more commonly known as "bearing splitter", is a must-have tool. The splitter can be adjusted to fit many different size bearings. The two-piece design allows pressure to be equally distributed around the bearing. This helps to prevent damage to the cage or rollers **(see illustrations)**.

2.95 The socket has a hub that slides into the axle housing allowing the socket tangs to fully engage into the locknut or adjustment nut. This prevents slipping while your loosing or tightening the nut

2.96 Typical bearing splitter, its adjustable and available in different sizes

2.97 The splitter grips the bearing from bottom of the bearing preventing damage to the bearing

2.98 Here's a relatively inexpensive bottle-jack type press

Hydraulic press

Hydraulic presses range from simple, bottle-jack type **(see illustration)** to the sophisticated and expensive 10 ton hydraulic ram press. Ideally, all press work should be done on a press where the workpiece is easily moved and secure. Even the bottle-jack type can be expensive and take up a lot of bench or floor space, so many do-it-yourselfers think they're out of the question. Before you make that decision, check out how much your local machine shop charges for press work - it may only take the savings from a couple of jobs to pay for it.

Special suspension-related tools

Coil spring compressor

Because of the high spring pressures and the possibility of injury involved, this is one area that we recommend professional help if you do not have the experience to safely complete the job. The coil spring tool has a set of

jaws on a threaded rod **(see illustration)**. The top set of jaws are inserted through the center of the spring and locked on the upper half of the spring. A plate or second set of jaws are then inserted at the bottom of the spring. As the rod is tightened, the jaws are drawn together and the spring is compressed.

Universal shock nut and strut nut removers

Replacing worn shock absorbers is not a cut-and-dry situation either, especially if your vehicle is equipped with MacPherson strut suspension or shock absorber/coil spring assemblies. Special sockets and tool sets are available to make removing the shock absorber nut easier **(see illustrations)**. The problem here is, the shock absorber shaft wants to turn along with the nut, so a method of holding the shaft and turning the nut is needed, and that's exactly what the tool does for you.

Strut spring compressor

Now that you've got the strut assembly off the car another special tool is necessary to complete the job. A MacPherson strut spring compressor is needed to separate the spring from the shock absorber **(see illustration)**.

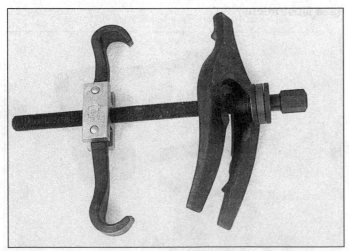

2.99 A typical aftermarket internal spring compressor tool: The hooked arms grip the upper coils of the spring, the plate is inserted below the lower coils, and when the nut on the threaded rod is turned, the spring is compressed

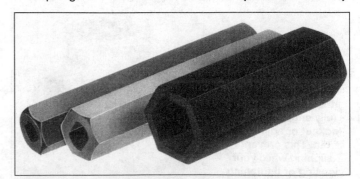

2.100 Removing shock absorbers is much easier when you use a special tool such as this universal shock nut set - use either one of the two different size stem sockets inside the double-ended shock-nut socket to remove the shock nut

2.101 This Lisle universal strut nut remover makes disassembly and assembly of a MacPherson strut easier - the double ended drivers remove the top nut on most all strut applications and work especially well on nuts that are recessed

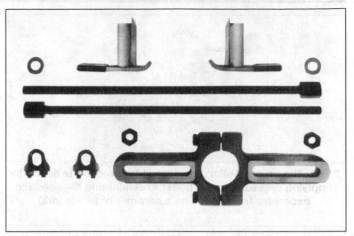

2.102 A MacPherson strut spring compressor is needed to disassemble and assemble the strut - this compressor is universal and safely handles all strut springs

2.103 The unique tool consists of a press and several adapters to fit various ball joints and U-joints

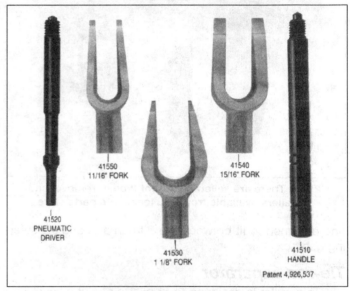

2.104 A great leverage tool for separating parts such as tie-rod ends steering linkages and knuckles

Because of the high spring pressures and the possibility of injury involved, we recommend professional help if you do not have the experience to safely complete the job.

Ball joint press tool

This tool is great for removing and installing ball joints and u-joints. The adapter fits over the ball joint and as pressure is applied, the ball joint is removed or installed without any damage to the control arm or drive shaft **(see illustration)**.

Picklefork set

Pickleforks are great for separating parts such as steering linkage joints, steering knuckles and bushings **(see illustration)**. Care should be taken when using these tools, as they can damage grease boots and gouge metal quickly.

Special steering tools

Pitman arm puller

These pullers are similar to a regular two-jaw puller. The only real difference is in the thickness of the arms **(see illustration)**. The Pitman arm puller jaws are heavy duty

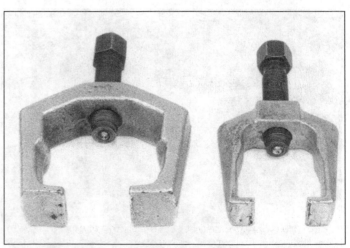

2.105 The Pitman arm pullers have heavy duty arms to take the pressure required to separate Pitman arms

2.106 The tie-rod puller is designed to remove the tie rod by applying pressure on the ballstud eliminating the need for excessive force (such as a hammer or pickle fork)

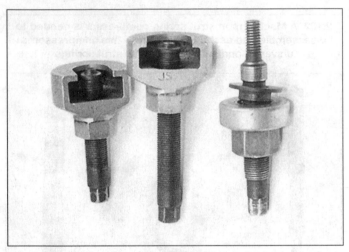

2.108 There are several different type of remover and installers available from you local auto parts store

and designed to fit between the Pitman arm and the steering gearbox.

Tie-rod separator

This puller is designed to fit over a steering knuckle or drag link, and apply pressure to the ballstud (see illustra-

2.109 You'll need a special puller to remove the power steering pump pulley on some vehicles

2.107 This small puller should be used to separate the tie-rod end from the steering knuckle. Note that - as a safety measure - the castellated nut has been loosened, but NOT removed; that way, if the ballstud pops out of the knuckle with a lot of force, there's no danger of the tie-rod end flying out violently enough to cause injury

tions). This prevents possible damage to the tie-rod, boot or knuckle and makes removal much easier, especially in a tight or confined location.

Power steering pulley remover/installer

Many of today's power steering pumps are mounted in such a way that the pulley must be removed to gain access to the pump mounting bolts and front seal. If you try to use a typical puller, you may find them impossible to get into such a small area or you'll damage the thin stamped, or plastic pulley and pump. The pulley tools mount on the center of the pulley and require very little effort to remove or install the pulley (see illustrations).

Steering wheel puller tool

The steering wheel is bolted to the steering shaft in the center of the steering column. To remove the steering

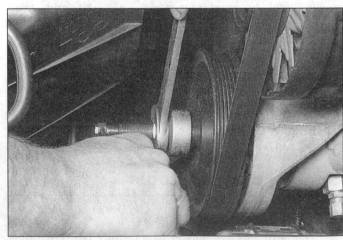

2.110 You'll also need a special pulley installer tool to press the pulley onto the pump shaft; NEVER hammer a pulley onto the shaft - you could damage the pump

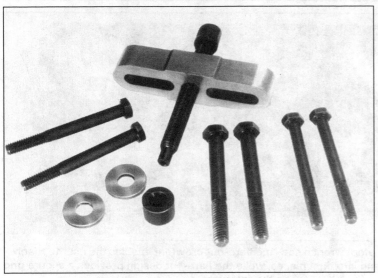

2.111 Steering wheel puller is universal and fits most makes and models

2.112 The puller must bolt to the steering wheel hub; pulling from anywhere else will damage the steering wheel

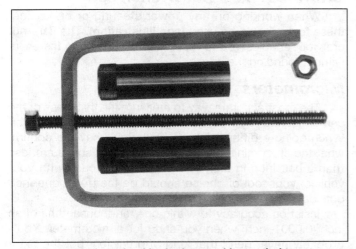

2.113 A steering wheel lock plate tool is needed to remove the lock-plate from a GM steering column

2.114 The lock plate tool depresses, and holds, the lock-plate so you can remove the retaining ring

wheel, loosen and remove the retaining nut or bolt. The second step is where the steering wheel removal tool comes in handy **(see illustrations)**. The steering wheel has a tapered shaft, and once the retaining nut is tightened, the steering wheel is pressed onto the steering shaft and will not come off without a puller. The puller is bolted to the steering wheel and the threaded rod in the center of the puller is tightened and the steering wheel is removed from the shaft.

Steering wheel lock plate removal tool

Let's say that you need to replace a turn-signal switch. It should be no problem, right? We have a steering wheel puller, just jerk the steering wheel off and dive in! Wrong! Under the steering wheel is a lock-plate on many vehicles, retained by a locking ring. A lock-plate depressing tool is needed to push to lock-plate down so you can get the lock-ring off **(see illustrations)**.

Front end alignment flex ratchet set

Sometimes trying to reach alignment bolts seem impossible or you need to be double jointed. This tool is double jointed so you don't have to be, making many tight spots easy to work with **(see illustration)**.

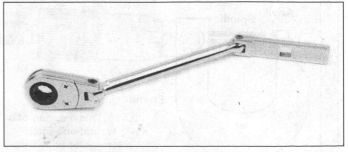

2.115 Here's a tool the pros use to make those impossible jobs seem so easy. This ratchet set should only be purchased if you're planning on doing front-end work

2.116 The steering wheel holder is one of those extra hand tools that allow you to do a better job

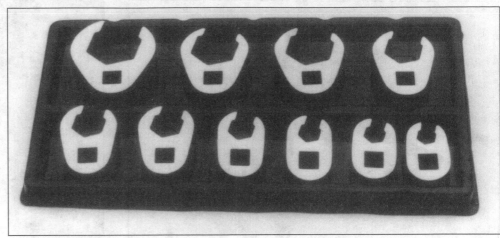

2.117 Flare-nut crowfoot wrench set - the flare-nut crowfoot enables the user to reach around obstructions and into tight places, while the flare-nut design provides a secure grip on the fastener

Steering wheel holder

The steering wheel holder grips the steering wheel, allowing for greater ease and accuracy when trying to set the toe adjustments or centering the steering wheel **(see illustration).**

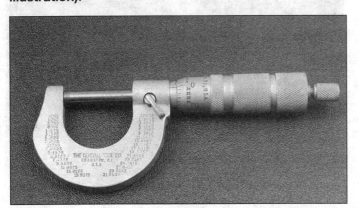

2.118 The one-inch micrometer is an essential precision measuring device for determining the adjustment shim thickness for ring gear backlash and pinion depth

Crowsfoot flare-nut wrench set

When working on any power steering or brake line, these tools should be used **(see illustration)**. The flare-nut crowfoot enables you to reach and securely grip fasteners without rounding the edges.

Micrometers

The most accurate way to measure the thickness of the carrier bearing or pinion bearing shims is with a micrometer. When doing a differential overhaul, you'll have to confirm what the shim minimum thickness is. Most shims are less than a half-inch in diameter. If this is the case with your vehicle, your tool of choice should be the trusty one-inch outside micrometer **(see illustration)**.

Insist on accuracy to within one ten-thousandths of an inch (0.0001-inch) when you shop for a micrometer. You'll probably never need that kind of precision, but the extra decimal place will help you decide which way to round off a close measurement.

High-quality micrometers have a range of one inch. If you plan to work on a wide variety of vehicles, you'll need a 1 to 2-inch micrometer.

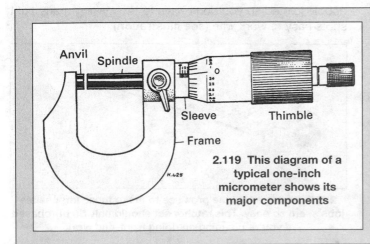

2.119 This diagram of a typical one-inch micrometer shows its major components

How to read a micrometer

The outside micrometer is without a doubt the most widely used precision measuring tool. It can be used to make a variety of highly accurate measurements without much possibility of error through misreading, a problem associated with other measuring instruments, such as vernier calipers.

Like any slide caliper, the outside micrometer uses the "double contact" of its spindle and anvil **(see illustration)** touching the object to be measured to determine that object's dimensions. Unlike a caliper, however, the micrometer also features a unique precision screw adjustment which can be read with a great deal more accuracy than calipers.

Why is this screw adjustment so accurate? Because years ago toolmakers discovered that a screw with 40 precision machined threads to the inch will advance one-fortieth (0.025) of an inch with each complete turn. The screw threads on the spindle revolve inside a fixed nut concealed by a sleeve.

On a one-inch micrometer, this sleeve is engraved longitudinally with exactly 40 lines to the inch, to correspond with the number of threads on the spindle. Every fourth line is made longer and is numbered one-tenth inch, two-tenths, etc. The other lines are often staggered to make them easier to read.

The thimble (the barrel which moves up and down the sleeve as it rotates) is divided into 25 divisions around the circumference of its beveled edge and is numbered from zero to 25. Close the micrometer spindle till it touches the anvil: You should see nothing but the zero line on the sleeve next to the beveled edge of the thimble. And the zero line of the thimble should be aligned with the horizontal (or axial) line on the sleeve. Remember: Each full revolution of the spindle from zero to zero advances or retracts the spindle one-fortieth or 0.025-inch. Therefore, if you rotate the thimble from zero on the beveled edge to the first graduation, you will move the spindle 1/25th of 1/40th, or 1/25th of 25/1000, which equals 1/1000th, or 0.001-inch.

Remember: Each numbered graduation on the sleeve represents 0.1-inch, each of the other sleeve graduations represents 0.025-inch and each graduation on the thimble represents 0.001-inch. Remember those three and you're halfway there.

For example: Suppose the 4 line is visible on the sleeve. This represents 0.400-inch. Then suppose there are an additional three lines (the short ones without numbers) showing. These marks are worth 0.025-inch each, or 0.075-inch. Finally, there are also two marks on the beveled edge of the thimble beyond the zero mark, each good for 0.001-inch, or a total of 0.002-inch. Add it all up and you get 0.400 plus 0.075 plus 0.002, which equals 0.477-inch.

Some beginners use a "dollars, quarters and cents" analogy to simplify reading a micrometer. Add up the bucks and change, then put a decimal point instead of a dollar sign in front of the sum!

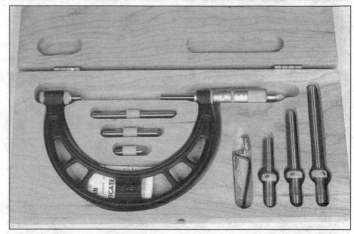

2.121 Avoid micrometer "sets" with interchangeable anvils - they're awkward to use when measuring little parts and changing the anvils is a hassle

Digital micrometers **(see illustration)** are easier to read than conventional micrometers, are just as accurate and are finally starting to become affordable. If you're uncomfortable reading a conventional micrometer (see sidebar), then get a digital.

Unless you're not going to use them very often, stay away from micrometers with interchangeable anvils **(see illustration)**. In theory, one of these beauties can do the work of five or six single-range micrometers. The trouble is, they're awkward to use when measuring little parts, and changing the anvils is a hassle.

Dial indicators

The dial indicator **(see illustrations)** is another measuring mainstay. You'll use this tool for measuring ring gear backlash and carrier runout (warpage). Make sure the dial indicator you buy is graduated in 0.001-inch increments.

Buy a dial indicator set that includes a flexible fixture

2.120 Digital micrometers are easier to read than conventional micrometers, are just as accurate and are finally starting to become affordable

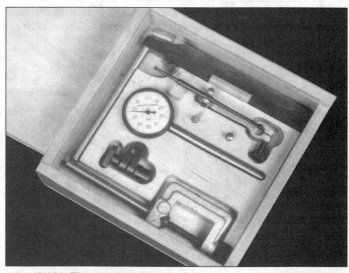

2.122 The dial indicator is the only way to measure carrier runout

2.123 Measure the backlash and ring gear runout with a dial indicator

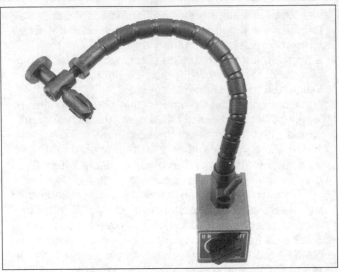

2.124 Get an adjustable, flexible fixture like this one, and a magnetic base, to ensure maximum versatility from your dial indicator

and a magnetic stand **(see illustration).** If the model you buy doesn't have a magnetic base, buy one separately. Make sure the magnet is plenty strong. If a weak magnet comes loose and the dial indicator takes a tumble on a concrete floor, you can kiss it good-bye. Make sure the arm that attach the dial indicator to the flexible fixture is sturdy and the locking clamps are easy to operate.

Calipers

Vernier calipers **(see illustration on opposite page)** aren't quite as accurate as a micrometer, but they're handy for quick measurements and they're relatively inexpensive. Most calipers have inside and outside jaws, so you can measure the inside diameter of a hole, or the outside diameter of a part.

Better-quality calipers have a dust shield over the geared rack that turns the dial to prevent small metal particles from jamming the mechanism. Make sure there's no play in the moveable jaw. To check, put a thin piece of metal between the jaws and measure its thickness with the

metal close to the rack, then out near the tips of the jaws. Compare your two measurements. If they vary by more than 0.001-inch, look at another caliper - the jaw mechanism is deflecting.

If your eyes are going bad, or already are bad, vernier calipers can be difficult to read. Dial calipers **(see illustration)** are a better choice. Dial calipers combine the measuring capabilities of micrometers with the convenience of dial indicators. Because they're much easier to read quickly than vernier calipers, they're ideal for taking quick measurements when absolute accuracy isn't necessary. Like conventional vernier calipers, they have both inside and outside jaws which allow you to quickly determine the diameter of a hole or a part. Get a six-inch dial caliper, graduated in 0.001-inch increments.

The latest calipers **(see illustration)** have a digital LCD display that indicates both inch and metric dimensions. If you can afford one of these, it's the hot setup.

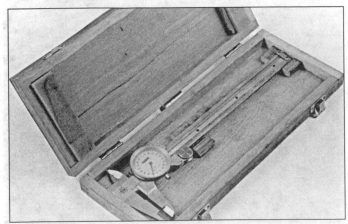

2.126 Dial calipers are a lot easier to read than conventional vernier calipers, particularly if your eyesight isn't as good as it used to be!

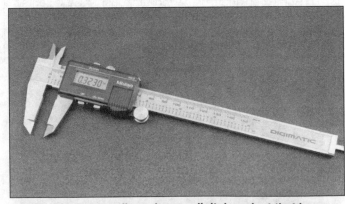

2.127 The latest calipers have a digital readout that is even easier to read than a dial caliper - another advantage of digital calipers is that they have a small microchip that allows them to convert instantaneously from inch to metric dimensions

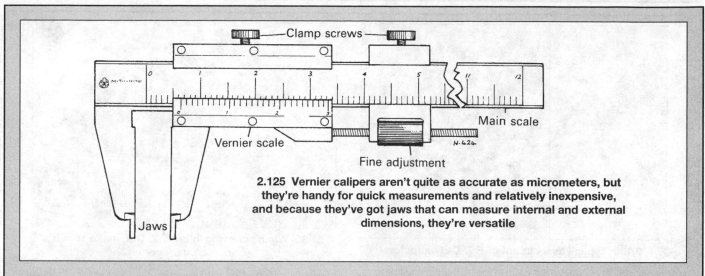

2.125 Vernier calipers aren't quite as accurate as micrometers, but they're handy for quick measurements and relatively inexpensive, and because they've got jaws that can measure internal and external dimensions, they're versatile

How to read a vernier caliper

On the lower half of the main beam, each inch is divided into ten numbered increments, or tenths (0.100-inch, 0.200-inch, etc.). Each tenth is divided into four increments of 0.025-inch each. The vernier scale has 25 increments, each representing a thousandth (0.001) of an inch.

First read the number of inches, then read the number of tenths. Add to this 0.025-inch for each additional graduation. Using the English vernier scale, determine which graduation of the vernier lines up exactly with a graduation on the main beam. This vernier graduation is the number of thousandths which are to be added to the previous readings.

For example, let's say:

1) *The number of inches is zero, or 0.000-inch;*
2) *The number of tenths is 4, or 0.400-inch;*
3) *The number of 0.025's is 2, or 0.050-inch; and*
4) *The vernier graduation which lines up with a graduation on the main beam is 15, or 0.015-inch.*
5) *Add them up:*
   0.000
   0.400
   0.050
   0.015
6) *And you get:*
   0.46-inch

That's all there is to it!

How to remove broken fasteners

Sooner or later, you're going to break off a bolt inside its threaded hole. There are several ways to remove it. Before you buy an expensive extractor set, try some of the following cheaper methods first.

First, regardless of which of the following methods you use, be sure to use penetrating oil. Penetrating oil is a special light oil with excellent penetrating power for freeing dirty and rusty fasteners. But it also works well on tightly torqued broken fasteners.

If enough of the fastener protrudes from its hole and if it isn't torqued down too tightly - you can often remove it with vise-grips or a small pipe wrench. If that doesn't work, or if the fastener doesn't provide sufficient purchase for pliers or a wrench, try filing it down to take a wrench, or cut a slot in it to accept a screwdriver **(see illustration).**

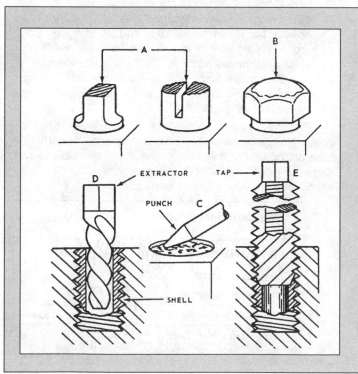

2.128 There are several ways to remove a broken fastener

A	File it flat or slot it	D	Use a screw extractor (like an E-Z-Out)
B	Weld on a nut	E	Use a tap to remove the shell
C	Use a punch to unscrew it		

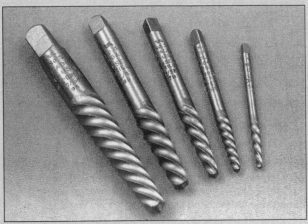

2.129 Typical assortment of E-Z-Out extractors

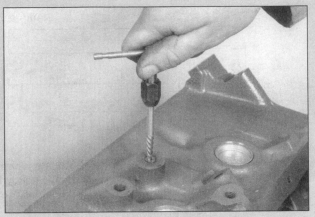

2.130 When screwing in the E-Z-Out, make sure it's centered properly

If you still can't get it off - and you know how to weld - try welding a flat piece of steel, or a nut, to the top of the broken fastener. If the fastener is broken off flush with - or below - the top of its hole, try tapping it out with a small, sharp punch. If that doesn't work, try drilling out the broken fastener with a bit only slightly smaller than the inside diameter of the hole. For example, if the hole is 1/2-inch in diameter, use a 15/32-inch drill bit. This leaves a shell which you can pick out with a sharp chisel.

If THAT doesn't work, you'll have to resort to some form of screw extractor, such as E-Z-Out (see illustration). Screw extractors are sold in sets which can remove anything from 1/4-inch to 1-inch bolts or studs. Most extractors are fluted and tapered high-grade steel. To use a screw extractor, drill a hole slightly smaller than the OD of the extractor you're going to use (Extractor sets include the manufacturer's recommendations for what size drill bit to use with each extractor size). Then screw in the extractor (see illustration) and back it - and the broken fastener - out. Extractors are reverse-threaded, so they won't unscrew when you back them out.

A word to the wise: Even though an E-Z-Out will usually save your bacon, it can cause even more grief if you're careless or sloppy. Drilling the hole for the extractor off-center, or using too small, or too big, a bit for the size of the fastener you're removing will only make things worse. So be careful!

How to repair broken threads

Warning: *Never attempt to repair the threads of the caliper mounting bolt holes, torque plate mounting bolt holes, stripped-out holes in two-piece calipers, wheel cylinder mounting bolt holes or any other critical component. Instead, replace the part with a new one.*

Sometimes, the internal threads of a nut or bolt hole can become stripped, usually from overtightening. Stripping threads is an all-too common occurrence, especially when working with aluminum parts, because aluminum is so soft that it easily strips out. Overtightened spark plugs are another common cause of stripped threads.

Usually, external or internal threads are only partially stripped. After they've been cleaned up with a tap or die, they'll still work. Sometimes, however, threads are badly damaged. When this happens, you've got three choices:

1) *Drill and tap the hole to the next suitable oversize and install a larger diameter bolt, screw or stud.*
2) *Drill and tap the hole to accept a threaded plug, then drill and tap the plug to the original screw size. You can also buy a plug already threaded to the original size. Then you simply drill a hole to the specified size, then run the threaded plug into the hole with a bolt and jam nut. Once the plug is fully seated, remove the jam nut and bolt.*
3) *The third method uses a patented thread repair kit like Heli-Coil or Slimsert. These easy-to-use kits are designed to repair damaged threads in spark plug holes, straight-through holes and blind holes. Both are available as kits which can handle a variety of sizes and thread patterns. Drill the hole, then tap it with the special included tap. Install the Heli-Coil (see illustration) and the hole is back to its original diameter and thread pitch.*

Regardless of which method you use, be sure to proceed calmly and carefully. A little impatience or carelessness during one of these relatively simple procedures can ruin your whole day's work and cost you a bundle if you wreck an expensive gearbox or housing.

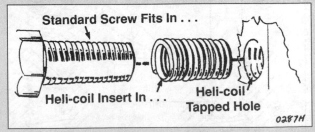

Standard Screw Fits In . . .

Heli-coil Insert In . . .

Heli-coil Tapped Hole

0287H

2.131 To install a Heli-Coil, drill out the hole, tap it with the special included tap and screw in the Heli-Coil

3 Troubleshooting

Contents

Symptom	Section

Clutch

Clutch fails to release 1
Clutch pedal travels to floor - no pressure
   or very little resistance 2
Clutch pedal stays on floor 3
Clutch slips (engine speed increases with
   no increase in vehicle speed 4
Grabbing (chattering) as clutch is engaged 5
High pedal effort 6
Rattling or clicking noise from clutch area 7
Squeal or rumble with clutch fully disengaged
   (pedal depressed) 8
Squeal or rumble with clutch fully engaged
   (pedal released) 9

Manual transmission/transaxle

Clicking noise in turns 10
Clunk on acceleration or deceleration 11
Hard to shift or unable to shift 12
Knocking noise at low speeds 13
Leaks lubricant 14
Noise most pronounced when turning 15
Noisy in all gears 16
Noisy in neutral with engine running 17
Noisy in one particular gear 18
Slips out of gear 19

Automatic transmission/transaxle

Engine will start in gears other than Park or Neutral 20
Fluid leakage 21
General shift mechanism problems 22
Transmission/transaxle fluid brown or has
   burned smell 23
Transmission/transaxle slips, shifts roughly, is noisy
   or has no drive in forward or reverse gears 24
Transmission/transaxle will not downshift with
   accelerator pedal pressed to the floor 25

Symptom	Section

Transfer case

Lubricant leaks from the vent or output shaft seals 26
Noisy or jumps out of four-wheel drive Low range 27
Transfer case is difficult to shift into the desired range 28
Transfer case noisy in all gears 29

Front driveaxles

Clicking noise in turns 30
Knock or clunk when the transmission/transaxle
   is under initial load 31
Oil leak at inner end of driveaxle 32
Shudder or vibration during acceleration 33
Vibration at highway speeds 34

Driveshaft

Knock or clunk when the transmission is under initial
   load (just after transmission is put into gear) 35
Metallic grinding sound consistent with vehicle speed 36
Oil leak at front of driveshaft 37
Vibration 38

Rear axle

Chattering noise on turns 39
Clunking noise under drive and coast conditions 40
Clunking or groaning noise as vehicle comes to a stop 41
Clunking or knocking noise under rough
   road conditions 42
Knocking or clicking noise as rear wheels rotate 43
Low pitch noise at low speeds 44
Noise changes under different road conditions 45
Noise changes under drive, float or coast conditions 46
Noise is most evident on turns 47
Noise is the same under all conditions 48
Noise lowers as vehicle speed lowers 49
Oil leakage 50
Vibration 51

Haynes suspension, steering and driveline manual

Symptom	Section
Suspension and steering systems	
Abnormal or excessive tire wear	52
Abnormal noise at the front end	53
Cupped tires	54
Erratic steering when braking	55
Excessive pitching and/or rolling around corners or during braking	56
Excessive play or looseness in steering system	57
Excessive tire wear on inside edge	58

Symptom	Section
Excessive tire wear on outside edge	59
Hard steering	60
Poor returnability of steering to center	61
Rattling or clicking noise in steering gear	62
Shimmy, shake or vibration	63
Suspension bottoms	64
Tire tread worn in one place	65
Vehicle pulls to one side	66
Wander or poor steering stability	67
Wheel makes a thumping noise	68

This Chapter provides an easy reference guide to the common problems which may occur in the driveline, suspension and steering systems used on the most popular foreign and domestic automobiles and light duty trucks on the road today. The symptoms, their possible causes and the corrective action necessary to restore proper operation are grouped under headings denoting various components or systems, such as Clutch, Driveaxle, Suspension and steering system, etc.

Remember that successful troubleshooting is not a mysterious "black art" practiced only by professional mechanics. It is simply the result of the right knowledge combined with an intelligent, systematic approach to the problem. Always work by a process of elimination, starting with the simplest solution and working through to the most complex - and never overlook the obvious.

Finally, always establish a clear idea of why a problem has occurred and take steps to ensure that it doesn't happen again. Remember, failure of a small component can often be indicative of potential failure or incorrect functioning of a more important component or system.

Clutch

1 Clutch fails to release

Probable cause	Corrective action
• Faulty clutch disc or pressure plate.	Replace clutch disc and pressure plate.
• Faulty release lever or release bearing.	Repair or replace release lever. Replace release bearing.
• Faulty hydraulic release system component or air in the system.	Replace the clutch master cylinder or release cylinder. Repair or replace the hydraulic lines. Bleed air from system.

2 Clutch pedal travels to floor - no pressure or very little resistance

Probable cause	Corrective action
• Hydraulic release system leaking or air in the system.	Replace or repair leaking hydraulic system component. Bleed air from system.
• Broken release bearing or fork.	Replace release bearing or fork.

3 Clutch pedal stays on floor

Probable cause	Corrective action
• Broken release bearing or fork.	Replace release bearing and/or release fork.
• Hydraulic release system leaking or air in the system.	Repair or replace leaking component. Bleed air from system.
• Clutch pedal or linkage binding.	Inspect clutch pedal and linkage, repair as necessary. Make sure the proper pedal stop (bumper) is installed.

4 Clutch slips (engine speed increases with no increase in vehicle speed)

Probable cause	Corrective action
• Clutch disc worn.	Replace clutch disc and pressure plate.
• Pressure plate defective (weak diaphragm springs).	Replace clutch disc and pressure plate.
• Clutch plate is oil soaked by leaking rear main seal or transmission input shaft seal.	Repair leaking seal. Clean flywheel with non-petroleum based cleaner and replace clutch disc and pressure plate.
• Pressure plate not seated.	Remove clutch components and determine reason pressure plate not seating properly.
• Clutch disc overheated.	Replace clutch disc and pressure plate. Resurface flywheel.
• Piston stuck in bore of clutch release cylinder, preventing clutch from fully engaging.	Repair or replace clutch release cylinder.
• Clutch out-of-adjustment.	Adjust clutch pedal freeplay.

5 Grabbing (chattering) as clutch is engaged

Probable cause	Corrective action
• Oil on clutch disc lining, burned or glazed facings.	Replace clutch disc and pressure plate. Resurface flywheel.
• Worn or loose engine or transmission/transaxle mounts.	Tighten loose fasteners. Replace defective engine or transmission/transaxle mounts.
• Worn splines on clutch disc hub.	Replace clutch disc and pressure plate.
• Warped pressure plate or flywheel.	Replace clutch disc and pressure plate. Resurface or replace flywheel.

6 High pedal effort

Probable cause	Corrective action
• Piston binding in bore of release cylinder.	Overhaul or replace release cylinder.
• Pressure plate faulty.	Replace clutch disc and pressure plate.
• Incorrect master or release cylinder installed.	Install correct master or release cylinder.
• Binding clutch pedal or linkage.	Inspect clutch pedal and linkage and repair as necessary. Replace clutch pedal bushings.
• Release bearing binding on transmission/transaxle bearing retainer.	Inspect release bearing and bearing retainer. Remove any burrs, nicks or corrosion. Clean and lubricate bearing retainer. Replace release bearing and/or bearing retainer, if necessary.

7 Rattling or clicking noise from clutch area

Probable cause	Corrective action
• Release fork loose.	Secure release fork to bellhousing. Replace fork, if necessary.
• Faulty clutch release bearing.	Replace clutch release bearing.
• Low engine idle speed.	Adjust engine idle speed to specifications.

8 Squeal or rumble with clutch fully disengaged (pedal depressed)

Probable cause	Corrective action
• Worn or defective clutch release bearing.	Replace release bearing.
• Worn or defective pressure plate springs or diaphragm fingers.	Replace clutch disc, pressure plate and release bearing.

9 Squeal or rumble with clutch fully engaged (pedal released)

Probable cause	Corrective action
• Release bearing binding on transmission/transaxle bearing retainer.	Inspect release bearing and bearing retainer. Remove any burrs, nicks or corrosion. Clean and lubricate bearing retainer. Replace release bearing and/or bearing retainer, if necessary.

Manual transmission/transaxle

Note: *Although, in most cases, the corrective action necessary to remedy the symptoms described below is beyond the scope of this manual, the information should be helpful in isolating the cause of the condition so that the owner can communicate clearly with a professional mechanic.*

10 Clicking noise in turns

Probable cause	Corrective action
• Worn or damaged outboard CV joint.	Overhaul or replace CV joints.

11 Clunk on acceleration or deceleration

Probable cause	Corrective action
• Loose or defective engine or transmission/transaxle mounts.	Tighten fasteners and/or replace mounts.
• Worn differential pinion shaft in case (transaxle only).	Requires transaxle disassembly.
• Worn side gear shaft counterbore in differential case (transaxle only).	Requires transaxle disassembly.
• Worn or damaged driveaxle inboard CV joints.	Overhaul or replace CV joints.

12 Hard to shift or unable to shift

Probable cause	Corrective action
• Clutch not releasing completely.	See **Clutch** symptoms.
• Shift linkage loose or worn.	Secure shift linkage, replace components as necessary.
• Faulty shift lever assembly, rods or linkage.	Repair transmission/transaxle as necessary.
• Faulty transmission/transaxle.	Repair transmission/transaxle as necessary.

13 Knocking noise at low speeds

Probable cause	Corrective action
• Worn driveaxle constant velocity (CV) joints.	Overhaul or replace CV joints.
• Worn side gear shaft counterbore in differential case (transaxle only).	Requires transaxle disassembly.

14 Leaks lubricant

Probable cause	Corrective action
• Excessive amount of lubricant.	Drain lubricant and refill to the proper level.
• Loose or broken input gear shaft bearing retainer.	Remove transmission/transaxle and secure bearing retainer. Replace bearing retainer, if damaged.
• Leaking seal, gasket or O-ring.	Most transmission/transaxle seals can be replaced without disassembling the transmission/transaxle. The exceptions would be seals or gaskets located between case components.

15 Noise most pronounced when turning

Probable cause	Corrective action
• Differential gear noise (transaxle only).	Requires transaxle disassembly.

16 Noisy in all gears

Probable cause	Corrective action
• Insufficient or incorrect lubricant.	Drain and refill with correct lubricant. If the noise is still present, the bearings are probably damaged.
• Damaged or worn bearings.	Remove transmission/transaxle for repair.
• Worn or damaged input gear shaft and/or output gear shaft.	Remove transmission/transaxle for repair.

17 Noisy in neutral with engine running

Probable cause	Corrective action
• Worn or damaged clutch release bearing.	Replace clutch release bearing.
• Worn or damaged input gear bearing.	Remove transmission/transaxle for repair.
• Worn or damaged main drive gear bearing.	Remove transmission/transaxle for repair.
• Worn or damaged counter shaft bearings or excessive countershaft endplay.	Remove transmission/transaxle for repair.

18 Noisy in one particular gear

Probable cause	Corrective action
• Damaged or worn constant mesh gears.	Remove transmission/transaxle for repair.
• Damaged or worn synchronizers.	Remove transmission/transaxle for repair.
• Bent reverse fork.	Remove transmission/transaxle for repair.
• Damaged fourth speed gear or output gear.	Remove transmission/transaxle for repair.
• Worn or damaged reverse idler gear or idler bushing.	Remove transmission/transaxle for repair.

19 Slips out of gear

Probable cause	Corrective action
• Worn or improperly adjusted linkage.	Adjust shift linkage. Replace components as necessary.
• Shift linkage does not work freely, binds.	Lubricate shift linkage. Replace components as necessary.
• Worn shift fork.	Replace shift fork (requires at least partial transmission/transaxle disassembly).
• Input gear bearing retainer broken or loose.	Remove transmission/transaxle and secure bearing retainer. Replace bearing retainer, if damaged.
• Transmission/transaxle loose on engine.	Tighten fasteners to proper torque specifications.
• Misalignment of the transmission/transaxle and the engine (foreign material between clutch cover and engine housing).	Remove transmission/transaxle and clean mounting surfaces of engine and transmission/transaxle or clutch cover.

Automatic transmission/transaxle

Note: *Although, in most cases, the corrective action necessary to remedy the symptoms described below is beyond the scope of this manual, the information should be helpful in isolating the cause of the condition so that the owner can communicate clearly with a professional mechanic.*

20 Engine will start in gears other than Park or Neutral

Probable cause	Corrective action
• Neutral start switch out of adjustment or malfunctioning.	Adjust or replace neutral start switch.
• Shift linkage loose or misadjusted.	Adjust shift linkage.

21 Fluid leakage

Automatic transmission/transaxle fluid is a deep red color. Fluid leaks should not be confused with engine oil, which can easily be blown onto the transmission/transaxle by air flow.

To pinpoint a leak, first remove all built-up dirt and grime from the transmission/transaxle housing with degreasing agents and/or steam cleaning. Then drive the vehicle at low speeds so air flow will not blow the leak far from its source. Raise the vehicle and determine where the leak is coming from. Common areas of leakage are:

a) Pan
b) Dipstick tube
c) Oil cooler lines
d) Speedometer/speed sensor
e) Output shaft seal (rear-wheel drive)
f) Driveaxle oil seals (front-wheel drive)

22 General shift mechanism problems

Probable cause	Corrective action

• Shift linkage loose or misadjusted. Common problems which may be attributed to poorly adjusted linkage are:
  a) *Engine starting in gears other than Park or Neutral.*
  b) *Indicator on shifter pointing to a gear other than the one actually being used.*
  c) *Vehicle moves when in Park.*
	Check and adjust the shift linkage.

23 Transmission/transaxle fluid brown or has a burned smell

Probable cause

- Transmission/transaxle fluid overheated.

Corrective action

Change fluid and filter. Flush oil cooler and replace fluid in torque converter, if possible. If internal damage to the clutches or band have occurred, remove the transmission/transaxle for repair.

24 Transmission/transaxle slips, shifts roughly, is noisy or has no drive in forward or reverse gears

There are many probable causes for the above problems. Before taking the vehicle to a transmission repair shop perform the following:

a) *Check the level and condition of the fluid. Correct the fluid level as necessary or change the fluid and filter if needed.*

b) *Adjust the Throttle Valve cable (if equipped).*
c) *Check the operation of the vacuum modulator (if equipped). Check the supply source for proper vacuum.*

If the problem persists, have a transmission specialist diagnose the cause.

25 Transmission/transaxle will not downshift with accelerator pedal pressed to the floor

Probable cause

- Throttle valve cable or vacuum modulator defective or misadjusted.

- Problem with Powertrain Control Module.

Corrective action

Adjust or replace throttle valve cable or vacuum modulator.

Many modern transmission/transaxles are electronically controlled. This type of problem may be caused by a malfunction in the control unit, sensor, solenoid or circuit. Take the vehicle to a dealer service department or a competent automatic transmission shop for diagnosis.

Transfer case

Note: *Although, in most cases, the corrective action necessary to remedy the symptoms described below is beyond the scope of this manual, the information should be helpful in isolating the cause of the condition so that the owner can communicate clearly with a professional mechanic.*

26 Lubricant leaks from the vent or output shaft seals

Probable cause

- Transfer case is overfilled.

- Vent is clogged or jammed closed.

- Output shaft seal incorrectly installed or damaged.

Corrective action

Drain to the proper level.

Clear or replace the vent.

Replace the seal and check contact surfaces for nicks and scoring.

27 Noisy or jumps out of four-wheel drive Low range

Probable cause

- Transfer case not fully engaged.

- Shift linkage loose, worn or binding.

- Shift fork cracked, inserts worn or fork binding on the rail.

Corrective action

Stop the vehicle, shift into Neutral and then engage 4L.

Tighten, repair or lubricate linkage as necessary.

Disassemble and repair as necessary.

28 Transfer case is difficult to shift into the desired range

Probable cause

Corrective action

- Speed may be too great to permit engagement.

Stop the vehicle and shift into the desired range.

- Shift linkage loose, bent or binding.

Check the linkage for damage or wear and replace or lubricate as necessary.

- If the vehicle has been driven on a paved surface for some time, the driveline torque can make shifting difficult.

Stop and shift into two-wheel drive on paved or hard surfaces.

- Insufficient or incorrect grade of lubricant.

Drain and refill the transfer case with the specified lubricant.

- Worn or damaged internal components.

Disassembly and overhaul of the transfer case may be necessary.

29 Transfer case noisy in all gears

Probable cause

Corrective action

- Insufficient or incorrect grade of lubricant.

Drain and refill with the correct lubricant.

Front driveaxles

30 Clicking noise in turns

Probable cause

Corrective action

- Worn or damaged outboard CV joint.

Replace the CV joint.

31 Knock or clunk as transmission/transaxle is under initial load

Probable cause

Corrective action

- Damaged inner or outer CV joints.

Replace CV joints.

- Loose or disconnected front suspension components or engine mounts.

Tighten all suspension and engine mount fasteners. Replace worn or damaged suspension components or engine mounts.

32 Oil leak at inner end of driveaxle

Probable cause

Corrective action

- Defective driveaxle seal.

Replace the driveaxle seal.

33 Shudder or vibration during acceleration

Probable cause

Corrective action

- Worn or damaged inboard or outboard CV joints.

Replace CV joint.

- Sticking inboard CV joint assembly.

Replace CV joint.

34 Vibration at highway speeds

Probable cause

Corrective action

- Out of balance front wheels and/or tires.

Balance wheels/tires.

- Out of round front tires.

Replace tires.

- Worn CV joint(s).

Replace CV joint.

Driveshaft

35 Knock or clunk when the transmission is under initial load (just after transmission is put into gear)

Probable cause	Corrective action
• Loose or disconnected rear suspension components.	Check all mounting bolts, nuts and bushings and tighten them to the specified torque.
• Loose driveshaft bolts.	Inspect all bolts and nuts and tighten them to the specified torque.
• Worn or damaged universal joint bearings.	Check for wear and replace as necessary.

36 Metallic grinding sound consistent with vehicle speed

Probable cause	Corrective action
• Worn or damaged universal joint bearings.	Check for wear and replace as necessary.

37 Oil leak at front of driveshaft

Probable cause	Corrective action
• Defective transmission or transfer case output shaft oil seal.	Replace the output shaft oil seal. While this is done, check the slip-yoke for burrs or a rough condition which may be damaging the seal. Burrs can be removed with crocus cloth or a fine whetstone.
• Defective driveshaft slip-yoke.	Replace the driveshaft slip-yoke.

38 Vibration

Note: *Before assuming that the driveshaft is at fault, make sure the tires are perfectly balanced and perform the following test.*

a) *Install a tachometer inside the vehicle to monitor engine speed as the vehicle is driven. Drive the vehicle and note the engine speed at which the vibration (roughness) is most pronounced. Now shift the transmission to a different gear and bring the engine speed to the same point.*

b) *If the vibration occurs at the same engine speed (rpm) regardless of which gear the transmission is in, the driveshaft is NOT at fault since the driveshaft speed varies.*

c) *If the vibration decreases or is eliminated when the transmission is in a different gear at the same engine speed, refer to the following probable causes.*

Probable cause	Corrective action
• Bent or dented driveshaft.	Inspect and replace as necessary.
• Undercoating or built-up dirt, etc. on the driveshaft.	Clean the shaft thoroughly and recheck.
• Worn universal joint bearings.	Check for wear and replace as necessary.
• Driveshaft and/or companion flange out of balance.	Check for missing weights on the shaft. Remove the driveshaft and reinstall 180-degrees from original position, then retest. Have the driveshaft professionally balanced if the problem persists.

Rear axle

Note: *Before attempting to diagnose a rear axle noise, be certain the noise is actually coming from the rear axle and not from the tires, engine, transmission, wheel bearings, etc. Road test the vehicle on a variety of road surfaces under drive (acceleration), float (steady speed), and coast (deceleration) conditions. Determine under which conditions the noise is most prominent and compare it with the following symptoms.*

39 Chattering noise on turns

Probable cause

- Limited-slip lubrication contaminated or clutch plates worn.

Corrective action

Flush and replace limited-slip differential with proper lubricant. Replace limited-slip unit.

40 Clunking noise under drive and coast conditions

Probable cause

- Worn or damaged differential cross shaft or case.

Corrective action

Replace cross shaft and/or case.

41 Clunking or groaning noise as vehicle comes to a stop

Probable cause

- Driveshaft slip-yoke binding.

Corrective action

Clean and lubricate driveshaft slip-yoke.

42 Clunking or knocking noise under rough road conditions

Probable cause

- Excessive axleshaft endplay.

Corrective action

Replace axleshafts, C-clips, differential cross shaft and/or case.

43 Knocking or clicking noise as rear wheels rotate

Probable cause

- Rear axle bearings loose, worn or damaged.

Corrective action

Replace rear axle bearings.

44 Low pitch noise at low speeds

Probable cause

- Pinion or differential side bearings worn or damaged.

Corrective action

Overhaul differential.

45 Noise changes under different road conditions

Probable cause

- Road noise.
- Tire noise.

Corrective action

No corrective procedures available.

Inspect tires and check tire pressures.

46 Noise changes under drive, float or coast conditions

Probable cause

- Ring-and-pinion gear noise.

Corrective action

Replace ring-and-pinion.

47 Noise is most evident on turns

Probable cause	Corrective action
• Differential side gears worn or damaged.	Overhaul differential.

48 Noise is the same under all conditions

Probable cause	Corrective action
• Road noise.	No corrective procedures available.
• Tire noise.	Inspect tires and check tire pressures.
• Front wheel bearings loose, worn or damaged.	Replace front wheel bearings.
• Rear axle bearings worn or damaged.	Replace rear axle bearings.

49 Noise lowers as vehicle speed lowers

Probable cause	Corrective action
• Tire noise.	Inspect tires and check tire pressures.

50 Oil leakage

Probable cause	Corrective action
• Pinion seal damaged.	Replace pinion seal.
• Axleshaft oil seals damaged.	Replace axleshaft oil seals.
• Differential inspection cover leaking.	Tighton the bolts or replace the gasket as required.

51 Vibration

See probable causes under *Driveshaft*. Proceed under the guidelines listed for the driveshaft. If the problem persists, check the rear axle bearings by raising the rear of the vehicle and spinning the rear wheels by hand. Listen for evidence of rough (noisy) bearings. Remove the axles and inspect the bearings, if necessary.

Suspension and steering systems

Note: *Before attempting to diagnose the suspension and steering systems, perform the following preliminary checks:*

a) *Check the tires for wrong pressure and uneven wear.*
b) *Check the steering universal joints from the column to the steering gear or rack and pinion for loose connectors or wear.*
c) *Check the front and rear suspension components and the steering gear or rack and pinion assembly for loose or damaged parts.*
d) *Check for out-of-round or out-of-balance tires, bent rims and loose and/or rough wheel bearings.*

52 Abnormal or excessive tire wear

Probable cause	Corrective action
• Wheel alignment out-of-specifications.	Align front wheels.
• Sagging or broken springs.	Replace weak or broken springs.
• Tire out-of-balance.	Balance tires.
• Worn shock absorber or strut damper.	Replace defective shock/strut.
• Overloaded vehicle.	Correct as necessary.
• Tires not rotated regularly.	Rotate or replace tires.

53 Abnormal noise at the front end

Probable cause	Corrective action
• Lack of lubrication at balljoints and tie-rod ends.	Lubricate suspension and steering components.
• Damaged shock absorber or strut mounting.	Repair damaged mount.
• Worn control arm bushings or tie-rod ends.	Replace control arm bushings or tie-rod ends.
• Loose stabilizer bar.	Secure stabilizer bar mounts. Replace bushings.
• Loose wheel nuts.	Tighten wheel lug nuts to proper torque specifications.
• Loose suspension bolts.	Tighten suspension bolts to proper torque specifications. Replace fasteners as necessary.

54 Cupped tires

Probable cause	Corrective action
• Front wheel or rear wheel alignment out-of-specifications.	Align front wheels.
• Worn shock absorber or strut dampers.	Replace shocks/struts.
• Wheel bearings worn.	Inspect, repack and adjust wheel bearing preload.
• Excessive tire or wheel runout.	Replace tire/wheel.
• Worn balljoints.	Replace balljoints.

55 Erratic steering when braking

Probable cause	Corrective action
• Wheel bearings loose or worn.	Inspect, repack and adjust wheel bearing preload.
• Broken or sagging springs.	Replace springs.
• Leaking wheel cylinder or caliper.	Overhaul or replace wheel cylinder or caliper.
• Warped rotors or drums.	Replace or resurface rotors or drums.

56 Excessive pitching and/or rolling around corners or during braking

Probable cause	Corrective action
• Loose stabilizer bar.	Tighten stabilizer bar mounts. Replace bushings.
• Worn shock absorber, strut dampers or mountings.	Replace shocks/struts. Repair damaged mounts.
• Broken or sagging springs.	Replace springs.
• Overloaded vehicle.	Correct as necessary.

57 Excessive play or looseness in steering system

Probable cause	Corrective action
• Wheel bearings worn.	Inspect, repack and adjust wheel bearing preload.
• Steering linkage components loose or worn.	Replace steering linkage components as necessary.
• Steering gear loose.	Tighten steering gear mounts. Replace bushings, if necessary.

58 Excessive tire wear on inside edge

Probable cause

- Inflation pressures incorrect.
- Front end alignment incorrect (toe-out).
- Loose or damaged steering components.

Corrective action

Adjust tire pressures.

Align front wheels.

Replace steering components as necessary.

59 Excessive tire wear on outside edge

Probable cause

- Inflation pressures incorrect.
- Excessive speed in turns.
- Front end alignment incorrect (excessive toe-in).
- Suspension arm bent or twisted.

Corrective action

Adjust tire pressures.

Correct as necessary.

Align front wheels.

Replace suspension arm.

60 Hard steering

Probable cause

- Lack of lubrication at balljoints, tie-rod ends and rack and pinion assembly.
- Front wheel alignment out-of-specifications.
- Low tire pressures.

Corrective action

Lubricate suspension and steering components.

Align front wheels.

Adjust tire pressures.

61 Poor returnability of steering to center

Probable cause

- Lack of lubrication at balljoints and tie-rod ends.
- Binding in balljoints.
- Binding in steering column.
- Lack of lubricant in steering gear assembly.
- Front wheel alignment out-of-specifications.

Corrective action

Lubricate suspension and steering components.

Lubricate or replace balljoints.

Check for interference with other components. Replace the steering column universal joint. Replace the steering column.

Fill steering gear with proper lubricant.

Align front wheels.

62 Rattling or clicking noise in steering gear

Probable cause

- Steering gear loose.
- Steering gear defective.

Corrective action

Tighten steering gear mounts. Replace bushings, if necessary.

Replace steering gear.

63 Shimmy, shake or vibration

Probable cause

- Tire or wheel out-of-balance or out-of-round.
- Loose or worn wheel bearings.
- Worn tie-rod ends.

Corrective action

Balance or replace tire or wheel.

Inspect wheel bearings. Replace defective bearings, repack and adjust bearing preload.

Replace tie-rod ends.

63 Shimmy, shake or vibration (continued)

Probable cause	Corrective action
• Worn lower balljoints.	Replace lower balljoints.
• Excessive wheel runout.	Replace bent wheel.
• Blister or bump on tire.	Replace defective tire.

64 Suspension bottoms

Probable cause	Corrective action
• Overloaded vehicle.	Correct as necessary.
• Worn shock absorber or strut dampers.	Replace shocks/struts.
• Incorrect, broken or sagging springs.	Replace springs.

65 Tire tread worn in one place

Probable cause	Corrective action
• Tires out-of-balance.	Balance tires.
• Damaged or buckled wheel.	Inspect and replace if necessary.
• Defective tire.	Replace tire.

66 Vehicle pulls to one side

Probable cause	Corrective action
• Mismatched or uneven tires.	Replace tires.
• Broken or sagging springs.	Replace springs.
• Wheel alignment out-of-specifications.	Align front wheels.
• Front brake dragging.	Inspect and adjust front brakes.

67 Wander or poor steering stability

Probable cause	Corrective action
• Mismatched or uneven tires.	Replace tires.
• Lack of lubrication at balljoints and tie-rod ends.	Lubricate suspension and steering components.
• Worn shock absorber or strut assemblies.	Replace shocks/struts.
• Loose stabilizer bar.	Tighten stabilizer bar mounts. Replace bushings.
• Broken or sagging springs.	Replace springs.
• Wheels out of alignment.	Align front wheels.

68 Wheel makes a thumping noise

Probable cause	Corrective action
• Blister or bump on tire.	Replace tire.
• Improper shock absorber or strut damper action.	Replace shock/strut.

4 Routine maintenance

Contents

	Section
Automatic transmission/transaxle fluid and filter change..	5
Automatic transmission/transaxle fluid level check	13
Brake and clutch fluid level checks	3
Chassis lubrication	11
Clutch freeplay check and adjustment	12
Differential lubricant change	16
Differential lubricant level check	10
Front wheel bearing check, repack and adjustment (4WD models)	18
Introduction	1

	Section
Maintenance schedule	2
Manual transmission/transaxle lubricant change	14
Manual transmission/transaxle lubricant level check	8
Power steering fluid level check	6
Suspension, steering and driveline check	7
Tire and tire pressure checks	4
Transfer case lubricant change (4WD models)	15
Transfer case lubricant level check (4WD models)	9
Wheel bearing check, repack and adjustment (RWD models)	17

1 Introduction

The key objective of this Chapter is to establish a preventive maintenance program to minimize the chances of driveline, suspension and steering system trouble or component failure. This is done by maintaining a constant awareness of the condition of the entire driveline, suspension and steering systems and by correcting defects or replacing worn-out parts before they become serious problems. The majority of the operations outlined in this Chapter are nothing more than inspections of the various components that make up the systems.

Secondly, this preventive maintenance program should cause most, if not all, maintenance and repairs to take place by intent, hopefully eliminating all "unscheduled maintenance." By doing so, all system repairs that may become necessary can be performed when and where you choose. Not only will it save you money by catching worn-out or damaged components before they cause serious damage, it'll also save you money by avoiding repair shop bills, tow truck fees etc.

The following maintenance schedule is very simple, as are the checking procedures that go with it. The maintenance intervals are based on normal operating conditions for a normal passenger vehicle or light truck. If a particular vehicle is subjected to more severe usage, the intervals should be shortened and the inspection procedures performed more often. Vehicles that tow trailers, carry heavy loads, are driven in a lot of stop-and-go traffic or in mountainous regions are going to wear out parts and need service more often than vehicles that enjoy mostly traffic-free highway trips. Mileage isn't always a factor - the important thing is to perform the inspections routinely. If your vehicle is operated under severe conditions, it would be a good idea to perform the inspections twice as frequently as outlined below, until you develop a feel for the rate at which your vehicle requires maintenance.

The maintenance schedule is basically an outline for the actual inspection and maintenance procedures. The individual procedures are then described in detail. Some of the items on the schedule may be listed at a shorter interval than recommended by some manufacturers, and some of the items listed may not even appear on the maintenance schedules of some manufacturers, but that's OK - it's better to be safe than sorry.

If you're experiencing a specific problem with any system, refer to Chapter 3, *Troubleshooting*, for the proper course of action to take.

2 Maintenance schedule

The following maintenance intervals are based on the assumption that the vehicle owner will be doing the maintenance or service work, as opposed to having a dealer service department do the work. Although the time/mileage intervals are loosely based on the manufacturers recommendations, most have been shortened to ensure, for example, that such items as lubricants and fluids are checked/changed at intervals that promote maximum engine/driveline service life. Also, subject to the preference of the individual owner interested in keeping his or her vehicle in peak condition at all times, and with the vehicle's ultimate resale in mind, many of the maintenance procedures may be performed more often than recommended in the following schedule. We encourage such owner initiative.

When the vehicle is new it should be serviced initially by a factory authorized dealer service department to protect the factory warranty. In many cases the initial maintenance check is done at no cost to the owner (check with your dealer service department for more information).

Every 250 miles or weekly, whichever comes first

Check the brake and clutch fluid levels (Section 3)
Check the tires and tire pressures (Section 4)
Check the automatic transmission/transaxle fluid level (Section 13)

Every 3000 miles or 3 months, whichever comes first

All items listed above, plus . . .
Check the power steering fluid level (Section 6)

Every 6000 miles or 6 months, whichever comes first

Inspect the suspension and steering components (Section 7)

Check the manual transmission/transaxle lubricant level (Section 8)
Check the transfer case lubricant level (Section 9)
Check the differential lubricant level (Section 10)
Lubricate the chassis components (Section 11)*
Check and adjust the clutch freeplay (Section 12)

Every 24,000 miles or 24 months, whichever comes first

All items listed above, plus . . .
Change the automatic transmission/transaxle fluid and filter (Section 5)**
Change the manual transmission/transaxle lubricant (Section 14)
Change the transfer case lubricant (Section 15)
Change the differential lubricant (Section 16)

Every 30,000 miles or 30 months, whichever comes first

All items listed above, plus . . .
Check and repack the wheel bearings (Sections 17 or 18)**

*   This item is affected by "severe" operating conditions, as described below. If the vehicle is operated under severe conditions, perform all maintenance indicated with an asterisk (*) at 3000 mile/three-month intervals. Severe conditions exist if you mainly operate the vehicle . . .

   in dusty areas
   towing a trailer
   idling for extended periods
   driving at low speeds when outside temperatures remain below freezing and most trips are less than four miles long

** Perform this procedure every 15,000 miles if operated under one or more of the following conditions:
   in heavy city traffic where the outside temperature regularly reaches 90-degrees F or higher
   in hilly or mountainous terrain
   frequent trailer pulling

Typical engine compartment maintenance components (rear-wheel drive model)

1   Brake fluid reservoir
2   Clutch fluid reservoir
3   Power steering fluid reservoir
4   Serpentine drivebelt

Typical front underside maintenance components (rear-wheel drive models)

1   Idler arm grease fitting
2   Stabilizer bar
3   Front wheel bearings
4   Lower balljoint grease fitting
5   Transmission
6   Lower balljoint grease fitting
7   Shock absorber
8   Tie-rod end grease fitting

Typical front underside maintenance components (four-wheel drive models)

1 Shock absorber
2 Tie-rod
3 Driveaxle boot
4 Transfer case
5 Front differential
6 Transmission

Typical rear underside maintenance components (rear-wheel drive models)

1 Rear shock absorber
2 Driveshaft universal joint
3 Rear drum brake
4 Differential check/fill plug

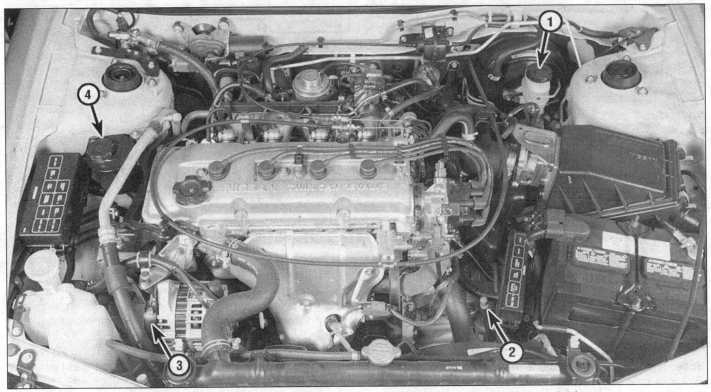

Typical engine compartment maintenance components (front-wheel drive models)

1 Brake fluid reservoir
2 Automatic transaxle fluid dipstick
3 Drivebelt
4 Power steering fluid reservoir

Typical front underside maintenance components (front-wheel drive models)

1 Disc brake caliper
2 Suspension strut and spring
3 Automatic transaxle drain plug
4 Driveaxle boot
5 Steering gear boot
6 Control arm bushing

3.3a On early models with a cast iron brake master cylinder, use a screwdriver to pry off the retainer and remove the cover to check the fluid level

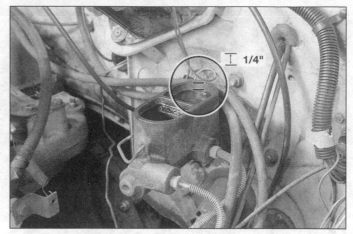

3.3b Maintain the brake fluid level 1/4-inch below the top edge

3 Brake and clutch fluid level checks (every 250 miles or weekly)

Refer to illustrations 3.3a, 3.3b, 3.3c and 3.3d

Note: *The following are fluid level checks to be done on a 250 mile or weekly basis. Additional fluid level checks can be found in specific maintenance procedures which follow. Regardless of intervals, be alert to fluid leaks under the vehicle which would indicate a fault to be corrected immediately.*

1   Fluids are an essential part of the brake and clutch systems. Because the fluids gradually become depleted and/or contaminated during normal operation of the vehicle, they must be periodically replenished. **Note:** *The vehicle must be on level ground when fluid levels are checked.*

2   The brake master cylinder is mounted on the upper left of the engine compartment firewall. The clutch cylinder, used on manual transmission/transaxle models with a hydraulic clutch release system, is mounted next to the master cylinder.

3   Early models may be equipped with a cast iron brake master cylinder while later models may use an aluminum cylinder with a plastic reservoir. The translucent plastic reservoir allows the fluid inside to be checked without removing the cover or cap. The hydraulic clutch system is a sealed unit and it shouldn't be necessary to add fluid under most conditions. Be sure to wipe the top of either reservoir cover with a clean rag to prevent contamination of the brake and/or clutch system before removing the cover **(see illustrations)**.

4   Most systems use brake fluid conforming to DOT 3 specifications. When adding fluid, pour it carefully into the reservoir to avoid spilling it on surrounding painted surfaces. Be sure the specified fluid is used, since mixing different types of brake fluid can cause damage to the system. **Warning:** *Brake fluid can harm your eyes and damage painted surfaces, so use extreme caution when handling or pouring it. Do not use brake fluid that has been standing open or is more than one year old. Brake fluid absorbs moisture from the air. Excess moisture can cause a danger-*

3.3c On later models with a plastic reservoir, the brake fluid level should be kept at the top of the slotted window - never let it drop below the MIN mark; flip up the reservoir cover to add fluid

3.3d The clutch fluid level should be kept near the full mark - never let it drop below the MIN mark; remove the reservoir cover to add fluid

ous loss of brake performance.

5   At this time, the fluid and master cylinder can be inspected for contamination. The system should be drained and refilled if deposits, dirt particles or water droplets are seen in the fluid.

6   After filling the reservoir to the proper level, make sure the cover or cap is on tight to prevent fluid leakage.

7   The brake fluid level in the master cylinder will drop slightly as the pads at the front wheels wear down during normal operation. If the master cylinder requires repeated additions to keep it at the proper level, it's an indication of leakage in the brake system, which should be corrected immediately. Check all brake lines and connections.

8   If, upon checking the master cylinder fluid level, you discover one or both reservoirs empty or nearly empty, the brake system should be bled.

4   Tire and tire pressure checks (every 250 miles or weekly)

Refer to illustrations 4.2, 4.3, 4.4a, 4.4b and 4.8

1   Periodic inspection of the tires may spare you the inconvenience of being stranded with a flat tire. It can also provide you with vital information regarding possible problems in the steering and suspension systems before major damage occurs.

2   The original tires on this vehicle are equipped with 1/2

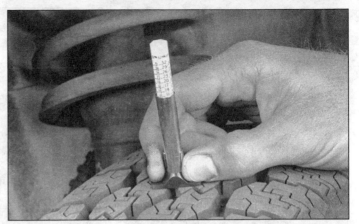

4.2 Use a tire tread depth indicator to monitor tire wear - they are available at auto parts stores and service stations and cost very little

inch wear bands that will appear when tread depth reaches 1/16-inch, but they don't appear until the tires are worn out. Tread wear can be monitored with a simple, inexpensive device known as a tread depth indicator **(see illustration)**.

3   Note any abnormal tread wear **(see illustration)**. Tread pattern irregularities such as cupping, flat spots and more wear on one side than the other are indications of front end alignment and/or balance problems. If any of these conditions are noted, take the vehicle to a tire shop or service station to correct the problem.

UNDERINFLATION

CUPPING

Cupping may be caused by:
● Underinflation and/or mechanical irregularities such as out-of-balance condition of wheel and/or tire, and bent or damaged wheel.
● Loose or worn steering tie-rod or steering idler arm.
● Loose, damaged or worn front suspension parts.

OVERINFLATION

INCORRECT TOE-IN OR EXTREME CAMBER

FEATHERING DUE TO MISALIGNMENT

4.3 This chart will help you determine the condition of the tires, the probable cause(s) of abnormal wear and the corrective action necessary

4.4a If a tire loses air on a steady basis, check the valve stem core first to make sure it's snug (special inexpensive wrenches are commonly available at auto parts stores)

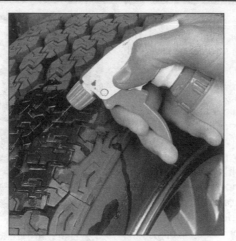

4.4b If the valve stem core is tight, raise the corner of the vehicle with the low tire and spray a soapy water solution onto the tread as the tire is turned slowly - leaks will cause small bubbles to appear

4.8 To extend the life of the tires, check the air pressure at least once a week with an accurate gauge (don't forget the spare!)

4   Look closely for cuts, punctures and embedded nails or tacks. Sometimes a tire will hold air pressure for a short time or leak down very slowly after a nail has embedded itself in the tread. If a slow leak persists, check the valve stem core to make sure it's tight **(see illustration)**. Examine the tread for an object that may have embedded itself in the tire or for a "plug" that may have begun to leak (radial tire punctures are repaired with a plug that's installed in a puncture). If a puncture is suspected, it can be easily verified by spraying a solution of soapy water onto the puncture area **(see illustration)**. The soapy solution will bubble if there's a leak. Unless the puncture is unusually large, a tire shop or service station can usually repair the tire.

5   Carefully inspect the inner sidewall of each tire for evidence of brake fluid leakage. If you see any, inspect the brakes immediately.

6   Correct air pressure adds miles to the lifespan of the tires, improves mileage and enhances overall ride quality. Tire pressure cannot be accurately estimated by looking at a tire, especially if it's a radial. A tire pressure gauge is essential. Keep an accurate gauge in the vehicle. The pressure gauges attached to the nozzles of air hoses at gas stations are often inaccurate.

7   Always check tire pressure when the tires are cold. Cold, in this case, means the vehicle has not been driven over a mile in the three hours preceding a tire pressure check. A pressure rise of four to eight pounds is not uncommon once the tires are warm.

8   Unscrew the valve cap protruding from the wheel or hubcap and push the gauge firmly onto the valve stem **(see illustration)**. Note the reading on the gauge and compare the figure to the recommended tire pressure shown on the placard on the driver's side door pillar. Be sure to reinstall the valve cap to keep dirt and moisture out of the valve stem mechanism. Check all four tires and, if necessary, add enough air to bring them up to the recommended pressure.

9   Don't forget to keep the spare tire inflated to the specified pressure (refer to your owner's manual or the tire sidewall).

5 Automatic transmission/ transaxle fluid level check (every 250 miles or weekly)

Refer to illustrations 5.3 and 5.6

1   The automatic transmission/transaxle fluid level should be carefully maintained. Low fluid level can lead to slipping or loss of drive, while overfilling can cause foaming and loss of fluid.

2   With the parking brake set, start the engine, then move the shift lever through all the gear ranges, ending in Park. The fluid level must be checked with the vehicle level and the engine running at idle. **Note:** *Incorrect fluid level readings will result if the vehicle has just been driven at high speeds for an extended period, in hot weather in city traffic, or if it has been pulling a trailer. If any of these conditions apply, wait until the fluid has cooled (about 30 minutes).*

3   With the transmission/transaxle at normal operating temperature, remove the dipstick from the filler tube. The dipstick is usually located at the rear of the engine compartment **(see illustration)**.

4   Wipe the fluid from the dipstick with a clean rag and push it back into the filler tube until the cap seats.

5   Pull the dipstick out again and note the fluid level.

6   If the fluid is warm, the level should be between the two dimples **(see illustration)**. If it's hot, the level should be in the crosshatched area, near the FULL line. If additional fluid is required, add it directly into the tube using a funnel. It takes about one pint to raise the level from the bottom of the crosshatched area to the FULL line with a hot transmission/transaxle, so add the fluid a little at a time and keep

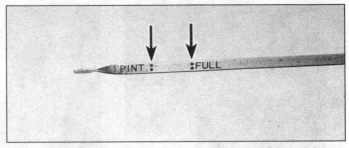

5.6 The automatic transmission/transaxle fluid must be maintained between the ADD mark and the FULL mark with the engine idling at operating temperature

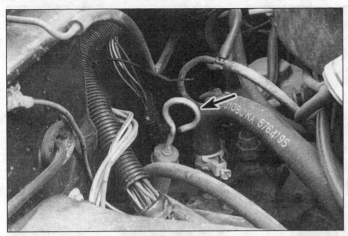

5.3 The automatic transmission/transaxle dipstick (arrow) is usually located at the rear of the engine compartment

checking the level until it's correct.

7   The condition of the fluid should also be checked along with the level. If the fluid at the end of the dipstick is a dark reddish-brown color, or if it smells burned, it should be changed. If you are in doubt about the condition of the fluid, purchase some new fluid and compare the two for color and smell.

6 Power steering fluid level check (every 3000 miles or 3 months)

Refer to illustrations 6.2a, 6.2b and 6.6

1   Unlike manual steering, the power steering system relies on fluid which may, over a period of time, require replenishing.

2   On most models, the fluid reservoir for the power steering pump is located on the pump body at the front of the engine **(see illustration)**. On some models (typically front-wheel drive models), the power steering fluid may be contained in a remote reservoir **(see illustration)**.

3   For the check, the front wheels should be pointed

6.2a The power steering fluid dipstick (arrow) is located in the power steering pump reservoir - turn the cap counterclockwise to remove it

straight ahead and the engine should be off.

4   Use a clean rag to wipe off the reservoir cap and the area around the cap. This will help prevent any foreign matter from entering the reservoir during the check.

5   Twist off the cap and check the temperature of the fluid at the end of the dipstick with your finger.

6   Wipe off the fluid with a clean rag, reinsert the dipstick, then withdraw it and read the fluid level. The fluid should be at the proper level, depending on whether it was checked hot or cold **(see illustration)**. Never allow the fluid level to

6.2b On some models, the power steering fluid reservoir may be attached to the firewall or inner fender panel

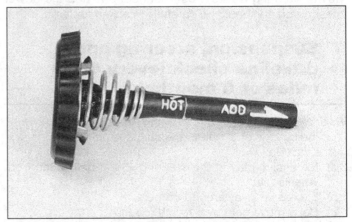

6.6 The marks on the power steering fluid dipstick indicate the safe fluid range

7.3 Grasp the bottom of the tire and force it in-and-out to check for excessive play in the wheel bearings

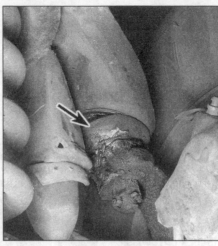

7.4a Inspect the balljoint boots for damage, indicated by leaking grease

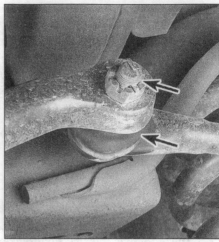

7.4b Check the steering linkage for torn boots and loose fasteners (arrows)

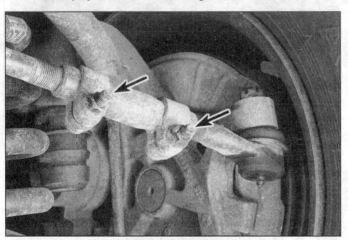

7.4c Make sure the steering tie-rod clamp nuts are tight (arrows)

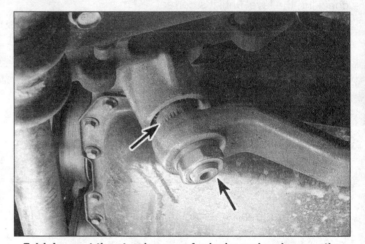

7.4d Inspect the steering gear for leaks and make sure the nut is tight (arrows)

drop below the lower mark on the dipstick.

7   If additional fluid is required, pour the specified type directly into the reservoir, using a funnel to prevent spills.

8   If the reservoir requires frequent fluid additions, all power steering hoses, hose connections and the power steering pump should be carefully checked for leaks.

7 Suspension, steering and driveline check (every 6000 miles or 6 months)

Refer to illustrations 7.3, 7.4a through 7.4e, 7.5, 7.7a, 7.7b and 7.8

1   Indications of a fault in the suspension or steering systems include:

a) Excessive play in the steering wheel before the front wheels react.

b) Excessive sway around corners.

c) Body movement over rough roads.

d) Binding at some point as the steering wheel is turned.

2   Raise the front of the vehicle periodically and visually check the suspension and steering components for wear. Make sure the vehicle is supported securely and cannot fall from the stands.

3   Check the wheel bearings. Do this by spinning the front wheels. Listen for any abnormal noises and watch to make sure the wheel spins true (doesn't wobble). Grab the top and bottom of the tire and pull in-and-out on it **(see illustration)**. Notice any movement which would indicate a loose wheel bearing assembly.

4   From under the vehicle, check for loose bolts, broken or disconnected parts and deteriorated rubber bushings on all suspension and steering components **(see illustrations)**. Look for fluid leaking from the steering gear assembly. Check the power steering hoses, belts and connections for leaks. Check the steering shaft universal joint for damage or wear, and on models equipped with rack and pinion steering, check the steering gear boots for damage **(see illustration)**.

5   Inspect the shock absorbers for fluid leaks, indicating the need for replacement **(see illustration)**. Never replace just one shock absorber; always replace them in sets.

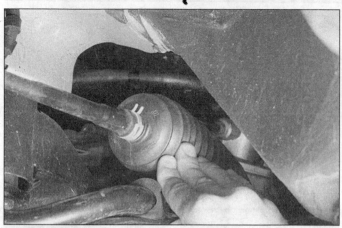

7.4e On models equipped with rack and pinion steering, check the steering boots for damage

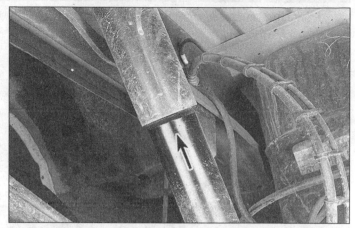

7.5 Check the shock absorbers for leaking fluid in this area (arrow)

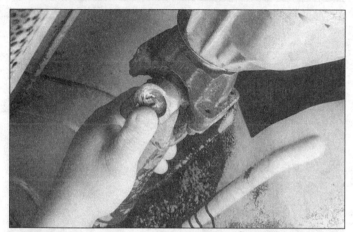

7.7a Twist the U-joint from side-to-side and check for side play, then force it back-and-forth to check for end play - if any play is found the U-joint should be replaced

7.7b If the driveshaft is equipped with a slip-yoke, grasp the driveshaft and force it up-and-down to check for excessive play at the slip-yoke

6   Have an assistant turn the steering wheel from side-to-side and check the steering components for free movement, chafing and binding. If the steering doesn't react with the movement of the steering wheel, try to determine where the slack is located.

7   On rear-wheel drive vehicles, inspect the driveshafts

7.8 On front-wheel drive models, check the driveaxle boots for cracks or leaking grease

for worn U-joints and for excessive play in the slip yoke and spline area (if equipped) (see illustrations).

8   On front-wheel drive vehicles, the driveaxle boots are very important because they prevent dirt, water and foreign material from entering and damaging the constant velocity joints. Oil and grease can cause the boot material to deteriorate prematurely, washing the boots with soap and water may prolong the life of the boots. Inspect the boots for tears and cracks (see illustration). If there is any evidence of leaking grease, replace the boot.

8 Manual transmission/ transaxle lubricant level check (every 6000 miles or 6 months)

Refer to illustration 8.2

1   The manual transmission/transaxle has a filler plug which must be removed to check the lubricant level. If the vehicle is raised to gain access to the plug, be sure to support it safely on jackstands - DO NOT crawl under a vehicle which is supported only by a jack! Be sure the vehicle is level or the check may be inaccurate.

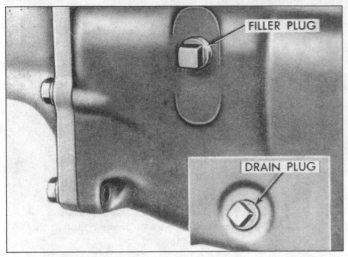

8.2 Typical filler and drain plug locations on a manual transmission/transaxle

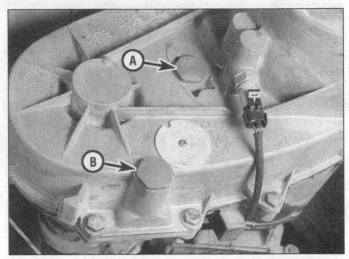

9.1 Typical 4WD transfer case filler plug (A) and drain plug (B) locations

2   Using the appropriate wrench, unscrew the plug from the transmission/transaxle **(see illustration)**; some models may require an Allen wrench.
3   Use your little finger to reach inside the housing to feel the lubricant level. The level should be at or near the bottom of the plug hole. If it isn't, add the recommended lubricant through the plug hole with a syringe or squeeze bottle.
4   Install and tighten the plug. Check for leaks after the first few miles of driving.

9   Transfer case lubricant level check (4WD models) (every 6000 miles or 6 months)

Refer to illustration 9.1
1   The transfer case lubricant level is checked by removing the upper plug located at the rear of the case **(see illustration)**.
2   After removing the plug, reach inside the hole. The lubricant level should be just at the bottom of the hole. If not, add the appropriate lubricant through the opening.

10   Differential lubricant level check (every 6000 miles or 6 months)

Refer to illustrations 10.2 and 10.3
Note: *4WD vehicles have two differentials - one in the center of each axle. 2WD vehicles have one differential - in the center of the rear axle. On 4WD vehicles, be sure to check the lubricant level in both differentials.*
1   The filler plug on all front and most rear differentials is a threaded metal type while the rear differential on some models may have a rubber press-in filler plug which must be removed to check the lubricant level. If the vehicle is raised to gain access to the plug, be sure to support it

10.2 On most differentials, the check/fill plug is removed by unscrewing it from the differential case using the appropriate-size open end wrench, 3/8-inch drive breaker bar or ratchet

safely on jackstands - DO NOT crawl under the vehicle when it's supported only by the jack. Be sure the vehicle is level or the check may not be accurate.
2   Remove the plug from the filler hole in the differential housing or cover **(see illustration)**.
3   The lubricant level should be at the bottom of the filler hole **(see illustration)**. If not, use a pump or squeeze bottle to add the recommended lubricant until it just starts to run out of the opening. On some models a tag is located in the area of the plug which gives information regarding lubricant type.
4   Install the plug securely into the filler hole.

11   Chassis lubrication (every 6000 miles or 6 months)

Refer to illustrations 11.1, 11.2a through 11.2h and 11.6
1   Obtain the necessary grease, tools, grease gun, etc.

10.3 Use your finger as a dipstick to check the lubricant level

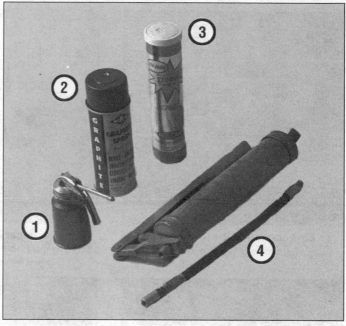

11.1 Materials required for chassis and body lubrication

1   *Engine oil* - Light engine oil in a can like this can be used for door and hood hinges
2   *Graphite spray* - Used to lubricate lock cylinders
3   *Grease* - Grease, in a variety of types and weights, is available for use in a grease gun. Check the Specification for your requirements
4   *Grease gun* - A common grease gun, shown here with a detachable hose and nozzle, is needed for chassis lubrication. After use, clean it thoroughly!

(see illustration). Occasionally plugs will be installed rather than grease fittings. If so, grease fittings will have to be purchased and installed.

2   Look under the vehicle and locate the grease fittings on the steering, suspension and driveline components **(see illustrations)**. They are normally found on the balljoints, tie-rod ends and universal joints. Some models with manual transmission/transaxles have grease fittings on the shift linkage and/or clutch linkage which should be lubricated whenever the chassis is lubricated.

3   For easier access under the vehicle, raise it with a jack and place jackstands under the frame. Make sure it's safely

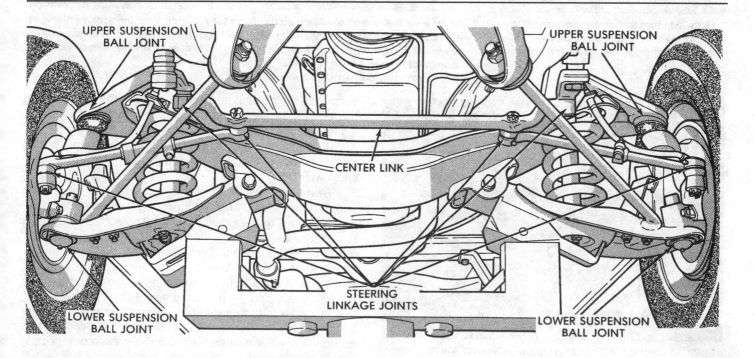

11.2a Typical lubrication points

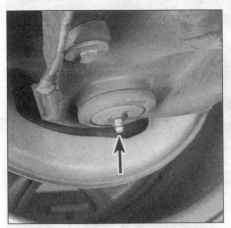

11.2b Location of the lower balljoint grease fitting (arrow)

11.2c The upper balljoint grease fitting (arrow) may be easier to access if the front wheel is removed

11.2d Be sure to grease each steering arm balljoint

11.2e Manual transmission/transaxle models may be equipped with lubrication fittings on the clutch linkage

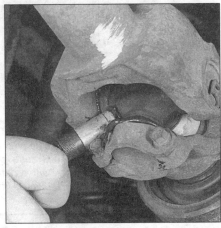

11.2f Rear-wheel drive vehicles may be equipped with grease fittings at the universal joints - apply one or two pumps of grease only, excessive lubrication may damage the seals

11.2g Check for a lubrication fitting at the slip-yoke - again, one or two pumps of grease is all that's necessary

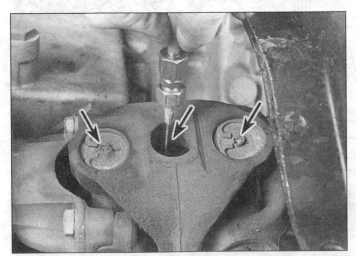

11.2h In addition to the conventional universal joints at each end of the driveshaft, 4WD models may be equipped with a constant velocity joint which requires a needle-like grease gun adapter to apply the lubrication

supported by the stands. If the wheels are to be removed at this interval for tire rotation or brake inspection, loosen the lug nuts slightly while the vehicle is still on the ground.

4   Before beginning, force a little grease out of the nozzle to remove any dirt from the end of the gun. Wipe the nozzle clean with a rag.

5   With the grease gun and plenty of clean rags, slide under the vehicle and begin lubricating the components.

6   Wipe one of the grease fitting nipples clean and push the nozzle firmly over it. Pump the gun until the component is completely lubricated. On balljoints and steering components, stop pumping when the rubber seal is firm to the touch **(see illustration)**. Do not pump too much grease into the fitting as it could rupture the seal. For all other suspension and steering components, continue pumping grease into the fitting until it oozes out of the joint between the two components. If it escapes around the grease gun nozzle, the nipple is clogged or the nozzle is not completely seated on the fitting. Resecure the gun nozzle to the fitting and try again. If necessary, replace the fitting with a new one.

7   Wipe the excess grease from the components and the grease fitting. Repeat the procedure for the remaining fittings.

8   Clean the fitting and pump grease into the driveline universal joints until the grease can be seen coming out of the contact points.

9   Also clean and lubricate the parking brake cable, along with the cable guides and levers. This can be done by smearing some of the chassis grease onto the cable and its related parts with your fingers.

12 Clutch freeplay check and adjustment (every 6,000 miles or 6 months)

Refer to illustrations 12.3 and 12.4

Note: *Clutch pedal freeplay adjustment generally applies to mechanically operated clutch systems only (cable or linkage). Hydraulic clutch release systems usually do not provide adjustment capabilities. In most cases the hydraulic systems are self-adjusting; in some cases, mechanical systems using a clutch cable are self-adjusting.*

1   Clutch pedal freeplay will change over time due to normal clutch disc and linkage wear.

2   Clutch pedal freeplay adjustment is an important maintenance item. Excessive freeplay can result in hard shifting and excessive transmission/transaxle wear. Insufficient freeplay can result in clutch slippage and premature clutch wear.

3   Press down lightly on the clutch pedal and with a ruler or tape, measure the distance the pedal moves freely before resistance is felt **(see illustration)**. Typically, the clutch pedal freeplay should measure approximately 1-inch. If there is little or no freeplay or if the freeplay is exces-

11.6 Wipe the dirt from the grease fittings before pushing the grease gun nozzle onto the fitting then pump the grease into the fitting until the rubber seal is firm to the touch

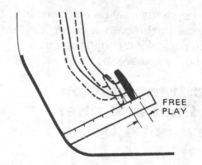

FREE PLAY

12.3 To determine the clutch pedal freeplay, depress the pedal, stop the moment clutch resistance is felt, then measure the distance (freeplay)

sive, it must be adjusted.

4   Locate the adjusting mechanism, usually an adjusting nut is located on the clutch fork rod at the clutch linkage **(see illustration)**. Some cable operated clutch systems may have a nut on the cable casing at the firewall. Some

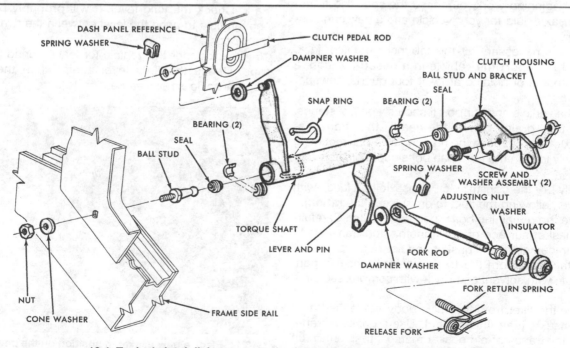

DASH PANEL REFERENCE
SPRING WASHER
CLUTCH PEDAL ROD
DAMPNER WASHER
CLUTCH HOUSING
BALL STUD AND BRACKET
SEAL
SNAP RING
BEARING (2)
BEARING (2)
SEAL
BALL STUD
SPRING WASHER
SCREW AND WASHER ASSEMBLY (2)
ADJUSTING NUT
WASHER
INSULATOR
TORQUE SHAFT
LEVER AND PIN
FORK ROD
DAMPNER WASHER
FORK RETURN SPRING
NUT
CONE WASHER
FRAME SIDE RAIL
RELEASE FORK

12.4 Typical clutch linkage - note the adjusting nut at the clutch fork rod

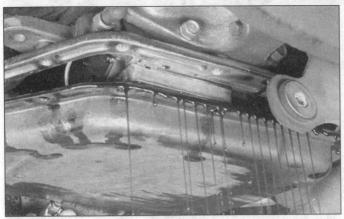

13.6 With the front bolts in place but loose, pull the rear of the pan down to drain the fluid

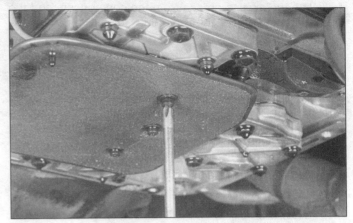

13.9 Remove the filter screws

hydraulic clutch release systems may be equipped with an adjustable master cylinder pushrod, look under the instrument panel at the clutch pedal for an adjusting nut.

5   Turn the adjusting nut to achieve the specified freeplay.

13 Automatic transmission/ transaxle fluid and filter change (every 24,000 or 24 months)

Refer to illustrations 13.6, 13.9 and 13.11

1   At the specified intervals, the transmission/transaxle fluid should be drained and replaced. Since the fluid will remain hot long after driving, perform this procedure only after the engine has cooled down completely.

2   Before beginning work, purchase the specified transmission/transaxle fluid for your vehicle and a new pan gasket and filter.

3   Other tools necessary for this job include a floor jack, jackstands to support the vehicle in a raised position, a drain pan capable of holding at least four quarts, newspapers and clean rags.

4   Raise the vehicle and support it securely on jackstands.

5   Place the drain pan underneath the transmission/transaxle pan. Remove the rear and side pan mounting bolts, but only loosen the front pan bolts approximately four turns.

6   Carefully pry the transmission/transaxle pan loose with a screwdriver, allowing the fluid to drain **(see illustration)**.

7   Remove the remaining bolts, pan and gasket. Carefully clean the gasket surface of the transmission/transaxle to remove all traces of the old gasket and sealant.

8   Drain the fluid from the transmission/transaxle pan, clean the pan with solvent and dry it with compressed air, if available.

9   Remove the filter from the valve body inside the transmission/transaxle **(see illustration)**. Use a gasket scraper to remove any traces of old gasket material that remain on the valve body. **Note:** *Be very careful not to gouge the delicate aluminum gasket surface on the valve body.*

10   Install a new gasket and filter. On many replacement filters, the gasket is attached to the filter to simplify installation.

11   Make sure the gasket surface on the transmission/transaxle pan is clean, then install a new gasket on the pan **(see illustration)**. Put the pan in place against the transmission/transaxle and, working around the pan, tighten each bolt a little at a time. Do not over-tighten the bolts and crush the gasket, Typically the transmission/transaxle bolts are tightened to approximately 10 to 12 Ft-lbs.

12   Lower the vehicle and add approximately 2 to 3 quarts of the specified type of automatic transmission/transaxle fluid through the filler tube (see Section 13).

13   With the transmission/transaxle in Park and the parking brake set, run the engine at a fast idle, but don't race it.

14   Move the gear selector through each range and back to Park. Check the fluid level. It will probably be low. Add enough fluid to bring the level between the two dimples on the dipstick.

15   Check under the vehicle for leaks during the first few trips. Check the fluid level again when the transmission/transaxle is hot (see Section 13).

13.11 Place a new gasket in position on the pan and install the bolts to hold it in place

14 Manual transmission/transaxle lubricant change (every 24,000 or 24 months)

1   This procedure should be performed after the vehicle has been driven so the lubricant will be warm and therefore will flow out of the transmission/transaxle more easily. Raise the vehicle and support it securely on jackstands.
2   Move a drain pan, rags, newspapers and wrenches under the transmission/transaxle.
3   Remove the transmission/transaxle drain plug at the bottom of the case and allow the lubricant to drain into the pan **(see illustration 8.2)**.
4   After the lubricant has drained completely, reinstall the plug and tighten it securely.
5   Remove the fill plug from the side of the transmission/transaxle case. Using a hand pump, syringe or funnel, fill the transmission/transaxle with the specified lubricant until it begins to leak out through the hole. Reinstall the fill plug and tighten it securely.
6   Lower the vehicle.
7   Drive the vehicle for a short distance, then check the drain and fill plugs for leakage.

15 Transfer case lubricant change (4WD models) (every 24,000 or 24 months)

1   This procedure should be performed after the vehicle has been driven so the lubricant will be warm and therefore will flow out of the transfer case more easily.
2   Raise the vehicle and support it securely on jackstands.
3   Remove the filler plug from the case **(see illustration 9.1)**.
4   Remove the drain plug from the lower part of the case and allow the lubricant to drain completely.
5   After the case is completely drained, carefully clean and install the drain plug. Tighten the plug securely.

6   Fill the case with the specified lubricant until it is level with the lower edge of the filler hole.
7   Install the filler plug and tighten it securely.
8   Drive the vehicle for a short distance and recheck the lubricant level. In some instances a small amount of additional lubricant will have to be added.

16 Differential lubricant change (every 24,000 or 24 months)

Refer to illustrations 16.3, 16.4a, 16.4b, 16.4c and 16.6
1   This procedure should be performed after the vehicle has been driven so the lubricant will be warm and therefore will flow out of the differential more easily.
2   Raise the vehicle and support it securely on jackstands. If the differential has a bolt-on cover at the rear, it is usually easiest to remove the cover to drain the lubricant (which will also allow you to inspect the differential. If there's no bolt-on cover, look for a drain plug at the bottom of the differential housing. If there's not a drain plug and no cover, you'll have to remove the lubricant through the filler plug hole with a suction pump. If you'll be draining the lubricant by removing the cover or a drain plug, move a drain pan, rags, newspapers and wrenches under the vehicle.
3   Remove the filler plug from the differential (see Section 10). If a suction pump is being used, insert the flexible hose. Work the hose down to the bottom of the differential housing and pump the lubricant out **(see illustration)**. If you'll be draining the lubricant through a drain plug, remove the plug and allow the lubricant to drain into the pan, then reinstall the drain plug.
4   If the differential is being drained by removing the cover plate, remove the bolts on the lower half of the plate. Loosen the bolts on the upper half and use them to keep the cover loosely attached. Allow the oil to drain into the pan, then completely remove the cover **(see illustrations)**.
5   Using a lint-free rag, clean the inside of the cover and the accessible areas of the differential housing. As this is done, check for chipped gears and metal particles in the lubricant, indicating that the differential should be more

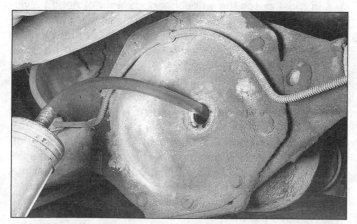

16.3 This is the easiest way to remove the lubricant: work the end of the hose to the bottom of the differential housing and draw out the old lubricant with a hand pump

16.4a Remove the bolts from the lower edge of the cover . . .

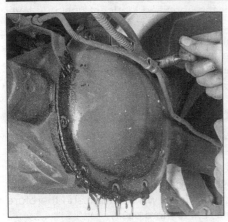

16.4b . . . then loosen the top bolts and allow the lubricant drain out

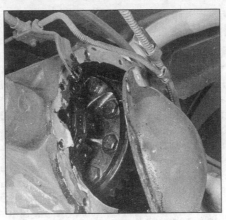

16.4c After the lubricant has drained, remove the remaining bolts and the cover

16.6 Carefully scrape the old gasket material off to ensure a leak-free seal

thoroughly inspected and/or repaired.

6   Thoroughly clean the gasket mating surfaces of the differential housing and the cover plate. Use a gasket scraper or putty knife to remove all traces of the old gasket (see illustration).

7   Apply a thin layer of RTV sealant to the cover flange, then press a new gasket into position on the cover. Make sure the bolt holes align properly.

8   Place the cover on the differential housing and install the bolts. Tighten the bolts securely.

9   Use a hand pump, syringe or funnel to fill the differential housing with the specified lubricant until it's level with the bottom of the plug hole.

10   Install the filler plug and make sure it is secure.

17 Wheel bearing check, repack and adjustment (RWD models) (every 30,000 miles or 30 months)

Refer to illustrations 17.1, 17.6, 17.7, 17.8, 17.9, 17.10, 17.12, 17.15, 17.19 and 17.22

Note: *Many front wheel drive vehicles are equipped with rear wheel bearings that may be checked, repacked and adjusted in the same manner as described in this Section.*

1   In most cases the wheel bearings will not need servicing until the brake pads are changed. However, the bearings should be checked whenever the front of the vehicle is raised for any reason. Several items, including a torque wrench and special grease, are required for this procedure (see illustration).

2   With the vehicle securely supported on jackstands, spin each wheel and check for noise, rolling resistance and freeplay.

3   Grasp the top of each tire with one hand and the bottom with the other. Move the wheel in-and-out on the spindle. If there's any noticeable movement, the bearings should be checked and then repacked with grease or replaced if necessary.

4   Remove the wheel.

5   Remove the brake caliper and hang it out of the way on a piece of wire. A wood block can be slid between the brake pads to keep them separated, if necessary.

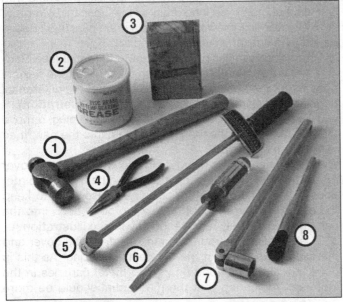

17.1 Tools and materials needed for front wheel bearing maintenance

1   **Hammer** - *A common hammer will do just fine*
2   **Grease** - *High-temperature grease that is formulated for front wheel bearings should be used*
3   **Wood block** - *If you have a scrap piece of 2x4, it can be used to drive the new seal into the hub*
4   **Needle-nose pliers** - *Used to straighten and remove the cotter pin in the spindle*
5   **Torque wrench** - *This is very important in this procedure; if the bearing is too tight, the wheel won't turn freely - if it's too loose, the wheel will "wobble" on the spindle. Either way, it could mean extensive damage*
6   **Screwdriver** - *Used to remove the seal from the hub (a long screwdriver is preferred)*
7   **Socket/breaker bar** - *Needed to loosen the nut on the spindle if it's extremely tight*
8   **Brush** - *Together with some clean solvent, this will be used to remove old grease from the hub and spindle*

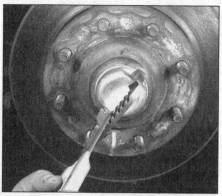

17.6 On some models its possible to use large pliers to grasp the grease cap securely and work it out of the hub - on others, you' have to use a hammer and chisel to detach it

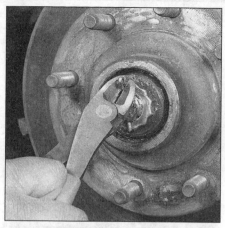

17.7 Use diagonal wire cutters to remove the cotter pin

17.8 Remove the nut lock and washer

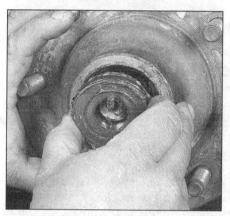

17.9 Pull the hub out to dislodge the outer wheel bearing

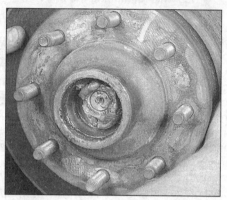

17.10 Reinstall the hub nut, then grasp the hub securely and pull out sharply to dislodge the inner bearing and seal against the back of the nut

17.12 Slide the inner bearing and seal assembly off the spindle

6   Remove the dust cap using large pliers or by prying it out of the hub using a screwdriver or hammer and chisel **(see illustration)**.

7   Straighten the bent ends of the cotter pin, then pull the cotter pin out of the nut lock **(see illustration)**. Discard the cotter pin and use a new one during reassembly.

8   Remove the nut lock, nut and washer from the end of the spindle **(see illustration)**.

9   Pull the hub/disc assembly out slightly, then push it back into its original position. This should force the outer bearing off the spindle enough so it can be removed **(see illustration)**.

10   Temporarily reinstall the hub/disc assembly and spindle nut. Dislodge the inner bearing and seal by grasping the assembly and pull out sharply **(see illustration)**.

11   Once the bearing and seal are free, remove the hub/disc assembly from the spindle.

12   Remove the inner wheel bearing and seal from the spindle, noting how the seal is installed **(see illustration)**.

13   Use solvent to remove all traces of the old grease from the bearings, hub and spindle. A small brush may prove helpful; however make sure no bristles from the brush embed themselves inside the bearing rollers. Allow the parts to air dry.

14   Carefully inspect the bearings for cracks, heat discoloration, worn rollers, etc. Check the bearing races inside the hub for wear and damage. If the bearing races are defective, the hubs should be taken to a machine shop with the facilities to remove the old races and press new ones in. Note that the bearings and races come as matched sets and old bearings should never be installed on new races.

15   Use high-temperature front wheel bearing grease to

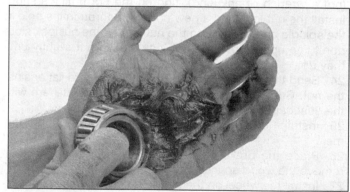

17.15 Work the grease completely into the bearing rollers - special bearing packing tools that work with a common grease gun are available inexpensively from auto parts stores

17.19 Use a block of wood and a hammer to tap the inner bearing seal evenly into the hub

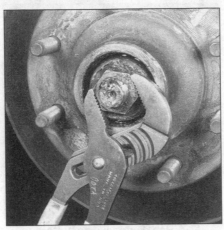

17.22 Seat the bearings by spinning the hub while tightening the hub nut

18.6a Remove the dust cap on vehicles with automatic locking hubs

pack the bearings. Work the grease completely into the bearings, forcing it between the rollers, cone and cage from the back side **(see illustration)**.

16 Apply a thin coat of grease to the spindle at the outer bearing seat, inner bearing seat, shoulder and seal seat.

17 Put a small quantity of grease inboard of each bearing race inside the hub. Using your finger, form a dam at these points to provide extra grease availability and to keep thinned grease from flowing out of the bearing.

18 Place the grease-packed inner bearing into the rear of the hub and put a little more grease outboard of the bearing.

19 Place a new seal over the inner bearing and tap the seal evenly into place with a hammer and blunt punch until it's flush with the hub **(see illustration)**.

20 Carefully place the hub assembly onto the spindle and push the grease-packed outer bearing into position.

21 Install the washer and spindle nut. Tighten the nut only slightly (no more than 12 ft-lbs of torque).

22 Spin the hub in a forward direction while tightening the spindle nut to approximately 20 ft-lbs to seat the bearings and remove any grease or burrs which could cause excessive bearing play later **(see illustration)**.

23 Loosen the spindle nut 1/4-turn, then using your hand (not a wrench of any kind), tighten the nut until it's snug. Install the nut lock and a new cotter pin through the hole in the spindle and the slots in the nut lock. If the nut lock slots don't line up, remove the nut lock and turn it slightly until they do.

24 Bend the ends of the cotter pin until they're flat against the nut. Cut off any extra length which could interfere with the dust cap.

25 Install the dust cap, tapping it into place with a hammer.

26 Place the brake caliper near the rotor and carefully remove the wood spacer. Install the caliper.

27 Install the wheel on the hub and tighten the lug nuts.

28 Grasp the top and bottom of the tire and check the bearings in the manner described earlier in this Section.

29 Lower the vehicle.

18 Front wheel bearing check, repack and adjustment (4WD models) (every 30,000 miles or 24 months)

Refer to illustrations 18.6a, 18.6b, 18.7a, 18.7b, 18.7c, 18.8, 18.9a, 18.9b, 18.10, 18.12, and 18.24.

Caution: *This procedure is generic for most full-size domestic 4WD pick-up and sport-utility models, consult the Haynes Automotive Repair Manual on your vehicle for specific torque specifications and bearing adjustment procedures. Some models may be equipped with a sealed hub and bearing assembly which is not serviceable. Other models may be equipped with a self-adjusting locknut assembly.*

Note: *On some models, this procedure requires a special socket (available at most automotive parts stores) to remove the inner and outer wheel bearing locking nuts.*

1 In most cases the front wheel bearings will not need servicing until the brake shoes or pads are changed. However, the bearings should be checked whenever the front of the vehicle is raised for any reason. Several items, including a torque wrench and special grease, are required for this procedure **(see illustration 17.1)**.

2 With the vehicle securely supported on jackstands, spin each wheel and check for noise, rolling resistance and freeplay.

3 Grasp the top of each tire with one hand and the bottom with the other. Move the wheel in and out on the spindle. If there's any noticeable movement, the bearings should be checked and then repacked with grease, or replaced if necessary.

4 Remove the wheel.

5 On disc brake vehicles, fabricate a wood block which can be slid between the brake pads to keep them separated, remove the caliper and hang it out of the way on a piece of wire.

6 Remove the dust cap on vehicles equipped with automatic locking hubs. On vehicles with manual locking hubs,

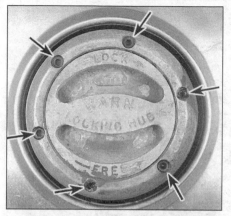

18.6b Remove the six hex-head capscrews securing the hub body on vehicles with manual locking hubs

18.7a Remove the inner snap ring from the axle shaft . . .

18.7b . . . remove the outer snap ring securing the hub clutch . . .

18.7c . . . pull the hub clutch assembly from the rotor/brake drum assembly

18.8 On vehicles with automatic locking hubs, detach the inner snap ring from the axle shaft and remove the drive hub and spring

detach the six hex-head capscrews securing the hub and remove the hub body **(see illustrations)**.

7   On vehicles equipped with manual locking hubs remove the inner and outer snap-rings then remove the front hub clutch assembly **(see illustrations)**. **Note:** *The front drive hubs consist of two sub assemblies, the hub body and the hub clutch assembly - DO NOT attempt to* *disassemble the sub assemblies.*

8   On vehicles equipped with automatic locking hubs remove the inner snap-ring **(see illustration)**. Then remove the drive hub and spring.

9   Using a special locknut removal socket, remove the

18.9a Remove the wheel bearing outer lock nut with the locknut removal socket (available at automotive parts stores)

18.9b Remove the locating washer and using the special socket, remove the inner locknut

18.10 Pull the rotor or brake drum assembly out slightly, then push it back in to disengage the outer wheel bearing and cup

18.12 Use a screwdriver or seal removal tool to pry out the grease seal

outer locknut, locating washer and inner locknut **(see illustrations)**.

10   Pull the rotor or brake drum assembly out slightly, then push it back in. This should force the outer bearing and cup off the spindle so it can be removed **(see illustration)**.

11   Pull the rotor or brake drum off the spindle.

12   Use a screwdriver or a seal puller tool to pry the seal out of the rear of the rotor or brake drum **(see illustration)**. Note how the seal is installed.

13   Remove the inner wheel bearing from the rotor or brake drum.

14   Use solvent to remove all traces of the old grease from the bearings, hub and spindle. A small brush may prove helpful; however make sure no bristles from the brush embed themselves inside the bearing rollers. Allow the parts to air dry.

15   Carefully inspect the bearings for cracks, heat discoloration, worn rollers, etc. Check the bearing races inside the hub for wear and damage. If the bearing races are defective, drive the bearing race out of the hub using a brass drift. Drive the new race into the hub using the appropriate size bearing driver (inexpensive bearing driver sets are available at most automotive parts stores). Note that the bearings and races are replaced as matched sets, used bearings should never be installed on new races.

16   Use only high-temperature front wheel bearing grease to pack the bearings. Inexpensive bearing packing tools are available at automotive parts stores, but not entirely necessary. If one is not available, pack the grease by hand completely into the bearings, forcing it between the rollers, cone and cage from the back side **(see illustration 17.15)**.

17   Apply a thin coat of grease to the spindle at the outer bearing seat, inner bearing seat, shoulder and seal seat.

18   Place a small quantity of grease inboard of each bearing race inside the hub. Using your finger, form a dam at these points to provide extra grease availability and to keep thinned grease from flowing out of the bearing.

19   Place the grease-packed inner bearing into the rear of the hub and put a little more grease outboard of the bearing.

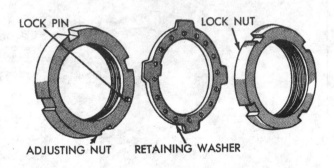

18.24 Engage the lock pin into one of the holes in the retaining washer

20   Place a new seal over the inner bearing and tap the seal evenly into place with a hammer and block of wood until it's flush with the hub.

21   Carefully place the rotor/brake drum assembly onto the spindle and push the grease-packed outer bearing into position.

22   Install the inner locknut with the locating peg facing outward. Then tighten the inner locknut securely (approximately 35 ft-lbs) while rotating the rotor or brake drum to seat the bearings.

23   Back off the inner locknut slightly to provide a slight amount of endplay, then tighten it until it's snug (hand-tight). Pull back-and-forth on the hub to check for endplay; if there's endplay, continue tightening the inner nut to just eliminate the endplay.

24   Install the locating washer, seating the inner locknut locating peg into one of the holes **(see illustration)**. If the peg does not align with a hole, back the nut off to the next nearest hole. Install the outer locknut and tighten it securely.

25   The remainder of the installation is the reverse of removal. **Note:** *Lightly grease the internal components of the front hub assembly with chassis grease. DO NOT pack the hubs full of grease or poor operation may occur.*

5 Driveline systems

Contents

	Section		Section
Axleshaft/driveaxle oil seal - replacement	15	General information	1
Axle assembly (front) - removal and installation	22	Hub/drum assembly and wheel bearings (full-floating axles) - removal, installation and adjustment	18
Axle assembly (rear) - removal and installation	20	Pinion oil seal - replacement	19
Carrier and bearings - removal and installation	24	Pilot bearing/bushing - inspection and replacement	5
Clutch components - removal, inspection and installation	3	Rear axle - check	13
Clutch - check	2	Rear axleshaft (semi-floating axles) - removal and installation	14
Clutch release bearing and lever - removal, inspection and installation	4	Rear axleshaft bearing (semi-floating axles) - replacement	16
Differential check and adjustments -	23	Rear axleshaft (full-floating axles) - removal installation and adjustment	17
Driveaxle boot replacement and CV joint inspection and overhaul	12	Ring and pinion - removal and installation	25
Driveaxles - removal and installation	10	Support bearing assembly - check and replacement	11
Driveshaft inspection	6	Universal joints - replacement	8
Driveshaft - removal and installation	7		
Driveshaft center bearing - check and replacement	9		
Front axleshaft and joint assembly (four wheel drive models) - removal component replacement and installation	21		

1 General information

Refer to illustrations 1.1a, 1.1b, 1.1c, 1.1d and 1.1e

The information in this Chapter deals with the compo-nents from the rear of the engine to the rear wheels, (except for the transmission/transaxle) commonly referred to as the driveline **(see illustrations)**. For the purposes of this Chap-ter, these components are grouped into four categories; clutch, driveshaft, driveaxle and rear axle assembly.

1.1a Typical front wheel drive vehicle driveline components

1	Outer CV joint	2	Inner CV joint	3	Intermediate shaft bearing

1.1b Typical four wheel drive vehicle driveline components (front)

1	Front differential	3	Front drive axle	5	Front leaf springs
2	Front driveshaft	4	Hub and bearings		

1.1c Typical rear driveline components - drop out type rear differential

1	Rear axle housing	3	Coil spring	5	Lower suspension arm	7	Differential
2	Shock absorber	4	Track bar	6	Upper suspension arm		

1.1d Typical driveline components - independent rear suspension type

1	Differential	3	Inner CV joint	5	Driveaxle
2	Driveshaft	4	Outer CV joint		

1.1e Typical rear driveline components with integral type differential

1	Stabilizer bar
2	Lower suspension arm pivot bolt/nut
3	Lower suspension arm-to-rear axle bolt/nut
4	Upper suspension arm pivot bolt/nut
5	Shock absorber

6	Shock absorber-to-rear axle bolt/nut
7	Coil spring
8	Differential cover
9	Differential
10	Rear driveshaft

Clutch

Refer to illustration 1.2

The manual transmission/transaxle uses a single dry plate type clutch. The clutch disc has a splined hub which allows it to slide along the splines of the transmission/trans-axle input shaft. The clutch and pressure plate are held in contact by spring pressure exerted by the diaphragm in the pressure plate **(see illustration)**.

The clutch release system is either mechanically or hydraulically operated. The mechanical release system consists of the clutch pedal, clutch cable and release bearing. The hydraulic release system consists of the clutch pedal, the clutch master cylinder, the clutch release cylinder, the hydraulic line between the master cylinder and release cylinder, and the clutch release bearing.

When pressure is applied to the clutch pedal to release the clutch, the clutch cable pulls the release lever and bearing on mechanical types. On hydraulic types, when pressure is applied to the clutch pedal to release the clutch, the clutch master cylinder transmits this movement to the clutch release cylinder, which moves the clutch release lever. As the lever pivots (both types), the shaft fingers push against the release bearing. The bearing pushes against the

fingers of the diaphragm spring of the pressure plate assembly, which in turn releases the clutch plate.

Terminology can be a problem regarding the clutch components because common names have in some cases changed from that used by the manufacturer. For example,

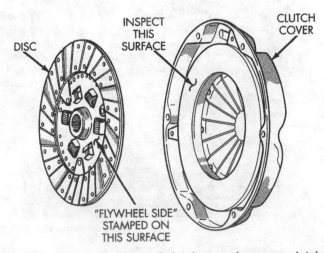

1.2 Typical clutch disc and clutch cover (pressure plate)

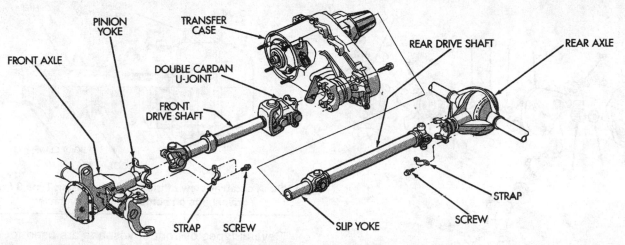

1.6 An exploded view of typical front and rear driveshafts on a 4WD model (the forward end of the Type 3 front driveshaft used on some new models has a CV joint instead of a slip yoke)

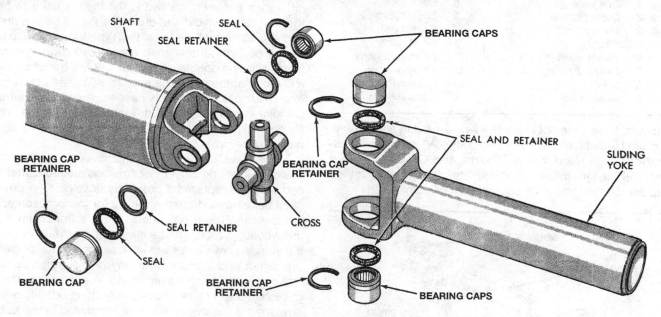

1.8a An exploded view of a typical single-cardan U-joint

the clutch release cylinder is sometimes referred to as a slave cylinder, the driven plate is also called the clutch plate or disc, the pressure plate assembly is also known as the clutch cover, and the clutch release bearing is sometimes called a throw-out bearing.

Driveshaft

Refer to illustrations 1.6, 1.8a, 1.8b, 1.8c, 1.9a, 1.9b and 1.9c

The driveshaft **(see illustration)** is a tube that transmits power from the transmission (2WD models) or the transfer case (4WD models) to the rear axle; on 4WD models, this driveshaft is referred to as the rear driveshaft. On 4WD models, another driveshaft also transmits power from the transfer case to the front axle; this driveshaft is referred to as the front driveshaft.

As the vehicle travels over irregular surfaces, the axles "float" up and down, so the driveshaft must be able to operate while constantly changing its angle relative to the transmission/transfer case and the axle. This is made possible by universal joints (U-joints) at each end of the driveshaft. As the angle between the transmission/transfer and the axle changes, so does the *length* between them, so the driveshaft must also be able to change its length. The necessary change in length is made possible by a sliding yoke, which is also referred to as a sleeve yoke or slip joint. An oil seal prevents leakage of fluid at this point and keeps dirt from entering the transmission. If leakage is evident at the front of the driveshaft, replace the oil seal.

There are two basic types of U-joint systems - single-cardan and double-cardan **(see illustrations)**. On later

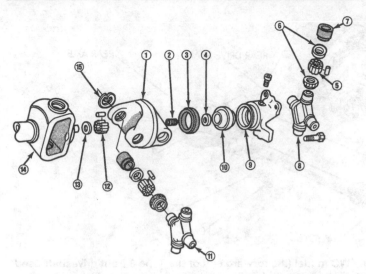

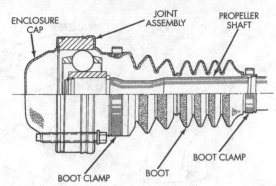

1.8c A cutaway view of the CV joint used on Type 3 front driveshafts on some late model vehicles

1.8b An exploded view of a typical double-cardan U-joint

1	Link yoke	8	Rear spider
2	Socket spring	9	Socket yoke
3	Socket ball retainer	10	Socket ball
		11	Front spider
4	Thrust washer	12	Needle bearings
5	Needle bearings	13	Thrust washer
6	Seal	14	Driveshaft yoke
7	Bearing cap	15	Retaining clip

models, three types of U-joints - single-cardan and double-cardan U-joints, and constant velocity (CV) joints **(see illustration)** - are used. Some U-joints and CV-joints are not repairable. On these vehicles, the driveshaft must be replaced if the joint or its boot is worn or damaged.

Several types of front driveshafts are used for 4WD models. For example Type 1 and Type 2 driveshafts **(see illustrations)** both use a single-cardan joint at the axle end and a double-cardan joint at the transfer case end. The only difference between the two is the means used to protect the slip yoke from dirt and dust. The Type 1 uses a dust cap and seal to protect the yoke; the Type 2 uses a rubber dust boot. The Type 3 driveshaft **(see illustration)** has no slip yoke. It uses a double-cardan joint at the transfer case and a CV joint at the front axle end. The CV joint contracts and expands, so no slip yoke is needed. A splined shaft on the CV joint allows the overall driveshaft length to be adjusted for optimal joint travel. The Type 4 driveshaft has a slip yoke built into the driveshaft.

The driveshaft assembly requires very little service. The universal joints on most newer models are lubricated for life and must be replaced if problems develop. The driveshaft must be removed from the vehicle for this procedure. Since the driveshaft is a balanced unit, it's important that no undercoating, mud, etc. be allowed to build up on it. When the vehicle is raised for service it's a good idea to clean the driveshaft and inspect it for any obvious damage. Also, make sure the small weights used to originally balance the driveshaft are in place and securely attached. Whenever the driveshaft is removed it must be reinstalled in the same relative position to preserve the balance.

Driveaxle

Refer to illustrations 1.11a and 1.11b

On front-wheel drive models power is transmitted from

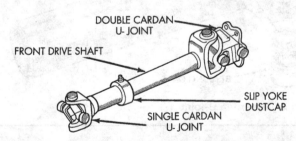

1.9a A Type 1 front driveshaft uses a dust cap for the slip yoke

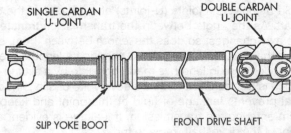

1.9b A Type 2 front driveshaft uses a dust boot for the slip yoke

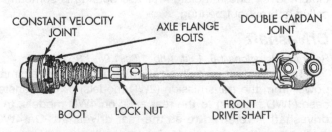

1.9c A Type 3 front driveshaft uses a constant velocity (CV) joint instead of a slip yoke at its forward end

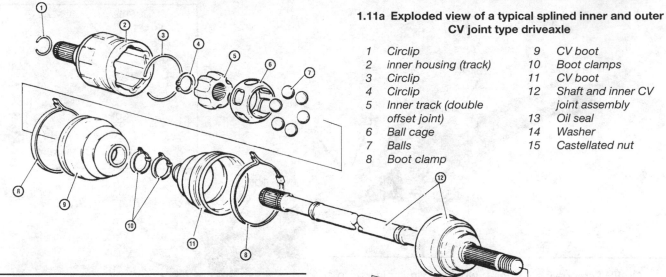

1	Circlip	9	CV boot
2	inner housing (track)	10	Boot clamps
3	Circlip	11	CV boot
4	Circlip	12	Shaft and inner CV joint assembly
5	Inner track (double offset joint)	13	Oil seal
6	Ball cage	14	Washer
7	Balls	15	Castellated nut
8	Boot clamp		

the trans-axle to the wheels through a pair of driveaxles. On rear-wheel drive independent rear suspension the power is transmitted from the transmission to the rear differential, to the rear wheels, through a pair of driveaxles (also called half-shafts) connected at the differential. The inner end of each driveaxle is either splined into the differential side gears (on most vehicles), or bolted to a flange connected to the transaxle or differential **(see illustrations)**. The outer ends of the driveaxles are either splined to the axle hubs and locked in place by a large nut or bolted to flanges connected to the hubs.

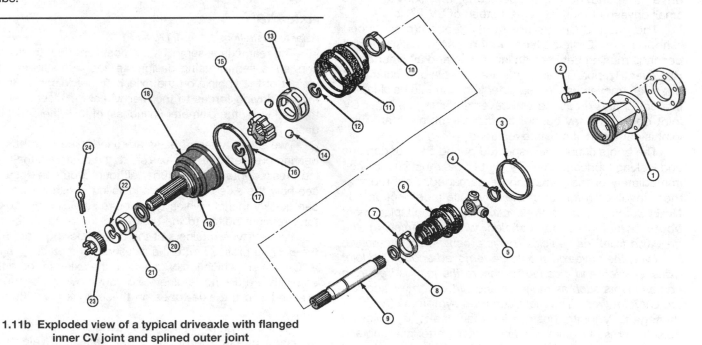

1.11b Exploded view of a typical driveaxle with flanged inner CV joint and splined outer joint

1	Inner housing (track)	7	Boot clamp	13	Cage	19	Wear sleeve
2	Mounting bolts	8	Spacer	14	Balls (6)	20	Washer
3	Boot clamp	9	Shaft	15	Cross (driver)	21	Hub nut
4	Snap ring	10	Boot clamp	16	Boot clamp	22	Spring washer
5	Tripod	11	CV boot	17	Circlip	23	Lock nut
6	CV boot	12	Spacer ring	18	Outer CV joint housing	24	Cotter pin

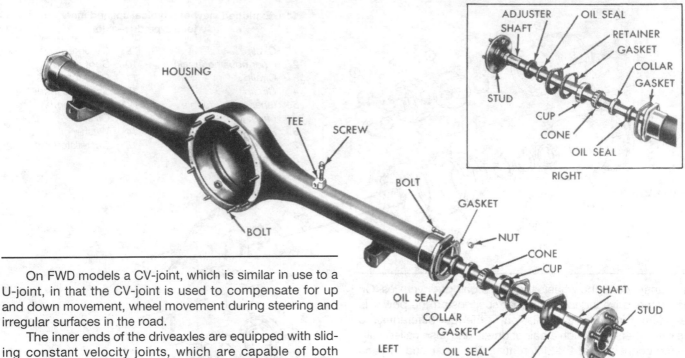

**1.17a Typical exploded view of a flanged type rear axle -
the axle bearing is pressed onto the axle**

On FWD models a CV-joint, which is similar in use to a U-joint, in that the CV-joint is used to compensate for up and down movement, wheel movement during steering and irregular surfaces in the road.

The inner ends of the driveaxles are equipped with sliding constant velocity joints, which are capable of both angular and axial motion. The inner joints are the "double-offset" type, which consists of ball bearings running between an inner race and an outer cage. They can also be disassembled, cleaned and inspected, but they must be replaced as a single unit if defective. On some rear-wheel drive independent rear suspensions, the axles look like small driveshafts with U-joints instead of CV-joints.

The outer CV joints are the "Rzeppa" (pronounced "sheppa") or "Birfield" type, which also consists of ball bearings running between an inner race and an outer cage. However, Rzeppa/Birfield joints are capable of angular - but not axial - movement. These outer joints should be cleaned, inspected and repacked whenever replacing an outer CV joint boot, but they cannot be disassembled. If an outer joint is damaged, it must be replaced.

The boots should be inspected periodically for damage and leaking lubricant. Torn CV joint boots must be replaced immediately or the joints can be damaged. Boot replacement involves removal of the driveaxle (see Section 10). **Note:** *Some auto parts stores carry "split" type replacement boots, which can be installed without removing the driveaxle from the vehicle. This is a convenient alternative; however, the driveaxle should be removed and the CV joint disassembled and cleaned to ensure the joint is free from contaminants such as moisture and dirt which will accelerate CV joint wear.* The most common symptom of worn or damaged CV joints, besides lubricant leaks, is a clicking noise in turns, a clunk when accelerating after coasting and vibration at highway speeds. To check for wear in the CV joints and driveaxle shafts, grasp each axle (one at a time) and rotate it in both directions while holding the CV joint housings, feeling for play indicating worn splines or sloppy CV joints. Also check the driveaxle shafts for cracks, dents and distortion.

Axleshaft

Refer to illustrations 1.17a, 1.17b and 1.18

The rear axle assemblies for most light duty trucks and cars are a semi-floating design, i.e. the axle supports the weight of the vehicle on the axleshaft in addition to transmitting driving forces to the rear wheels. These axles are referred to by the diameter (in inches) of the differential ring gear.

Two basic types of axles are used: One type has the bearing pressed onto the axle shaft; the other has the bearing pressed into the axle tube. Although you can't actually see how the axle bearings are installed in either type, you can quickly identify which type you're servicing by looking for a retainer bolted to the brake backing plate. Axles that use flanged-type retainers at the brake backing plate have the bearing pressed onto the axleshaft. The other type use a "C-washer" which locks to secure the axles to the differential. There are no retainers, so you know the bearing is pressed into the axle tubes on these units **(see illustrations)**.

Heavy duty models use full-floating axles. A full floating axleshaft doesn't carry any of the vehicle's weight; the weight is supported on the axle housing itself by roller bearings. Full-floating axles can be identified by the large hub projecting from the center of the wheel. The axle flange is secured by bolts on the end of the hub. **(see illustration)**.

The rear axle assembly is a hypoid, semi-floating type

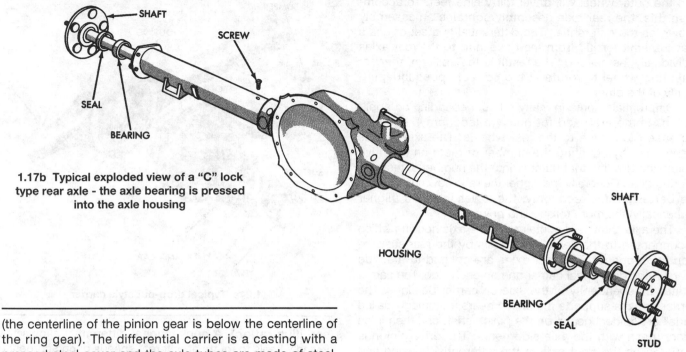

1.17b Typical exploded view of a "C" lock type rear axle - the axle bearing is pressed into the axle housing

(the centerline of the pinion gear is below the centerline of the ring gear). The differential carrier is a casting with a pressed steel cover and the axle tubes are made of steel, pressed and welded into the carrier.

Optional limited-slip rear axles are also available on some vehicles. This type of differential allows for normal operation until one wheel loses traction. There are several different types available: One uses small stout springs with "brake cones" which are a tapered machined cones; the next type uses a set of driven clutch's and large springs and the last unit utilizes multi-disc clutch packs and a speed sensitive engagement mechanism which locks both axleshafts together, applying equal rotational power to both wheels.

Differential

Refer to illustrations 1.24a, 1.24b, 1.24c, 1.25 and 1.26

The rear axles must be able to turn at different speeds to compensate for the difference in the distance the rear wheels travel when rounding a turn. For example on a twenty foot turn the inner wheel will travel thirty one feet

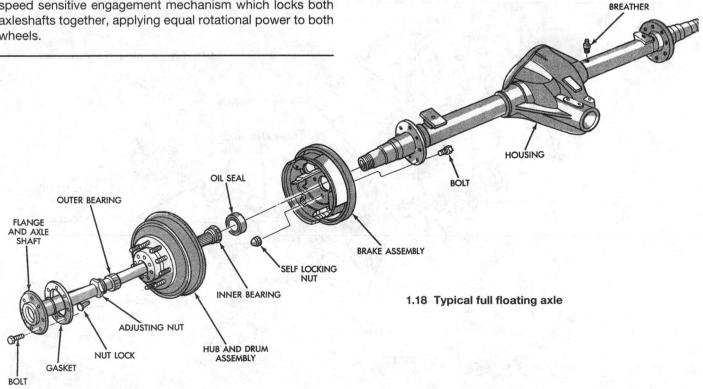

1.18 Typical full floating axle

and the outer wheel will travel thirty nine feet. To accomplish this, the rear axle assembly contains an assembly known as the differential. The differential is a set of gears that transmit torque from the driveshaft to the rear axles individually, as needed. The result is the ability of the rear axle, and wheel to rotate at the correct speed independently of the other.

Differentials come in many ratios, depending on application. Some are used for mileage (economy), some to increase the torque to the rear wheels. The axle ratio is determined by counting the number of teeth on the drive pinion and dividing that number into the number of teeth on the ring gear. Generally the higher the ratio (low numerically) the better the fuel economy; the lower the ratio (higher numerically) the more torque and power.

The axle shafts are contained in the axle housing which is connected to the vehicle's chassis by the rear suspension components. The rear axles are splined to the side gears in the differential carrier one on each side. The carrier rotates independently of the axle on carrier bearings. The carrier or case supports the pinion gears, commonly called "side" or "spider" gears, on the pinion shaft and the pinion gears mesh with the axle side gears. The drive pinion is mounted on the front side of the differential housing and drives the ring gear mounted on the flanged side of the carrier or case. Driven by the drive pinion gear, the differential ring gear transfers the power to the differential case and pinion gears which engage the differential side gears that drive the axle shafts. This setup allows one wheel to turn faster than the other during a turn.

Most vehicles use a conventional differential **(see illustrations)**. But some models have a limited slip differential

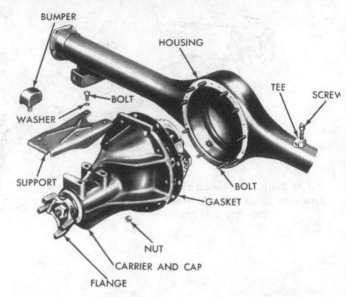

1.24a Typical drop-out style carrier

(see illustration). A conventional differential allows the driving wheels to rotate at different speeds during cornering (the outside wheel must turn faster than the inside wheel), while equally dividing the driving torque between the two wheels. In most normal driving situations, this function of the differential is adequate. However, the total driving torque can't be more than double the torque at the lower traction wheel. When traction conditions aren't the same for both driving wheels, some of the available traction is lost. On many vehicles an optional limited slip differential is

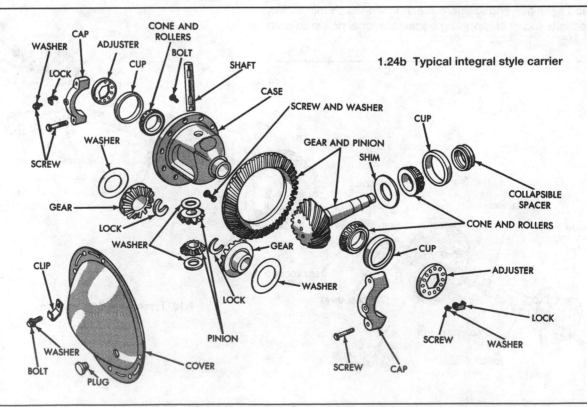

1.24b Typical integral style carrier

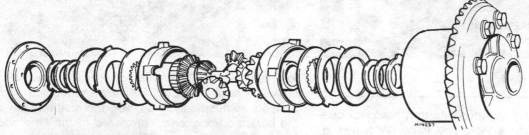

1.24c Typical limited slip differential

1.25 This tag on the differential housing contains information such as the part number and gear ratio

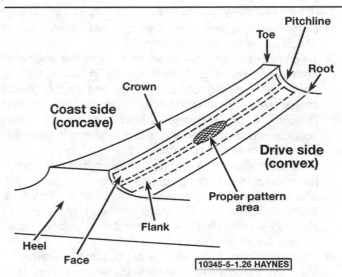

1.26 Typical ring gear tooth pattern

available for performance and poor traction conditions - i.e. mud, ice or snow. During normal conditions, its controlled internal friction is easily overcome, allowing the driving wheels to turn at different speeds, just like a conventional differential. However, during extremely slippery conditions, the limited slip differential allows the driving wheel with better traction condition to develop more driving torque than the other wheel. Thus the total driving torque can be significantly greater than with a conventional differential.

It is important in any repair to use the correct replacement parts for your vehicle. The manufacture provides this information on a stamped metal tag attached to one of the bolts that secures either the carrier or rear cover **(see illustration)**.

There are two preload adjustments necessary for all differentials, the drive pinion bearing preload and carrier bearing preload. There are also two gear adjustments necessary on all differentials. The first is the drive pinion depth; which is the point the drive pinion meshes with the ring gear to produce the best gear pattern and wear **(see illustration)**. The second adjustment is the backlash; this is the actual play between the drive pinion and the ring gear measured in thousands of an inch. Backlash provides clearance for lubrication and allows for heat expansion.

Frequently, when working on the suspension or steering system components, you may come across fasteners which seem impossible to loosen. These fasteners on the underside of the vehicle are continually subjected to water, road grime, mud, etc., and can become rusted or "frozen," making them extremely difficult to remove. In order to unscrew these stubborn fasteners without damaging them

(or other components), be sure to use lots of penetrating oil and allow it to soak in for a while. Using a wire brush to clean exposed threads will also ease removal of the nut or bolt and prevent damage to the threads. Sometimes a sharp blow with a hammer and punch will break the bond between a nut and bolt threads, but care must be taken to prevent the punch from slipping off the fastener and ruining the threads. Heating the stuck fastener and surrounding area with a torch sometimes helps too, but isn't recommended because of the obvious dangers associated with fire. Long breaker bars will increase leverage. Sometimes tightening the nut or bolt first will help to break it loose. Fasteners that require drastic measures to remove should always be replaced with new ones.

Since most of the procedures dealt with in this Chapter involve jacking up the vehicle and working underneath it, a good pair of jackstands will be needed. A hydraulic floor jack is the preferred type of jack to lift the vehicle, and it can also be used to support certain components during various operations. **Warning:** *Never, under any circumstances, rely on a jack to support the vehicle while working on it. Whenever any of the suspension or steering fasteners are loosened or removed they must be inspected and, if necessary, replaced with new ones of the same part number or of original equipment quality and design. Never attempt to heat or straighten any suspension or steering components. Instead, replace any bent or damaged part with a new one.*

Clutch

2 Clutch - check

1   Other than replacing components that have obvious damage, some preliminary checks should be performed to diagnose a clutch system failure.

   a) *Before proceeding, check and, if necessary, adjust clutch pedal freeplay and height.*
   b) *To check "clutch spin down time," run the engine at normal idle speed with the transaxle in Neutral (clutch pedal up - engaged). Disengage the clutch (pedal down), wait several seconds and shift the transaxle into Reverse. No grinding noise should be heard. A grinding noise would most likely indicate a problem in the pressure plate or the clutch disc.*
   c) *To check for complete clutch release, run the engine (with the parking brake applied to prevent movement) and hold the clutch pedal approximately 1/2-inch from the floor. Shift the transaxle between 1st gear and Reverse several times. If the shift is not smooth, component failure is indicated.*
   d) *Visually inspect the clutch pedal bushing at the top of the clutch pedal to make sure there is no sticking or excessive wear.*
   e) *Under the vehicle, verify that the clutch release lever is solidly mounted on the ball stud.*
   f) *Make sure that the hydraulic lines aren't leaking at either the master cylinder or the release cylinder. Bleed the system if necessary.*
   g) *Make sure that the clutch cable is not kinked or frayed. Replace the cable if necessary.*

3 Clutch components - removal, inspection and installation

Warning: *Dust produced by clutch wear and deposited on clutch components may contain asbestos, which is hazardous to your health. DO NOT blow it out with compressed air and DO NOT inhale it. DO NOT use gasoline or petroleum-based solvents to remove the dust. Brake system cleaner should be used to flush the dust into a drain pan. After the clutch components are wiped clean with a rag, dispose of the contaminated rags and cleaner in a labeled, covered container.*

Removal

Refer to illustrations 3.5 and 3.7

1   Access to the clutch components is normally accomplished by removing the transmission/transaxle, leaving the engine in the vehicle. **Note:** *On some front-wheel drive vehicles the engine/transaxle assembly must be removed together as a unit. If, of course, the engine is being removed*

for major overhaul, then the opportunity should always be taken to check the clutch for wear and replace worn components as necessary. However, the relatively low cost of the clutch components compared to the time and labor involved in gaining access to them warrants their replacement any time the engine or transaxle is removed, unless they are new or in near-perfect condition. The following procedures assume that the engine will stay in place.

2   Remove the transmission/transaxle from the vehicle. Support the engine while the transmission is out. On rear-wheel drive models a floor jack can be used to support the engine. On front-wheel drive models, an engine hoist or support fixture should be used to support it from above. If a jack is used underneath the engine, make sure a wood block is placed between the jack and oil pan to spread the load. **Caution:** *The pick-up for the oil pump is very close to the bottom of the oil pan. If the pan is bent or distorted in any way, engine oil starvation could occur.*

3   The release fork and release bearing can remain attached to the transmission/transaxle; however, you should inspect them while the transmission/transaxle is removed.

4   To support the clutch disc during removal, install a clutch alignment tool through the clutch disc hub.

5   Carefully inspect the flywheel and pressure plate for indexing marks. The marks are usually an X, an O or a white letter. If they cannot be found, scribe marks yourself so the pressure plate and the flywheel will be in the same alignment during installation (**see illustration**).

6   Slowly loosen the pressure plate-to-flywheel bolts. Work in a diagonal pattern and loosen each bolt a little at a time until all spring pressure is relieved.

3.5 Mark the relationship of the pressure plate to the flywheel (only necessary if you're going to re-use the same pressure plate)

3.7 Remove the pressure plate and the clutch disc together

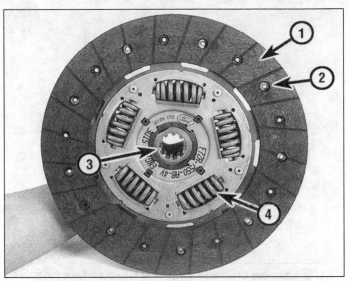

3.10 Examine the clutch disc for evidence of excessive wear, such as burned friction material (1), loose rivets (2), worn hub splines (3) and distorted damper cushions or springs (4)

7 Hold the pressure plate securely and completely remove the bolts. Grasp the pressure plate <u>and</u> clutch disc securely as the last bolt is removed or the clutch disc may drop out **(see illustration)**.

Inspection

Refer to illustrations 3.10, 3.12a and 3.12b

8 Ordinarily, when a problem occurs in the clutch, it can be attributed to wear of the clutch driven plate assembly (clutch disc). However, all components should be inspected at this time.

9 Inspect the flywheel for cracks, heat checking, score marks and other damage. If the imperfections are slight, a machine shop can resurface it to make it flat and smooth.

10 Inspect the lining on the clutch disc. There should be at least 1/16-inch of lining above the rivet heads. Check for loose rivets, distortion, cracks, broken springs and other obvious damage **(see illustration)**. As mentioned above, ordinarily the clutch disc is replaced as a matter of course, so if in doubt about the condition, replace it with a new one.

11 The release bearing should be replaced along with the clutch disc.

12 Check the machined surface and the diaphragm spring fingers of the pressure plate **(see illustrations)**. If the surface is grooved or otherwise damaged, replace the pressure plate assembly. Also check for obvious damage, distortion, cracking, etc. Light glazing can be removed with emery cloth or sandpaper. If a new pressure plate is indicated, new or factory rebuilt units are available.

Installation

Refer to illustration 3.13

13 Carefully wipe the flywheel and pressure plate machined surfaces clean. It's important that no oil or grease is on these surfaces or the lining of the clutch disc. Handle these parts only with clean hands. Position the clutch disc and pressure plate with the clutch held in place

NORMAL FINGER WEAR

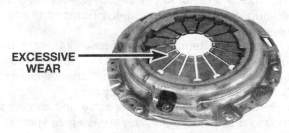

EXCESSIVE WEAR

EXCESSIVE FINGER WEAR

BROKEN OR BENT FINGERS

3.12a Replace the pressure plate if any of these conditions are noted

3.12b Examine the pressure plate friction surface for score marks, cracks and evidence of overheating (blue spots)

3.13 Center the clutch disc in the pressure plate with a clutch alignment tool, then tighten the pressure plate-to-flywheel bolts a little at a time, working around the pressure plate in a criss-cross pattern

with an alignment tool **(see illustration)**. Make sure it's installed properly (most replacement clutch plates will be marked "flywheel side" or something similar - if not marked, install the clutch disc with the damper springs or cushion toward the transaxle).

14   Tighten the pressure plate-to-flywheel bolts only finger tight, working around the pressure plate.

15   Center the clutch disc by ensuring the alignment tool is through the splined hub and into the recess in the crankshaft. Wiggle the tool up, down or side-to-side as needed to bottom the tool. Tighten the pressure plate-to-flywheel bolts a little at a time, working in a crisscross pattern to prevent distortion of the cover. After all of the bolts are snug, tighten them securely. Remove the alignment tool.

16   Using high-temperature grease, lubricate the inner groove of the release bearing. Also place grease on the release lever contact areas and the transaxle input shaft.

17   Install the clutch release bearing.

18   Install the transmission/transaxle and all components removed previously, tightening all fasteners securely.

4   Clutch release bearing and lever - removal, inspection and installation

Warning: *Dust produced by clutch wear and deposited on clutch components may contain asbestos, which is hazardous to your health. DO NOT blow it out with compressed air and DO NOT inhale it. DO NOT use gasoline or petroleum-based solvents to remove the dust. Brake system cleaner should be used to flush it into a drain pan. After the clutch components are wiped clean with a rag, dispose of the contaminated rags and cleaner in a labeled, covered container.*

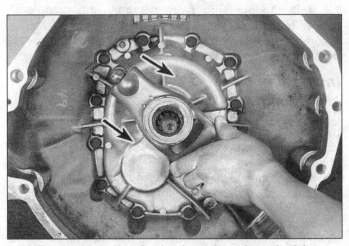

4.4 To disengage the clutch release lever from the fulcrum ball stud, pull the release lever in the direction of the arrows until it pops loose

Removal

Refer to illustration 4.4

1   Raise the vehicle and place it securely on jackstands.

2   Remove the transmission/transaxle.

3   Remove the bellhousing (if necessary).

4   Remove the clutch release lever from the ball stud, then remove the bearing from the lever **(see illustration)**.

Inspection

Refer to illustrations 4.5a, 4.5b and 4.5c

5   Hold the center of the bearing and rotate the outer portion while applying pressure **(see illustration)**. If the bearing doesn't turn smoothly or if it's noisy, replace it with a new one. Wipe the bearing with a clean rag and inspect it for damage, wear and cracks. The bearing is sealed for life, so don't immerse it in solvent or you'll ruin it. Inspect the friction surfaces and the bearing retainer on the release

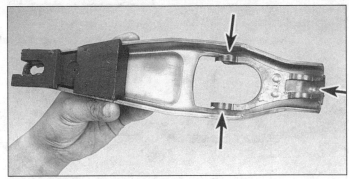

4.5b Inspect the ballstud socket (right arrow) for wear and make sure the retainer fingers (arrows) are still strong; if the socket is worn, or the fingers are weak, bent or broken, replace the release lever

4.5a To check the release bearing, hold it by the outer cage and rotate the inner race while applying a side load to it - the bearing should turn smoothly and easily; if it doesn't, replace it

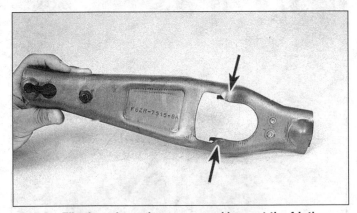

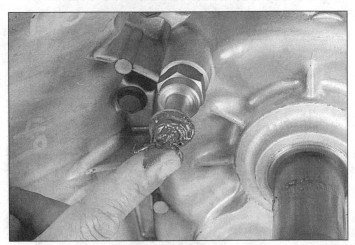

4.6a Lubricate the release lever fulcrum stud . . .

4.5c Flip the release lever over and inspect the friction surfaces (arrows) that push the release bearing; if they're excessively worn, replace the release lever

Installation

Refer to illustrations 4.6a, 4.6b, 4.6c, 4.6d, 4.6e, 4.7 and 4.8

lever **(see illustrations)**. If the friction surfaces are worn excessively, or the retainer fingers are weak, bent or broken, replace the release lever.

6   Lightly lubricate the release lever ball stud, the release lever and the release bearing with high temperature grease **(see illustrations)**.

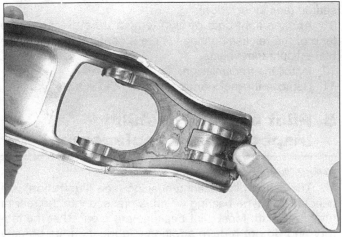

4.6b . . . the fulcrum ball stud socket . . .

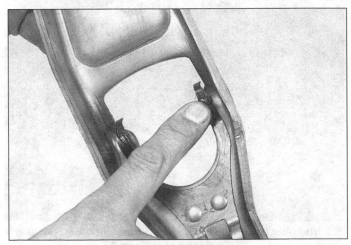

4.6c . . . the retainer fingers . . .

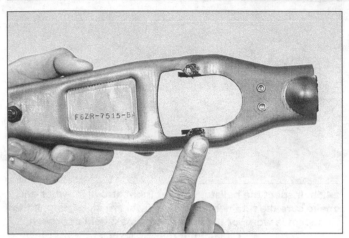

4.6d ... and the friction surfaces of the release lever

4.6e Lubricate the inner hub and the thrust side (where the release fingers contact it) of the release bearing with high-temperature grease

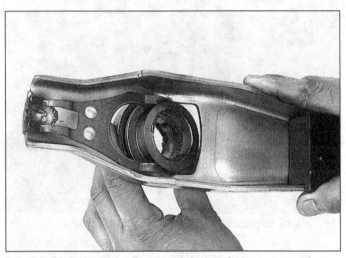

4.7 Make sure the fingers of the retainer are properly engaged with the groove in the release bearing

4.8 This is how the release bearing and release lever should look when properly installed; don't forget to lubricate the front bearing surface (arrows) of the release bearing

7 Attach the release bearing to the clutch lever (see illustration).

8 Push the release lever onto the ball stud until it's firmly seated (see illustration).

9 Apply a light coat of high temperature grease to the face of the release bearing, where it contacts the pressure plate diaphragm fingers.

10 Install the transmission.

11 Remove the jackstands and lower the vehicle.

5 Pilot bearing/bushing - inspection and replacement

Refer to illustrations 5.1, 5.5, 5.8, 5.9 and 5.10

1 The clutch pilot bearing/bushing (see illustration) is a needle roller type bearing which is pressed into the rear of the crankshaft. Most pilot bearings are greased at the factory and do not require additional lubrication. Its primary purpose is to support the front of the transmission/trans-

5.1 The pilot bearing (arrow) should turn smoothly and quietly if it does not replace it

5.5 Remove the old pilot bearing with a small slide-hammer, as shown

5.8 If you don't have a small slide-hammer, pack the cavity behind the pilot bearing with grease . . .

5.9 . . . then force out the bearing hydraulically with a steel rod or wood dowel slightly smaller in diameter than the bearing bore - when the hammer strikes the rod or dowel, the grease will transmit this force to the backside of the bearing and push it out

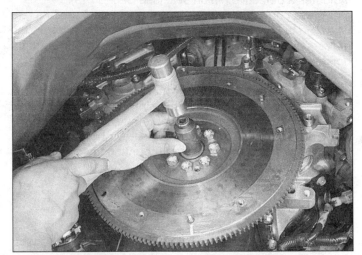

5.10 Use a large socket and hammer, or a soft-faced hammer, to install the new pilot bearing; make sure the bearing is fully seated

axle input shaft. The pilot bearing/bushing should be inspected whenever the clutch components are removed from the engine. Due to its inaccessibility, if you are in doubt as to its condition, replace it with a new one. **Note:** *If the engine has been removed from the vehicle, disregard the following steps which do not apply.*

2   Remove the transmission.

3   Remove the clutch components.

4   Inspect for any excessive wear, scoring, lack of grease, dryness or obvious damage. If any of these conditions are noted, the bearing/bushing should be replaced. A flashlight will be helpful to direct light into the recess.

5   The pilot bearing can be removed with a special puller and slide hammer **(see illustration)**, but if you don't have a suitable tool, the following alternative method works as well.

6   Obtain a solid steel bar or wood dowel which is slightly smaller in diameter than the bearing.

7   Check the bar or dowel for fit - it should just slip into the bearing with very little clearance.

8   Pack the bearing and the area behind it (in the crankshaft recess) with heavy grease **(see illustration)**. Pack it tightly to eliminate as much air as possible.

9   Insert the bar or dowel into the bearing bore and strike it sharply with a hammer **(see illustration)**; this will force the grease to the backside of the bearing and push it out. Remove the bearing and clean all grease from the crankshaft recess.

10   To install the new bearing or bushing, lightly lubricate the outside surface with lithium-based grease, then drive it into the recess with a large socket **(see illustration)**. If the bearing is equipped with a seal, make sure that the seal faces out, toward the transmission. Pilot bushings should be lubricated with a light film of high-temperature grease.

11   Install the clutch components, transmission and all other components removed previously, tightening all fasteners securely.

Driveshaft

6 Driveshaft inspection

1   Raise the rear of the vehicle and support it securely on jackstands. Block the front wheels to keep the vehicle from rolling off the stands.

2   Crawl under the vehicle and visually inspect the driveshaft(s). Look for any dents or cracks in the tubing. If any are found, the driveshaft must be replaced.

3   Check for oil leakage at the front and rear of the driveshaft. Leakage where the driveshaft enters the transmission extension housing indicates a defective extension housing seal. Leakage where the driveshaft enters the differential indicates a defective pinion seal (see Section 19).

4   While under the vehicle, have an assistant rotate a rear wheel so the driveshaft will rotate. As it does, make sure the universal joints are operating properly without binding, noise or looseness.

5   The universal joints can also be checked with the driveshaft motionless, by gripping your hands on either side of a joint and attempting to twist the joint. Any movement at all in the joint is a sign of considerable wear. Lifting up on the shaft will also indicate movement in the universal joints.

6   Finally, check the driveshaft mounting bolts at the ends to make sure they're tight.

7 Driveshaft - removal and installation

1   Disconnect the negative cable from the battery.

2   Raise the vehicle and support it securely on jackstands. Place the transmission in Neutral with the parking brake off.

Rear driveshaft

Removal

Refer to illustrations 7.3 and 7.4

3   On conventional driveshafts, make reference marks on the driveshaft and the pinion flange in line with each other **(see illustration)**. This is to make sure the driveshaft is reinstalled in the same position to preserve the balance. On models with a CV or double cardan joint, make reference marks on the driveshaft, the rear constant velocity U-joint assembly and the pinion flange.

4   On conventional driveshafts, remove the rear universal joint bolts and clamps **(see illustration)**. Turn the driveshaft (or wheels) as necessary to bring the bolts into the most accessible position. On CV models, remove the nuts and lockwashers attaching the rear constant velocity U-joint to the rear axle pinion flange **(see illustration 8.13)**.

5   On conventional driveshafts, tape the bearing caps to the cross to prevent the caps from coming off during removal.

6   Lower the rear of the driveshaft. Slide the front of the driveshaft out of the transmission extension housing.

7   Wrap a plastic bag over the transmission extension housing and hold it in place with a rubber band. This will prevent loss of fluid and protect against contamination while the driveshaft is out.

Installation

8   Remove the plastic bag from the transmission extension housing and wipe the area clean. Inspect the oil seal carefully. Procedures for replacement of this seal can be found in Chapter 7.

9   Slide the front of the driveshaft into the transmission.

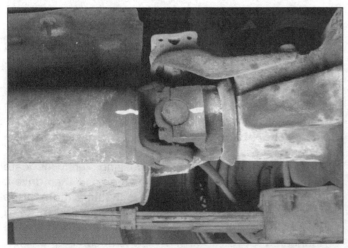

7.3 Always make reference marks on the driveshaft and pinion flange to insure that the driveshaft balance is maintained

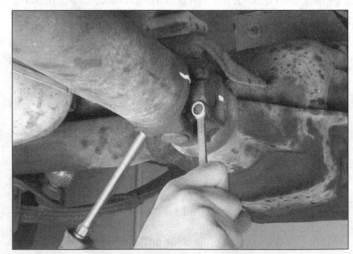

7.4 Use a large screwdriver to hold the flange while breaking the U-joint bolts loose

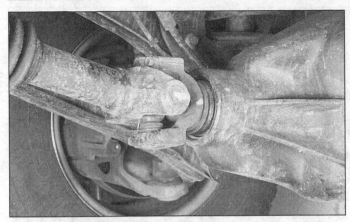

7.12a Before removing the front driveshaft from a 4WD
model, mark the U-joint and pinion yokes . . .

7.12b . . . and mark the transfer case output shaft flange and
the driveshaft U-joint flange

10 Raise the rear of the driveshaft into position, checking
to be sure the marks are in alignment. If not, turn the rear
wheels to match the pinion flange and the driveshaft.

11 Remove the tape securing the bearing caps and install
the clamps and bolts. Lower the vehicle and connect the
negative battery cable.

Front driveshaft (4WD models only)

Refer to illustrations 7.12a and 7.12b

12 Be sure to make alignment marks on the driveshaft
yokes, the axle pinion yoke and the transfer case output
flange **(see illustrations)**.

13 Remove the U-joint strap bolts **(see illustration 7.4)**.
Note: *The manufacturer recommends using new straps and
bolts upon installation.*

14 Remove the bolts from the transfer case yoke flange.

15 Remove the driveshaft.

16 Installation is the reverse of removal.

8 Universal joints - replacement

Conventional U-joints

Refer to illustrations 8.3a, 8.3b, 8.3c, 8.5a, 8.5b and 8.9

Note: *A press or large vise will be required for this procedure. It's not a good idea to repeatedly beat the driveshaft
with a hammer. If you don't have the equipment it may be a
good idea to take the driveshaft to a repair or machine shop
where the universal joints can be replaced for you, normally
at a reasonable charge. The following overhaul procedure
depicts a rear U-joint, but the procedure for rebuilding the
front U-joint is basically the same.*

1 Remove the driveshaft (see Section 7).

2 Apply penetrating oil to the bearing caps.

3 Remove the bearing cap retainers, if equipped **(see
illustrations)**. **Note:** *Some models have snap-rings on the*

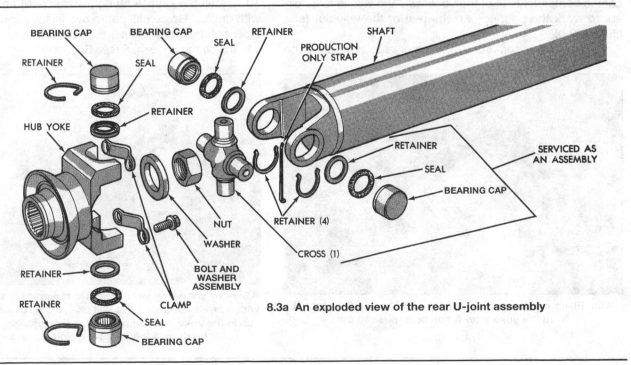

8.3a An exploded view of the rear U-joint assembly

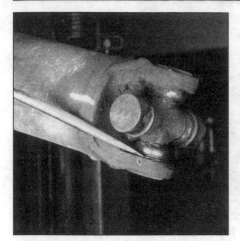

8.3b On some models remove the inner retainers from the U-joint by tapping them off with a screwdriver and hammer . . .

8.3c . . . or a pair of needle-nose pliers can be used to remove the outer retainers from the U-joint

8.5a To press the universal joint out of the driveshaft yoke, set it up in a vise with the small socket pushing the joint and bearing cap into the large socket

outer surface of the bearing caps, while on other models the snap-rings are seated in a groove on the inner surface of the cap. Additionally, on some models the bearing caps are retained by injected-plastic retaining rings - the plastic breaks as the joint is disassembled. Snap-rings will be installed during reassembly (they'll be included in the U-joint kit).

4   Supporting the driveshaft, place it in position on either an arbor press or on a workbench equipped with a vise.

5   Place a piece of pipe or a large socket, having an inside diameter slightly larger than the outside diameter of the bearing caps, over one of the bearing caps. Position a socket with an outside diameter slightly smaller than that of the opposite bearing cap against the cap **(see illustration)** and use the vise or press to force the bearing cap out (inside the pipe or large socket), stopping just before it comes completely out of the yoke. Use the vise or large pliers to work the bearing cap the rest of the way out **(see illustration)**.

6   Transfer the sockets to the other side and press the

opposite bearing cap out in the same manner.

7   Coat the new universal joint bearings with bearing grease. Position the spider in the yoke and partially install one bearing cap in the yoke. If the replacement spider is equipped with a grease fitting, be sure it's offset in the proper direction (toward the driveshaft).

8   Start the spider into the bearing cap and then partially install the other bearing cap. Align the spider and press the bearing caps into position, being careful not to damage the dust seals.

9   Install the bearing cap retainers or snap-rings. If difficulty is encountered in seating the retainers/snap-rings, strike the driveshaft yoke sharply with a hammer. This will spring the yoke ears slightly and allow the retainers to seat in their grooves **(see illustration)**.

10 Install the grease fitting (if equipped) and fill the joint with grease. Be careful not to overfill the joint, as this could blow out the grease seals.

11 Install the driveshaft (see Section 7).

8.5b Pliers can be used to grip the bearing cap to detach it from the yoke after it has been pushed out

8.9 If the snap-ring will not seat in the groove, strike the yoke with a brass hammer - this will relieve the tension that has set up in the yoke, and slightly spring the yoke ears (this should also be done if the joint feels tight when assembled)

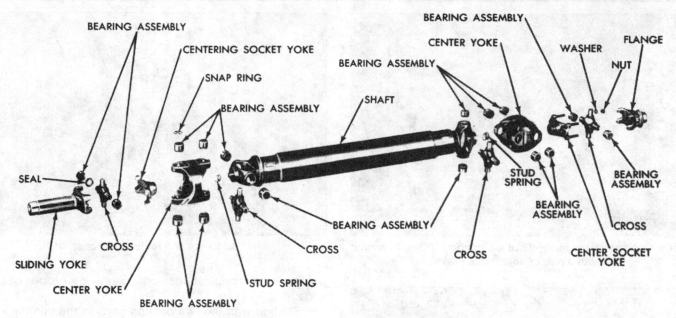

8.12a A typical double-cardan universal joint

Double-cardan universal joints

Refer to illustrations 8.12a, 8.12b, 8.13, 8.14a, 8.14b, 8.15a, 8.15b, 8.16a, 8.16b, 8.17, 8.18, 8.19, 8.20a, 8.20b, 8.22a, 8.22b, 8.22c, 8.22d and 8.22e

Note: *The following overhaul procedure depicts a rear double-cardan U-joint, but the procedure for rebuilding the front U-joint is basically the same. The only difference is that, instead of a companion flange bolted to the drive flange (as used at the differential end of the driveshaft), the front end uses a centering socket bolted to the slip yoke. So, overhauling the front U-joint is actually a little easier than the rear unit depicted here, because you only have to press out two front spider bearing caps; the other two can be replaced by simply unbolting the centering socket from the slip yoke.*

12 Remove the eight snap-rings that retain the bearing caps **(see illustrations)**. **Note:** *Some models have snap-rings on the outer surface of the bearing caps, while on other models the snap-rings are seated in a groove on the inner surface of the cap. Additionally, on some models the bearing caps are retained by injected-plastic retaining rings - the plastic breaks as the joint is disassembled. Snap-rings will be installed during reassembly (they'll be included in the U-joint kit).*

13 Mark the relationship of the companion flange, the center yoke and the driveshaft yoke **(see illustration)** or, if you're overhauling the front U-joint, the slip yoke, the centering socket yoke, the center yoke and the front driveshaft yoke. **Caution:** *The U-joints must be assembled with these*

8.12b Remove all the snap-rings

8.13 Marking the relationship of the parts to one another is essential, if you want to preserve dynamic balance and fit; if you're rebuilding the rear U-joint (the one depicted in these photos), mark the companion flange, the center yoke and the driveshaft yoke; if you're overhauling the front U-joint, mark the slip yoke, the centering socket yoke, the center yoke and the front driveshaft yoke.

8.14a Once you have pushed out a cap about 3/8-inch, twist it out with a pair of adjustable pliers . . .

8.14b . . . and remove the cap

8.15a Once you have removed the center yoke, you can remove the two bearing caps from the driveshaft yoke in the same fashion: Place a large socket on one side and a smaller socket on the opposite side, then push the cap on the right into the larger socket until it protrudes about 3/8-inch as described above, twist the cap off with a pair of adjustable pliers . . .

pieces in their original positions to provide proper clearance.

14 Starting with the two bearing caps in the companion flange end of the center yoke, position the U-joint in a bench vise as shown, with a smaller socket on one side, to push against the bearing cap and spider, and a larger socket on the other side, with an inside diameter large enough to allow the opposite bearing cap to protrude into it without interference **(see illustration 8.13)**. Tighten the jaws of the vise until the bearing protrudes about 3/8-inch out of the yoke. Loosen the vise jaws, rotate the U-joint 90-degrees so that you can get at the protruding bearing cap, work the cap out with a pair of adjustable pliers and remove the cap **(see illustrations)**. Then rotate the U-joint another 90-degrees, install the large and small socket again and press out and remove the opposite bearing cap (the one you just pushed in to drive out the first cap). Note that the companion flange spider bearings are being removed first. Then do the other two bearing caps for that spider. It's easier to start with the outermost spider, regardless of which

8.15b . . . and remove the cap; then flip the U-joint around and remove the other bearing cap the same way

8.16a Place the companion flange in a bench vise as shown and remove the dust cover and centering ball: Since you're not going to reuse it, you can knock the old dust cover off with a hammer and chisel

8.16b Once you've got the dust cover off, rotate the centering ball as shown, grab it with the bench vise jaws and carefully tap off the companion flange

8.17 Carefully tap a new centering ball and dust cover onto the companion flange with a large socket; just make sure you don't damage the bearing surface of the centering ball (Dana units simply use a circular, flat seal, not a dust cover, to protect the centering ball)

U-joint you're overhauling. In other words, if you're rebuilding the rear U-joint, remove the spider and bearing caps that attach the companion flange to the rear end of the center yoke, then remove the spider and bearing caps that attach the forward end of the center yoke to the rear driveshaft yoke. If you're rebuilding the front U-joint, remove the spider and caps that attach the slip yoke and center yoke to the front of the center yoke, then do the spider and caps that attach the rear end of the center yoke to the forward driveshaft yoke.

15   After you've got the center yoke off, remove the two bearing caps from the driveshaft yoke in the same fashion as described above **(see illustrations)**.

16   Place the companion flange in a bench vise as shown and remove the dust cover and centering ball **(see illustrations)**.

17   Install a new centering ball and dust cover on the com-

panion flange **(see illustration)**.

18   Install the new spider bearing in the driveshaft yoke and secure it with a couple of new bearing caps **(see illustration)**. Make sure the grease fitting on the spider faces *away* from the center yoke; if the spider is installed with the grease fitting facing toward the center yoke, it could create a clearance problem, and you won't be able to grease the spider. Don't forget to install a new seal with each new bearing cap **(see illustration 8.12a)**. **Note:** *The seals are integral with the caps on some rebuild kits.*

19   Place the driveshaft yoke in the bench vise and press the two caps into the yoke until they're flush with the driveshaft yoke **(see illustration)**.

20   Install the center yoke on the other two legs of the new spider and hold it in place with new bearing caps and seals

8.18 Install a new spider bearing into the driveshaft yoke and secure it with new bearing caps (be sure to install a new seal with each new bearing cap unless, of course, the seals are integral with the caps, as they are in the Dana kit used in these photos)

8.19 Place the driveshaft yoke in the bench vise and press two opposite caps into the yoke until they're flush with the driveshaft yoke, then install a pair of sockets and press the caps into the yoke until they're fully seated (tops of caps flush with snap-ring grooves)

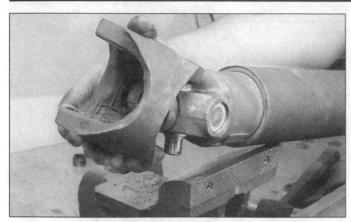

8.20a Install the center yoke on the other two legs of the new spider (make sure the marks you made on the center yoke and the driveshaft yoke are aligned) . . .

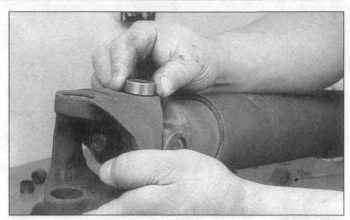

8.20b . . . and hold it in place with new bearing caps and seals place the driveshaft in the vise, press the two caps into the center yoke until they're flush with the yoke, install the two sockets, press the caps all the way in and install the snap-rings

(see illustrations). Make sure the marks you made on the center yoke and the driveshaft yoke are lined up. Place the driveshaft in the vise and press the two caps into the center yoke until they're flush with the yoke.

21 Install a pair of sockets between each opposing pair of bearing caps and press in the caps until they're fully seated (the tops of the bearing caps must be flush with the lower edges of the snap-ring grooves in the driveshaft yoke, center yoke and companion flange (or the driveshaft yoke and

center yoke, if you're rebuilding a front U-joint).

22 Install the centering spring **(see illustration)** and a new center seal on the centering stud and guide the centering stud into the centering ball. Connect the companion flange to the other end of the center yoke with a new spider **(see illustration)**. Make sure the grease fitting on the spider faces away from the center yoke **(see illustration)**; if the spider is installed with the grease fitting facing toward the center yoke, it could create a clearance problem, and you won't be able to grease the spider. Tap two bearing caps into the center yoke to hold everything together **(see illustration)**, press in the caps so they're flush with the center yoke, press the other two caps into the companion flange until they're flush with the companion flange, then press in all four caps with a pair of sockets until they're fully seated. Install new snap-rings **(see illustration)**. **Note:** *If difficulty is encountered in seating the snap-rings, strike the driveshaft yoke(s) sharply with a hammer. This will spring the yoke ears slightly and allow the snap-rings to seat in the groove* **(see illustration)**.

8.22a Install the centering spring and, if applicable, a new dust seal for the centering ball

8.22b Attach the companion flange to the center yoke with the other new spider

8.22c Make sure the grease fitting on the spider faces away from the center yoke (toward the companion flange) to avoid clearance problems, and to provide access to the fitting for lubrication

8.22d Tap the new bearing caps into place to hold the spider, then press in the caps the same way you did the caps for the other spider, with a pair of sockets

8.22e After the new bearing caps are fully seated (flush with the snap-ring grooves in the yokes), install new snap-rings

23 Disassembling and reassembling a front U-joint is a similar procedure, except that the front spider can be disconnected from the slip yoke by removing the centering socket.

9 Driveshaft center bearing - check and replacement

Refer to illustration 9.1

1 The center bearing can be checked in the same manner as the U-joints. Inspect the rubber center bearing mounts for damage and deterioration **(see illustration)**.
2 Remove the driveshaft assembly.
3 Mount the front driveshaft assembly in a vise and pull the bearing support and insulator away from the bearing.

4 Use a hammer to bend the slinger away from the bearing enough to allow sufficient clearance to install a bearing puller.
5 Remove the bearing with a puller.
6 Remove the slinger.
7 Always use an overhaul kit when reassembling the center bearing assembly.
8 To assemble the bearing, position the slinger, bearing assembly and retainer on the shaft.
9 Use a piece of proper size pipe to drive the parts onto the shaft until they bottom on the shoulder (be careful not to damage the shaft splines).
10 Place the slip yoke seal cap and seal on the center bearing splines for installation purposes.
11 Install the driveshaft assembly and crimp the seal cap tabs to the slip yoke.

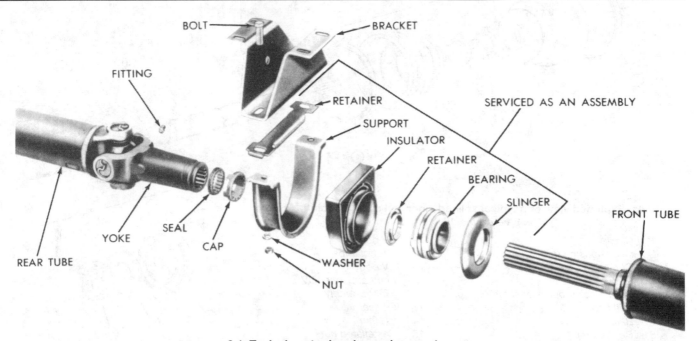

9.1 Typical center bearing and support

Driveaxles

10 Driveaxles - removal and installation

Front driveaxle

Splined type driveaxle

Removal

Refer to illustrations 10.4a, 10.4b, 10.4c, 10.5, 10.6, 10.9, 10.10a, 10.10b and 10.10c

1   Disconnect the cable from the negative terminal of the battery.
2   Loosen the front wheel lug nuts, raise the vehicle and support it securely on jackstands.
3   Remove the wheel.
4   If equipped, remove the cotter pin, the nut lock and the felt washer from the driveaxle/hub nut **(see illustrations)**.

5   Remove the driveaxle/hub nut and washer. To prevent the hub from turning, wedge a prybar between two of the wheel studs and allow the prybar to rest against the ground

10.4b Remove the cotter pin . . .

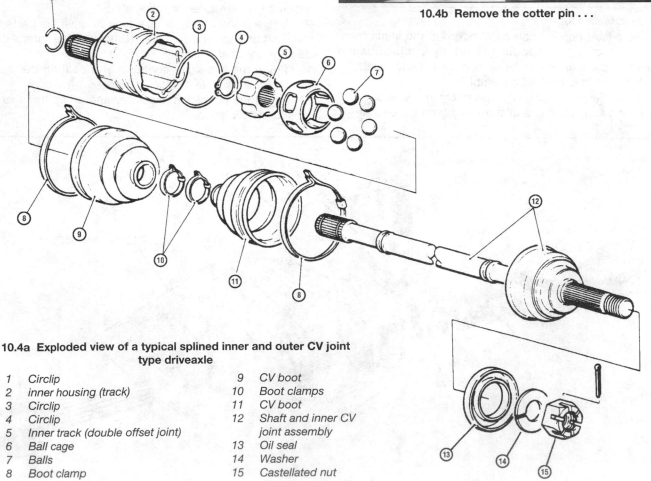

10.4a Exploded view of a typical splined inner and outer CV joint type driveaxle

1	Circlip	9	CV boot
2	inner housing (track)	10	Boot clamps
3	Circlip	11	CV boot
4	Circlip	12	Shaft and inner CV
5	Inner track (double offset joint)		joint assembly
6	Ball cage	13	Oil seal
7	Balls	14	Washer
8	Boot clamp	15	Castellated nut

10.4c . . . the bearing nut lock and the felt washer from the driveaxle/hub nut

10.5 Remove the driveaxle/hub nut and washer. To prevent the hub from turning, wedge a prybar between two of the wheel studs and allow the prybar to rest against the ground or the floorpan of the vehicle

or the floorpan of the vehicle **(see illustration)**.

6   If the driveaxle splines are "frozen," free them by tapping the end of the driveaxle with a soft-faced hammer or a hammer and a brass punch **(see illustration)**.

7   Remove the engine splash shields, if equipped. Place a drain pan underneath the transaxle to catch the lubricant that may spill out when the driveaxles are removed.

8   Disconnect the control arm from the steering knuckle.

9   Pull out on the steering knuckle and detach the driveaxle from the hub **(see illustration)**.

10   Carefully pry the inner CV joint out of the transaxle **(see illustration)** and remove the driveaxle assembly. If equipped with equal-length driveaxles and an intermediate shaft, the inner CV joint housing on one driveaxle may terminate at a support bracket. To detach the driveaxle assembly from the bracket, remove the retainer-to-bracket bolts, mark the relationship of the bearing to the support bracket and pull out the driveaxle assembly **(see illustrations)**. Insert a plug into the differential housing to prevent lubricant loss. Do not try to separate the bearing from the

10.6 If the driveaxle splines are "frozen," knock the driveaxle loose with a hammer and brass punch

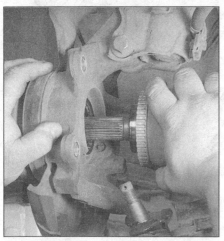

10.9 Pull out on the steering knuckle and detach the driveaxle from the hub

10.10a If the driveaxle is splined directly into the transaxle, carefully pry the inner CV joint out of the transaxle and remove the driveaxle assembly

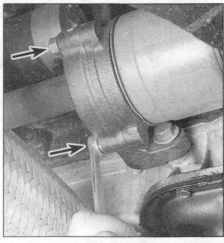

10.10b If one driveaxle is connected to an intermediate shaft, remove the retaining bolts (arrows) . . .

inner CV joint until you have the entire assembly on the bench (see Section 12).

Installation

11 Install a new driveaxle oil seal if necessary (see Section 15).

12 Installation is the reverse of the removal procedure with the following additional points.

a) *When installing the driveaxle, push the driveaxle in sharply to seat the retaining ring on the inner CV joint in its groove in the differential side gear.*

b) *Tighten the control arm balljoint-to-steering knuckle securely.*

c) *Tighten the driveaxle/hub nut securely, then install the nut lock and a new cotter pin. On models that use a self-locking nut (no cotter pin), be sure to install a new nut.*

d) *Install the wheel and lug nuts, lower the vehicle and tighten the lug nuts securely.*

e) *Check the transaxle or differential lubricant and add, if necessary, to bring it to the proper level.*

Flanged type driveaxle

Removal

Refer to illustrations 10.16a, 10.16b, 10.17a, 10.17b and 10.17c

13 Disconnect the cable from the negative terminal of the battery.

14 Loosen the front wheel lug nuts, raise the vehicle and support it securely on jackstands.

15 Remove the wheel.

10.10c ... mark the relationship of the bearing retainer to the support bracket and remove the driveaxle assembly

16 Remove the bolts **(see illustrations)** from the inner CV joint and detach it from the inner flange. **Note:** *The bolts used to retain the CV joints to the hubs vary depending on the manufacturer. There are several different types used, Torx, 12-point head, reverse Torx and Allen-head, etc.*

17 Remove the bolts from the outer CV joint **(see illustration)** and detach it from the outer flange (if equipped). On some models the outer CV joint may be splined to he hub, remove the axle nut and press the stub axle out of the wheel hub with a puller **(see illustration)**. If necessary, prevent the hub from turning by reinstalling two lug bolts and wedging a large screwdriver or prybar between them.

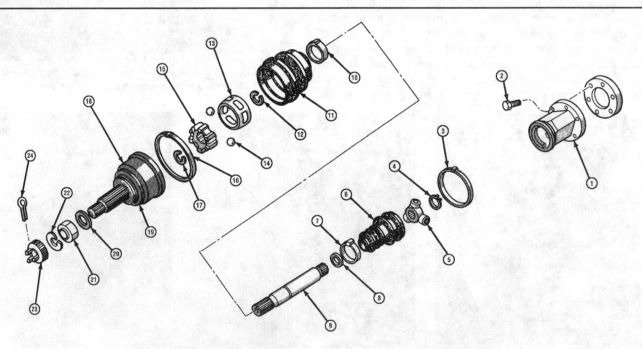

10.16a Exploded view of a typical driveaxle with flanged inner CV joint

1	Inner housing (track)		4	Snap ring		7	Boot clamp
2	Mounting bolts		5	Tripod		8	Spacer
3	Boot clamp		6	CV boot		9	Shaft

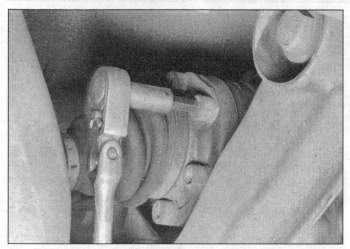

10.16b Remove the retaining bolts from the inner CV joint and detach the joint from the drive flange

10.17a Remove the retaining bolts from the outer CV joint and detach the joint from the drive flange

10.17b A two-jaw puller works well for pushing the stub axle from the hub

10.17c Brace a pry bar across two studs to prevent the hub from turning as the nut is loosened

Installation

18 Install a new driveaxle oil seal if necessary (see Section 15).
19 Installation is the reverse of the removal procedure.
20 Install the wheel and lug nuts, lower the vehicle and tighten the lug nuts securely
21 Check the transaxle or differential lubricant and add, if necessary, to bring it to the proper level.

Rear driveaxle

Splined type driveaxle

Removal

Refer to illustrations 10.24, 10.26a, 10.26b, 10.27a, 10.27b and 10.28

22 On models equipped with rear disc brakes, remove the rear caliper and disc.
23 On all models, remove the rear hub nut.
24 Remove the upper control arm nut and bolt **(see illustration)**.

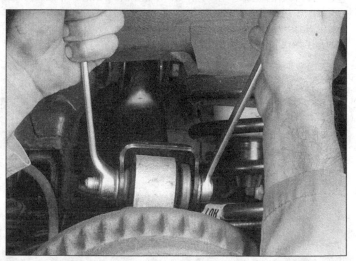

10.24 Removing the upper control arm nut

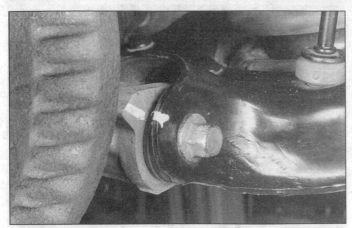

10.26a With the lower arm in the relaxed position, mark the position of the lower control arm to the knuckle

10.26b Remove the nut and bolt retaining the lower control arm to the knuckle

10.27a Use a puller to push the driveaxle from the hub - on models with drum brakes a special attachment that fits over the studs will be necessary, or this method may be used if the drum is removed first

10.27b Support the knuckle assembly with wire or rope

25 Use wire or rope to support the upper control arm.

26 With the lower arm in the relaxed position, reference mark the position of the lower control arm to the knuckle **(see illustration). Caution:** *Failure to mark this position will cause bushing "wind-up" on assembly and the wrong ride*

height. Remove the nut and bolt retaining the lower control arm-to-knuckle **(see illustration)**.

27 With the hub nut removed, use a two-jaw puller to push the driveaxle out of the hub **(see illustration)**. Allow the driveaxle to rest on the lower control arm. Support the knuckle assembly with rope or wire **(see illustration)**.

10.28 Pry the driveaxle loose from the differential housing

10.31 Always replace the circlip on the inner CV joint stub shaft (if equipped) before reinstalling the driveaxle

12.2a Pry up the retaining tabs on the boot clamps . . .

12.2b . . . then open the clamps and remove them from the boot

28   Pry the driveaxle loose from the differential **(see illustration)**. **Caution:** *Care must be taken not to damage the differential oil seal, the housing, the CV joint boots or the anti-lock brake sensor ring (if equipped).*
29   It's a good idea to replace the differential oil seal whenever the driveaxle is removed (see Section 15).
30   Insert a plug into the differential housing to prevent lubricant loss.

Installation
Refer to illustration 10.31
31   Install a new circlip on the inner end of the driveaxle **(see illustration)**. Do not bend or twist the circlip.
32   Remove the plug from the differential housing.
33   Lightly lubricate the driveaxle splines and carefully align the splines of the inner stub shaft with the splines in the differential.
34   Push the driveaxle into the differential until you feel the circlip engage with the groove in the differential side gear.
35   With the exception of using new hub nuts, the remainder of installation is the reverse of removal. Be sure to align the previously made matchmarks before tightening the lower control arm-to-knuckle bolts. Also, tighten the knuckle-to-control arm bolts securely.

Flanged or strapped type driveaxle

Removal
36   Scribe alignment marks on the CV or U-joint at both ends of the driveaxle, inline with marks made on the flange or yoke. This is ensure the drive axle is reinstalled in the same relative position.
37   Remove the bolts from the inner CV joint and detach it from the inner flange. On U-joint types remove the bolts and straps and detach it from the yoke. **Note:** *The bolts used to retain the CV or U-joints to the hub/yoke vary depending on the manufacturer. There are several different types used, Torx, 12-point head, reverse Torx and Allen-head, etc.*
38   Remove the bolts from the outer CV joint and detach it from the outer flange. On U-joint types remove the upper control arm nut and bolt. Use wire or rope to support the upper control arm.

Installation
39   Installation is the reverse of removal. Be sure to align the previously made matchmarks before tightening the CV joint or U-joint bolts securely.

11 Support bearing assembly - check and replacement

Note: *This procedure applies to models with an intermediate shaft only.*
1   Remove the driveaxle connected to the intermediate shaft (see Section 10)
2   Rotate the intermediate shaft and listen to the bearing. It should operate smoothly and quietly.
3   If the bearing is rough or noisy, unbolt the support bearing bracket from the engine block and remove the intermediate shaft, bearing and bracket. Take the assembly, a new bearing and new dust shields (if equipped) to an automotive machine shop. They will have the right tools to press off the old bearing and press on the new one.
4   Install the intermediate shaft assembly, tightening the bearing support-to-engine block bolts securely.
5   Install the driveaxle (see Section 10).

12 Driveaxle boot replacement and CV joint inspection and overhaul

Note: *If the CV joints must be overhauled (usually due to torn boots), explore all options before beginning the job. Complete rebuilt driveaxles are available on an exchange basis from local auto parts stores, which eliminates much time and work. Whichever route you choose to take, check on the cost and availability of parts before disassembling the vehicle.*
1   Remove the driveaxle (see Section 10).

Inner CV joint
Tripod type
Disassembly
Refer to illustrations 12.2a, 12.2b, 12.3, 12.4, 12.5 and 12.6
2   Remove the boot clamps **(see illustrations)**.

12.3 Once the boot is detached from the inner CV joint housing, the housing can be removed

12.4 Use a center punch to place marks (arrows) on the tripod and the driveaxle to ensure that they're properly reassembled

12.5 Remove the snap-ring from the groove in the end of the axleshaft

12.6 Drive the tripod joint from the axleshaft with a brass punch and hammer - make sure you don't damage the bearing surfaces or the splines on the shaft

12.8a If the tripod has a chamfer on one side, be sure to install it facing the correct direction, as noted on removal (some face inward, others face outward)

3   Pull the boot back from the inner CV joint, remove the retainer ring (if equipped) and slide the joint housing off **(see illustration)**.

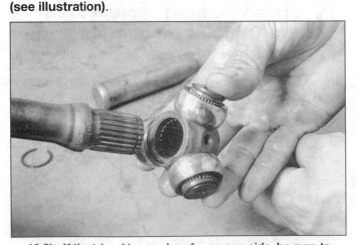

12.8b If the tripod has a chamfer on one side, be sure to install it facing the correct direction, as noted on removal (some face inward, others face outward)

4   Use a center punch to mark the tripod and axleshaft to ensure that they are reassembled properly **(see illustration)**.

5   Remove the snap-ring from the end of the axleshaft with a pair of snap-ring pliers **(see illustration)**.

6   Use a hammer and a brass punch to drive the tripod joint from the driveaxle **(see illustration)**. Note how the tripod is installed. Some models may have a chamfer on one side or the other. It is imperative that it be reinstalled facing the same direction.

Check

7   Clean all components with solvent to remove the grease, and check for cracks, pitting, scoring and other signs of wear.

Reassembly

Refer to illustrations 12.8a, 12.8b, 12.10a, 12.10b and 12.10c

8   Slide the clamps and boot onto the axleshaft. It's a good idea to wrap the axleshaft splines with tape to prevent damaging the boot **(see illustration)**. Place the tripod on the shaft and install the snap-ring **(see illustration)**. Apply

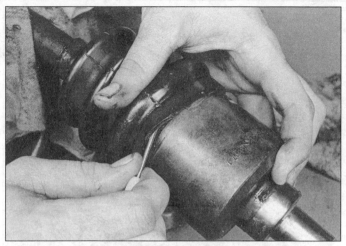

12.10a Equalize the pressure inside the boot by inserting a small, DULL screwdriver between the boot and the CV joint housing

12.10b To install the new clamps, bend the tang down . . .

12.10c . . . and tap the tabs down to hold it in place

12.11 Remove both boot clamps, then slide the boot down the axleshaft so it's out of the way

12.12 Mark the shaft, inner race, cage and outer race (housing) so they can be reassembled in the same relationship to each other

grease to the tripod assembly, the inside of the joint housing and the inside of the boot.

9   Slide the boot into place, making sure both ends seat in their grooves.

10   Adjust the length of the joint by positioning it mid-way through its travel, equalize the pressure in the boot **(see illustrations)**, then tighten and secure the boot clamps. Proceed to Step 30.

Double-offset (ball-and-cage) type

Disassembly

Refer to illustrations 12.11, 12.12, 12.13, 12.14 and 12.15

11   Remove both boot clamps and discard them. Slide the boot out of the way **(see illustration)**.

12   Mark the shaft, the inner race, the cage and the outer race (housing) so they can be reassembled in the same way **(see illustration)**.

13   Pry the wire ring bearing retainer from the housing **(see illustration)**.

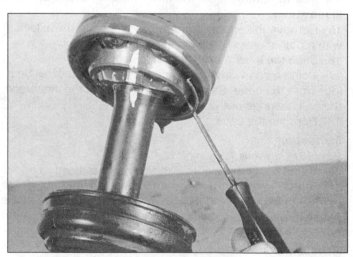

12.13 Pry the retainer from the housing with a small screwdriver

12.14 Slide the housing off the bearing assembly - some of the ball bearings may fall out when the race is removed, so be ready to catch them

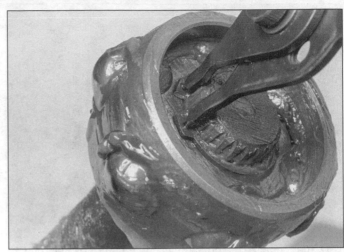

12.15 Remove the snap-ring from the groove in the axleshaft with a pair of snap-ring pliers

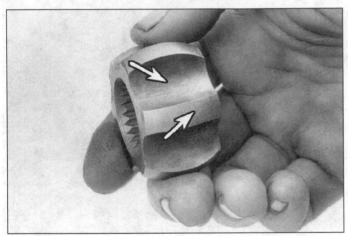

12.19a Inspect the inner race lands and grooves for pitting, score marks, cracks and other signs of wear and damage

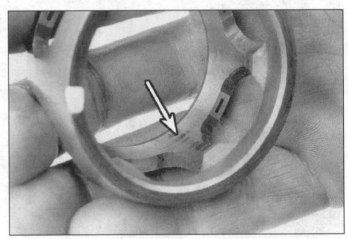

12.19b Inspect the cage for cracks, pitting and score marks (shiny, polished spots are normal and will not adversely affect CV joint performance)

14   Pull the housing off the inner bearing assembly **(see illustration)**.

15   Remove the snap-ring from the groove in the axleshaft with a pair of snap-ring pliers **(see illustration)**.

16   Slide the inner race off the axleshaft.

17   Using a screwdriver or piece of wood, pry the ball bearings from the cage. Be careful not to scratch the inner race, the ball bearings or the cage.

18   Remove the cage.

Inspection
Refer to illustrations 12.19a and 12.19b

19   Clean the components with solvent to remove all traces of grease. Inspect the cage and races for pitting, score marks, cracks and other signs of wear and damage **(see illustrations)**. Shiny, polished spots are normal and will not adversely affect CV joint performance.

Reassembly
Refer to illustration 12.25

20   Wrap the axleshaft splines with tape to avoid damaging the boot. Slide the small boot clamp and boot onto the axleshaft, then remove the tape. Slide the large boot clamp over the boot.

21   Install the cage on the axleshaft with the smaller diameter side of the cage facing toward the boot.

22   Install the inner race onto the axleshaft with the match-mark on the race (or the larger diameter side) aligned with the mark on the end of the axleshaft.

23   Install the snap-ring in the groove. Make sure it's completely seated by pushing on the inner race.

24   Move the cage up over the inner race, aligning the match marks. Press the ball bearings into the cage windows with your thumbs. If they won't stay in place, apply CV joint grease to hold them.

25   Fill the outer race and boot with CV joint grease (normally included with the new boot kit). Pack the inner race and cage assembly with grease, by hand, until grease is worked completely into the assembly **(see illustration)**.

26   Slide the inner race, balls and cage into the CV joint

12.25 Pack the inner race and cage assembly with grease, by hand, until grease is worked completely into the assembly (also note that the larger diameter side, or "bulge," is facing out)

12.34 After the old grease has been rinsed away and the cleaning solvent has been blown out with compressed air, rotate the outer joint through its full range of motion and inspect the bearing surfaces for wear or damage - if any of the balls, the race or the cage look damaged, replace the outer joint assembly

housing and install the wire ring bearing retainer.

27 Wipe any excess grease from the axle boot groove on the outer race. Seat the small diameter of the boot in the recessed area on the axleshaft. Push the other end of the boot onto the CV joint housing and position the joint midway through its travel.

28 Equalize the pressure in the boot by inserting a dull screwdriver between the boot and the outer race. Don't damage the boot with the tool.

29 Install the boot clamps.

All inner CV joints

30 Install a new circlip on the inner CV joint stub axle.

31 Install the driveaxle (see Section 10).

Outer CV joint

Refer to illustration 12.34

32 Remove the boot clamps and slide the boot back far enough to inspect the joint. If the axleshaft is rough and you're planning to re-use the same boot, wrap the axleshaft with tape to protect the boot from damage.

33 Thoroughly wash the outer CV joint in clean solvent and blow dry it with compressed air, if available. The outer joint on many driveaxles can't be disassembled, so it's difficult to wash away all the old grease and to rid the bearing of solvent once it's clean. But it's imperative that the job be done thoroughly, so take your time and do it right.

34 Bend the outer CV joint housing at an angle to the driveaxle to expose the bearings, inner race and cage. Inspect the bearing surfaces for signs of wear. If the joint is

worn, replace it (see step 38) **(see illustration)**.

35 If the boot is damaged but the joint is OK, remove the inner CV joint and boot (see Steps 2 through 6 for tripod and Steps 11 through 18 for double-offset). **Caution:** *The outer CV joint should only be removed if it is going to be replaced.* Proceed to the next step.

36 Slide the new outer boot onto the driveaxle. It's a good idea to wrap vinyl tape around the shaft splines to prevent damage to the boot. When the boot is in position, add the specified amount of grease (included in the boot replacement kit) to the outer joint and the boot (pack the joint with as much grease as it will hold and put the rest into the boot). Slide the boot on the rest of the way, equalize the pressure inside the boot and install the new clamps.

37 Slide on the inner boot and install the inner CV joint (see Steps 8 through 10 for tripod and 20 through 29 for double-offset).

38 If the bearings are damaged or worn, replace the joint, if possible (check with your local auto parts store to see if parts are available; some outer joints can't be removed). On some driveaxles, this can be done by knocking the joint from the shaft using a hammer and punch, or by pulling the joint off with a slide hammer. Install the new joint by first sliding the new boot and clamps onto the axleshaft. Then thread the old driveaxle/hub nut onto the end of the shaft and assemble the joint onto the shaft splines. Drive the joint into place. Be sure to use a new driveaxle/hub nut when installing the driveaxle.

Rear axle

13 Rear axle - check

1 On all models, many times a problem is suspected in an axle area when, in fact, it lies elsewhere. For this reason, a thorough check should be performed before assuming an axle problem.

2 The following noises are those commonly associated with axle diagnosis procedures:

a) *Road noise is often mistaken for mechanical faults. Driving the vehicle on different surfaces will show whether the road surface is the cause of the noise. Road noise will remain the same if the vehicle is under power or coasting.*

b) *Tire noise is sometimes mistaken for mechanical problems. Tires which are worn or low on pressure are particularly susceptible to emitting vibrations and noises. Tire noise will remain about the same during varying driving situations, where axle noise will change during coasting, acceleration, etc.*

c) *Engine and transmission/transaxle noise can be deceiving because it will travel along the driveline. To isolate engine and transmission/transaxle noises, make a note of the engine speed at which the noise is most pronounced. Stop the vehicle and place the transmission/transaxle in Neutral and run the engine to the same speed. If the noise is the same, the axle is not at fault.*

14 Rear axleshaft (semi-floating axles) - removal and installation

1 Raise the rear of the vehicle and support the axle housing securely on jackstands.

2 Remove the rear wheels and brake drums.

"C" lock type

Removal

Refer to illustrations 14.3, 14.4a, 14.4b, 14.5 and 14.7

3 Clean off all the dirt from around the differential housing cover, then progressively loosen the cover bolts and allow the lubricant to drain. The cover can be removed when the oil has stopped flowing out **(see illustration)**.

4 Turn the differential for access to the lock bolt. Remove the lock bolt and differential shaft **(see illustrations)**.

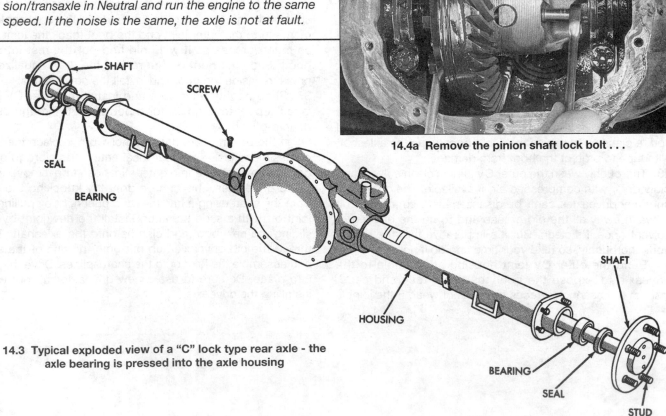

14.4a Remove the pinion shaft lock bolt . . .

14.3 Typical exploded view of a "C" lock type rear axle - the axle bearing is pressed into the axle housing

14.4b ... then carefully remove the pinion shaft from the differential carrier (don't turn the axleshafts or the carrier after the shaft has been removed, or the pinion gears may fall out)

14.5 Push the axle flange in, then remove the C-lock from the inner end of the axleshaft

14.7 Check the surface of the axle where the bearing rides for galling, pitting and hard spots (arrow)

5   Push the axle in and remove the C-lock from the inner end of the shaft (see illustration).

6   Remove the axle from the housing, being careful not to damage the bearing or seal.

7   Inspect the axle for galling, pitting and hard spots (see illustration). If these conditions are present, replace the axle and the bearing. Inspect the oil seal and replace it if necessary.

Installation

8   When installing the axle, first make sure that all the parts are clean.

9   Install the oil seal (if removed) and lubricate the lips with differential oil. Note that the lips must face in. Insert the axleshaft into the housing and engage the splines into the differential side gears. Install the C-locks and differential shaft. Apply thread locking compound to the threads and install the lock bolt.

10   The remainder of the installation is the reverse of removal. Make sure that the housing cover and differential gasket flange are clean. Apply RTV-type sealant to the cover and install it. Be sure to refill the differential with oil.

Flanged type

Removal

Refer to illustrations 14.11a, 14.11b and 14.12

11   Unscrew and remove the nuts which attach the retainer to the brake backing plate (see illustrations). A large hole is provided in the axle flange (where the drum attaches) for

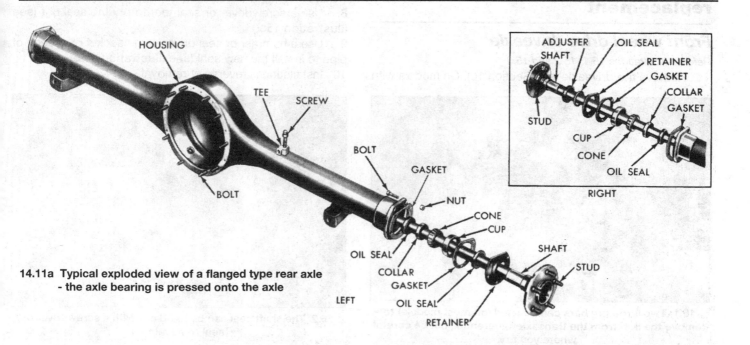

14.11a Typical exploded view of a flanged type rear axle - the axle bearing is pressed onto the axle

14.11b Unscrew the four brake backing plate nuts using a socket on an extension passing through the hole in the axle flange

14.12 Pull the axle from the housing using a slide hammer and axle flange adapter

access to the nuts with a socket, ratchet and extension. Rotate the axle to align this hole with each nut.

12 Attach a slide hammer to the wheel mounting studs and withdraw the axleshaft **(see illustration)**. Sometimes, if the axleshaft is not too tight, you can pull it out by hand, but be careful not to pull too hard or you'll pull the vehicle off of the jackstands.

Installation

13 Insert the axle shaft into the differential housing and engage the splines with the differential side gears. Tap the axleshaft in with a soft-faced hammer until it seats completely.

14 Install the retainer nuts, tightening the lower nuts first.

15 Install the brake drums and wheels and lower the vehicle to the ground.

15 Axleshaft/driveaxle oil seal - replacement

Front wheel drive driveaxle

Refer to illustrations 15.1, 15.2 and 15.3

1   Remove the driveaxle (see Section 10). On models with

a stub or differential shaft; unbolt the hub or use two pry-bars to gently pry the hub out of the transaxle **(see illustration)**.

2   Use a screwdriver or seal tool to pry the seal out **(see illustration)**.

3   Use a hammer or seal driver, large socket or section of pipe to install the new seal **(see illustration)**.

4   Installation is reverse of removal.

Rear wheel drive driveaxle

5   Raise the rear of the vehicle and support it on jackstands.

6   Drain the differential lubricant.

7   Separate the driveaxle from the differential (see Section 10) and suspend it out of the way. On models with a stub or differential shaft; Use two prybars, gently pry the differential output shaft out of the differential **(see illustration 15.1)**. **Note:** *Be ready to catch the shaft as it comes out.*

8   Use a screwdriver or seal tool to pry the seal out **(see illustration 15.2)**.

9   Use a hammer or seal driver, large socket or section of pipe to install the new seal **(see illustration 15.3)**.

10   Installation is reverse of removal.

15.1 Two large pry bars can be used (on most models) to remove the hub from the transaxle/differential, but be careful where you pry

15.2 The shaft seal can be pried out with a screwdriver or seal removal tool

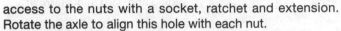

15.3 If you don't have a seal driver, a large socket or a piece of pipe with a diameter slightly less than that of the seal can be used to drive the seal into place

15.12a Use a seal removal tool to remove the old seal from the axle housing

15.12b If a seal removal tool isn't available, a prybar or even the end of the axle can be used to pry the seal out of the housing

15.13 Use a seal driver or a large socket to tap the new axleshaft oil seal into place

15.14 Remove the inner axle seal from the housing using a slide hammer and internal puller jaw attachment

Rear wheel drive axleshaft

Note: *See Section 1 for axleshaft identification.*

11   Remove the axleshaft (see Section 14).

"C" lock type

Refer to illustrations 15.12a, 15.12b and 15.13

12   Pry the oil seal out of the end of the axle housing with a seal removal tool, a large screwdriver or the inner end of the axleshaft **(see illustrations)**.

13   Apply high-temperature grease to the oil seal recess and tap the new seal evenly into place with a hammer and seal installation tool **(see illustration)**, large socket or piece of pipe so the lips are facing in and the metal face is visible from the end of the axle housing. When correctly installed, the face of the oil seal should be flush with the end of the axle housing then install the axleshaft (see Section 14).

Flanged type

Refer to illustrations 15.14 and 15.15

14   A slide hammer with a seal hook will be needed to remove the inner oil seal from the axle housing **(see illustration)**.

15.15 Lubricate the seal lip, then drive the seal into position with an installation tool or a piece of pipe with an outside diameter slightly smaller than the outside diameter of the seal

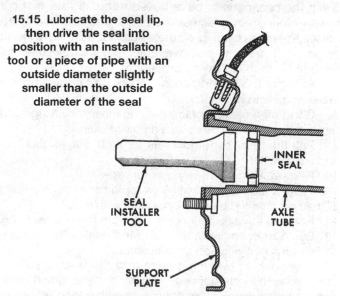

INNER SEAL

SEAL INSTALLER TOOL

AXLE TUBE

SUPPORT PLATE

15   Wipe the axle housing seal bore clean and drive in a new seal with a piece of pipe or a large socket and a hammer **(see illustration)**, then install the axleshaft (see Section 14).

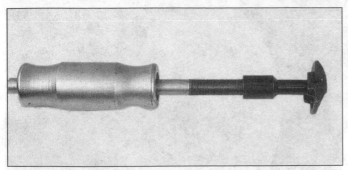

16.2a A typical axleshaft bearing removal tool (available at stores which carry automotive tools)

16 Rear axleshaft bearing (semi-floating axle) - replacement

"C" lock type

Note: *See Section 1 for axleshaft identification.*

Refer to illustrations 16.2a, 16.2b and 16.4

1   Remove the axleshaft (see Section 14) and the oil seal (see Section 15).

2   You'll need a bearing puller **(see illustrations)**, or you'll need to fabricate a similar tool.

3   Attach a slide hammer and pull the bearing out of the axle housing.

4   Clean out the bearing recess and drive in the new bearing with a bearing driver or a large socket **(see illustration)**. Lubricate the new bearing with gear lubricant. Make sure that the bearing is seated into the full depth of its recess.

5   Discard the old oil seal and install a new one (see Section 15), then install the axleshaft (see Section 14).

Flanged type

6   If the bearing is to be removed from the axle, first cut three or four grooves across the bearing collar with a sharp chisel. Press the bearing off the axle shaft using a hydraulic press. If a press is not available, the bearing may be removed with the following procedure.

7   Remove the bearing collar retainer by cutting off the lower edge with a chisel.

8   Grind off a section of the bearing inner race flange and remove the bearing rollers with a pair of pliers.

9   Pull the roller cage down as far as it will go, then cut and remove it.

10   Remove the roller bearing outer race.

11   Protect the seal surface by wrapping some tape around it, then remove the bearing inner race with a puller.

12   Remove the axleshaft outer seal from the retainer plate.

13   Begin reassembly by installing the retainer plate, complete with a new seal, on the axleshaft.

14   Press wheel bearing grease into the bearing rollers, then install the new bearing outer race, bearing and collar using a hydraulic press or drive it on with a long section of pipe the exact diameter of the inner bearing race.

15   Install a new rubber coated steel gasket over the axle housing flange studs, then install the brake assembly.

16.2b To remove the axle bearing, insert a bearing removal tool attached to a slide hammer through the center, pull the tool up against the back side and use the slide hammer to drive the bearing from the axle housing

16.4 A correctly-sized bearing driver must be used to drive the bearing into the housing

16   Apply a little grease to the outer diameter of the bearing outer race to prevent corrosion.

17   Lubricate the inner seal and install the seal into the axle housing.

18   Install a foam gasket over the axle housing flange studs, then carefully slide the axleshaft assembly into the housing until the splines on the shaft engage in the differential side gears. Seat the axleshaft against the mounting flange (see Section 14).

17 Rear axleshaft (full-floating axles) - removal, installation and adjustment

Refer to illustrations 17.2, 17.3, 17.6a and 17.6b

1   Loosen the rear wheel lug nuts. Raise the vehicle and place it securely on jackstands. Remove the rear wheel.

2   Remove the axleshaft flange bolts or nuts **(see illustration)**.

17.2 Remove the axleshaft flange bolts . . .

17.3 . . . and pull out the axleshaft

17.6a Before installing the axleshaft, apply a coat of silicone sealant to the mating surface of the hub . . .

17.6b . . . then place the new gasket in position on the hub

3   Pull out the axleshaft (see illustration). Some units may be equipped with tapered dowel pins which can "stick" to the flange and/or the hub and bearing assembly. If the axleshaft is stuck, rap it sharply in the center of the flange with a hammer to knock the dowels loose.

4   While the axleshaft is removed, inspect and, if necessary, replace the hub and bearing seals and bearings (see Section 18). This is also a good time to inspect the rear brake assembly and replace any parts that are worn.

5   Slip a new gasket over the end of the axleshaft and slide the axleshaft into the axle housing.

6   Before engaging the axleshaft splines with the differential, clean the gasket sealing area, apply silicone sealant to the gasket mating surface of the hub (see illustration) and place the new gasket in position (see illustration). Push the axleshaft into the axle housing until the axleshaft splines are fully engaged with the differential. If you have difficulty engaging the splines with the differential, have an assistant turn the other wheel slightly while you push on the axleshaft.

7   Install the flange bolts.

18 Hub/drum assembly and wheel bearings (full-floating axles) - removal, installation and adjustment

Warning: *The dust created by the brake system may contain asbestos, which is harmful to your health. Never blow it out with compressed air and don't inhale any of it. An approved filtering mask should be worn when working on the brakes. Do not, under any circumstances, use petroleum-based solvents to clean brake parts. Use brake system cleaner only!*

Removal

Refer to illustrations 18.3a, 18.3b, 18.3c, 18.3d, 18.4, 18.5a, 18.5b, 18.6, 18.7, 18.9a and 18.9b

1   Raise the vehicle, support it securely on jackstands and remove the rear wheels.

2   Remove the axleshaft (see Section 17).

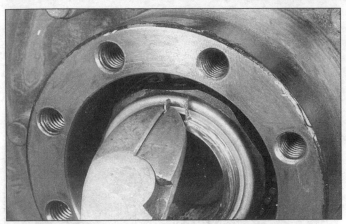

18.3a This nut lock must be removed from the keyway . . .

18.3b . . . before loosening the adjusting nut

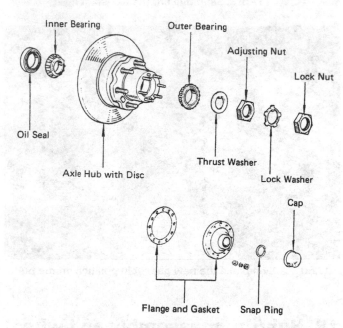

18.3c Typical axle hub and bearing assembly

3   Remove the nut lock (see illustration) and remove the adjusting nut (see illustration). There are several different types of nut locks or locknuts used on the full floating axles. Many use a variation of the same type, a lockring that requires a special tanged socket to remove. Other types used are a flat tabbed lock washer that is bent over a flat spot of the locknut, or a locknut that uses screws to secure the locknut to the lockring (see illustrations).
4   Remove the outer wheel bearing (see illustration).
5   Pull the hub and brake drum straight off the axle tube (see illustration). If necessary, separate the drum from the hub. If the hub and drum are stuck, lay them on the floor, tap the studs to knock the hub loose and separate the two (see illustration).
6   Place the hub assembly in a bench vise with the inner side facing toward you. Using a large screwdriver, pry bar or seal removal tool to pry out the oil seal (see illustration).
7   Remove the inner wheel bearing from the hub (see illustration).
8   Use solvent to wash the bearings, hub and axle tube. A small brush may prove useful; make sure no bristles from the brush embed themselves between the bearing rollers.

18.3d Remove the screws that secure the locknut to the lockring

18.4 Remove the outer bearing

18.5a To remove the hub and bearing assembly, you'll have to pull off the brake drum

18.5b If the hub and drum are stuck together (which they often are), dropping them on the floor will usually free them; if that doesn't work, tapping on the wheel studs will knock the hub loose from the drum

18.6 To remove the hub inner seal, place the hub assembly in a bench vise and pry the seal out with a seal removal tool (shown) or a screwdriver (it may not be possible to separate the hub from the drum on some models)

18.7 Remove the inner bearing from the hub

18.9a To replace a stud, knock it out with a brass hammer

Allow the parts to air dry.

9   Carefully inspect the bearings for cracks, wear and damage. Check the axle tube flange, studs, and hub splines for damage and corrosion. Check the bearing cups (races) for pitting or scoring. Worn or damaged components must be replaced with new ones. If necessary, tap the bearing cups from the hub with a hammer and brass drift and install new ones with the appropriate size bearing driver. This is also a good time to replace any damaged or worn wheel studs **(see illustrations)**.

10 Inspect the brake drum for scoring or damage and check the condition of the brake shoes and repair as necessary.

Installation and adjustment

Refer to illustrations 18.11, 18.12, 18.13a, 18.13b, 18.15, 18.16a and 18.16b

11 Pack the bearings with wheel bearing grease **(see illus-**

18.9b To install a stud, insert the stud through the hole in the flange, drop a spacer or several washers the height of the shoulder over the bolt, install the wheel lug nut and tighten the nut until the stud is pulled up through the flange

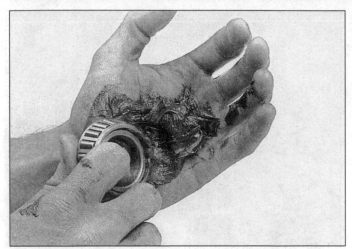

18.11 Work the grease completely into the rollers

18.12 Slide the hub onto the axle tube and install the outer bearing

18.13a Install a new adjusting nut

18.13b Tighten the adjusting nut securely while rotating the hub

tration). Work the grease completely into the bearings, forcing it between the rollers, cone and cage. Place the inner bearing into the hub and install the seal (with the seal lip facing into the hub). Using a seal driver or block of wood, drive the seal in until it's flush with the hub. Lubricate the seal with gear oil or grease.

12 Place the hub assembly on the axle tube, taking care not to damage the oil seals. Place the outer bearing into the hub **(see illustration).**

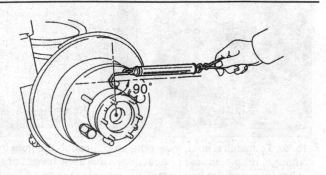

18.15 Use a spring scale to measure the hub bearing preload

18.16a Insert a new nut lock into the axle tube keyway . . .

18.16b ... then tap the nut lock into the keyway until it stops (the vertical fingers of the nut lock should be firmly seated against the nylon portion of the adjusting nut) (not all models are equipped with this type of lock)

19.4 Mark the relationship between the pinion and flange as shown

13 Install the new adjusting nut and tighten it to approximately 40 ft-lbs while turning the hub **(see illustrations)**.

14 Back off the adjusting nut until it is loose, then tighten it by hand while turning the hub.

15 Tighten the nut a little more, to ensure that there will be no endplay in the bearings. When properly adjusted, the force required to start the hub turning, when measured with a spring scale attached to one of the wheel studs, should be approximately 5 to 12 inch-lbs **(see illustration)**.

16 Lock the adjusting nut in position, install the locking wedge, lockring or tabbed washer securely **(see illustrations)**.

17 Install the axleshaft (see Section 17). Install the brake drum and adjust the brake shoes if necessary.

18 Install the rear wheel, remove the jackstands and lower the vehicle. Tighten the wheel lug nuts.

19 Pinion oil seal - replacement

Refer to illustrations 19.4, 19.5, 19.6a, 19.6b, 19.7 and 19.8

1 Raise the rear of the vehicle and place it securely on jackstands.

2 Remove the rear wheels and brake drums.

3 Mark the driveshaft and companion flange for ease of realignment during reassembly, then remove the driveshaft (see Section 7).

4 Mark the relationship between the pinion and companion flange **(see illustration)**.

5 Using an inch-pound torque wrench, measure and record the torque required to turn the pinion nut through several revolutions **(see illustration)**. This value, known as pinion bearing preload, will be used when the pinion flange is reinstalled.

6 Using a suitable tool, hold the companion flange and remove the pinion nut **(see illustration)**. Using a suitable

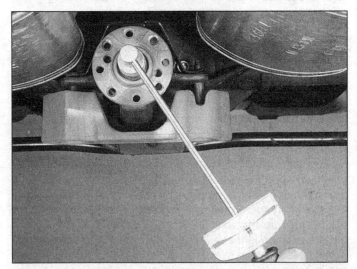

19.5 Using an inch-pound torque wrench, measure the pinion bearing preload (turning torque) and record this measurement

19.6a Some ingenuity or special tools will be needed to keep the flange from turning while you're loosening and removing the pinion nut. On this model, hold the flange with a punch jammed through a hole in the flange and wedged under the reinforcing rib on the side of the differential housing

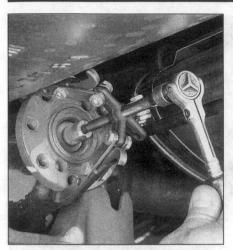

19.6b If necessary, use a small puller to separate the pinion flange from the pinion shaft

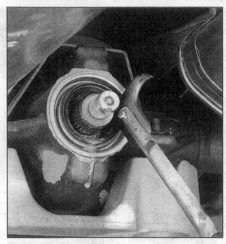

19.7 Pry out the old seal with a large screwdriver or seal removal tool

19.8 Using a seal driver or a large socket, tap the new pinion seal into place with the seal square to the bore

puller (see illustration), remove the companion flange.

7   Pry out the old seal with a seal removal tool (see illustration).

8   Clean the oil seal mounting surface, then tap the new seal into place, taking care to insert it squarely as shown (see illustration).

9   Inspect the splines on the pinion shaft for burrs and nicks. Remove any rough areas with a crocus cloth. Wipe the splines clean.

10   Install the companion flange, aligning it with the marks made during removal. Gently tap the flange on with a soft-faced hammer until you can start the pinion nut on the pinion shaft.

11   Using a suitable tool, hold the companion flange while tightening the pinion nut until all of the in and out free play is gone. Continue tightening, take frequent rotational torque measurements, using the inch-pound torque wrench, until the measurement recorded in Step five is reached. **Caution:** *If the measurement recorded in Step five was less than the manufacturer's recommended pinion bearing preload torque, continue tightening until the specified torque is reached. If it was more than specified, continue tightening until the recorded measurement is reached. Under no circumstances should the pinion nut be backed off to reduce pinion bearing preload. Increase the nut torque in small increments and check the preload after each increase.*

12   Reinstall the driveshaft, brake drums and wheels.

13   Check the differential oil level and fill as necessary.

14   Lower the vehicle and take a test drive to check for leaks.

20 Axle assembly (rear) - removal and installation

1   Loosen the rear wheel lug nuts, raise the vehicle and support it securely on jackstands placed underneath the frame. Remove the wheels.

2   Support the rear axle assembly with a floor jack placed underneath the differential.

3   Remove the shock absorber lower mounting nuts and compress the shocks to get them out of the way (see Chapter 6).

4   Disconnect the driveshaft from the differential pinion shaft yoke and hang the rear of the driveshaft from the underbody with a piece of wire (see Section 7).

5   Unbolt the stabilizer bar from the stabilizer bar link, if equipped.

6   Disconnect the parking brake cables from the equalizer. Disconnect the height sensing proportioning valve control rod from the axle housing (if equipped).

7   Disconnect the flexible brake hose from the junction block on the rear axle housing. Plug the end of the hose or wrap a plastic bag tightly around it to prevent excessive fluid loss and contamination.

8   On models equipped with leaf springs, either remove the U-bolts that attach the leaf springs to the differential housing or remove the spring eye-to-frame bracket mounting bolts (see Chapter 6).

9   On models equipped with coil springs, secure the coil springs with safety chains to prevent the springs from dropping out when the axle is lowered. Remove the suspension arm mounting bolts and pry the arms away from the housing.

10   Carefully lower the jack and move the axle assembly out from under the vehicle.

11   Installation is the reverse of the removal procedure. Be sure to bleed brake system.

Front axle (Four-wheel drive models)

21.5a Remove the spindle securing nuts . . .

21 Front axleshaft and joint assembly (four wheel drive models) - removal, component replacement and installation

Removal

Refer to illustrations 21.5a, 21.5b, 21.6, 21.7a, 21.7b

1   Loosen the lug nuts on the front wheels.
2   Raise the front of the vehicle, support it securely on jackstands and remove the front wheels.
3   Remove the front brake calipers.
4   Remove the locking hubs, wheel bearings and brake disc or drum.
5   Remove the nuts securing the spindle to the steering knuckle. Tap the spindle with a plastic or soft faced hammer to jar the spindle free (**see illustrations**).

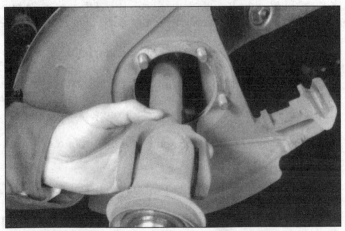

21.7a Withdraw the shaft and joint assembly from the axle housing

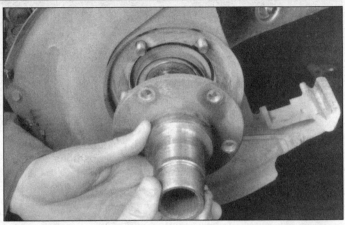

21.5b . . . then tap the spindle loose with a soft face hammer and remove it

21.6 Remove the spindle seat

6   Remove the spindle seat (**see illustration**).
7   Remove the shaft and joint assembly out of the carrier (**see illustrations**).

21.7b If necessary, tap the slinger off with a hammer - a press is needed to reinstall it

21.8a Mark the shaft and yoke with paint so they can be reassembled in the same relative positions

21.8b Remove the retaining rings with a screwdriver or other suitable tool

Shaft and universal joint assembly replacement

Refer to illustrations 21.8a, 21.8b and 21.9

8   Remove the snap-rings from the U-joint with a screwdriver **(see illustrations)**.

9   The remainder of the procedure is the same as that for the driveshaft universal joints, which is described in Section 8 **(see illustration)**.

Installation

10   Slide the shaft and joint assembly through the knuckle and engage the splines on the shaft in the carrier.

11   Install the splash shield. Install the spindle on the steering knuckle, then install and tighten the spindle nuts.

12   Repack the wheel bearings with grease and install the brake disc or drum. Adjust the wheel bearing preload and tighten the locknut (see Chapter 4). Install the locking hubs.

13   Install the brake caliper.

14   Install the front wheels and lower the vehicle.

15   Tighten the lug nuts securely.

22 Axle assembly (front) - removal and installation

Removal

1   Loosen the front wheel lug nuts, raise the front of the vehicle and support it securely on jackstands positioned under the frame rails. Remove the front wheels.

2   Unbolt the front brake calipers and hang them out of the way with pieces of wire - don't let the calipers hang by the brake hose.

3   Remove the brake pads, anchor plates and brake discs.

4   Mark the relationship of the front driveshaft to the front differential pinion shaft yoke, then disconnect the driveshaft from the yoke (see Section 7). Check for wear or damage to the straps and replace as necessary.

5   Disconnect the stabilizer bar link from the bracket on the axle (see Chapter 6).

21.9 Position a pair of sockets in a vise to push the bearing out of the joint - the small socket is smaller than the bearing; the large socket is large enough so the bearing will fit into it

6   Disconnect the tie-rod and center link from the steering knuckle arms. Position them out of the way and hang them with pieces of wire from the underbody.

7   Unbolt the lower ends of the shock absorbers from the axle.

8   Remove the steering damper (if equipped).

9   Remove the ABS sensor (if equipped).

10   Position a hydraulic jack under the differential. If two jacks are available, place one under each axle tube to balance the assembly.

11   On models equipped with coil springs, unbolt the lower and/or upper suspension arms from the axle.

12   On models equipped with leaf springs, either remove the U-bolts that attach the leaf springs to the differential housing or remove the spring eye-to-frame bracket mounting bolts (see Chapter 6).

13   Slowly lower the axle assembly to the ground.

Installation

14   Installation is the reverse of the removal procedure. When raising the axle into position, make sure the coil springs seat properly. Be sure to use new U-joint straps if the old ones were damaged. Tighten all mounting fasteners securely.

Differential

23 Differential - check and adjustments

Check

Refer to illustrations 23.3, 23.4a, 23.4b, 23.5, 23.7 and 23.8

1   Before beginning any differential repairs it's important to properly diagnosis the problem and make some basic checks before teardown begins (see Chapter 3).

2   Remove the driveshaft (see Section 7).

3   Using a dial or beam type inch-pound torque wrench and socket, rotate the pinion shaft and record the amount of force required to rotate the pinion. This is called pinion bearing preload **(see illustration)**. Typical pinion bearing preload is 15 to 25 in-lbs.

4   Remove the differential cover and drain the fluid (see Chapter 1). Attach a dial indicator to the carrier housing and place the dial indicator on the drive side of the ring gear. Zero the dial indicator and move the ring gear back and forth without turning the pinion gear. This is the backlash measurement and should be taken in several places **(see illustrations)**.

5   Remove the dial indicator from the tooth side of the ring gear and place it on the back side of the gear. Rotate the gear and check the dial indicator for movement; this is ring gear runout **(see illustration)**.

6   The last check is the wear pattern on the ring gear.

23.3 Using an inch pound dial type torque wrench rotate the pinion several revolutions noting the amount of force required to turn the differential - this is pinion bearing preload

23.4a Mount the dial indicator on the axle housing and place the indicator tip against the drive side of the gear (integral carrier shown)

23.4b Zero the dial indicator and move the ring gear back and forth without turning the gear. This is the backlash measurement and should be taken in several places (drop-out style carrier shown)

23.5 Place the dial indicator tip against the back of the ring gear and rotate the carrier through several revolutions, noting the reading on the dial indicator - this is ring gear runout

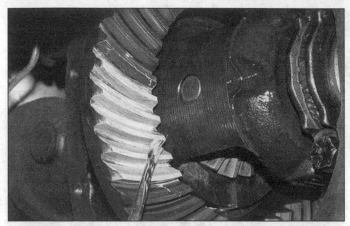

23.7 Using a clean brush, coat the teeth of the ring gear with gear marking compound until a small section of the gear is completely covered. Then rotate the ring gear several times in both directions and check the pattern

7   Clean the gear and apply gear marking compound or white grease to both sides of several gear teeth **(see illustrations)**.

8   Rotate the ring gear several revolutions in one direction and then in the opposite direction. Look at the area where the compound or grease has been rubbed off. This is the wear area or contact point between the ring and pinion **(see illustration)**. A correct pattern will be centered on the gear tooth.

Adjustments

Drive pinion depth

Note: *Pinion depth is most often adjusted by a shim placed under the rear pinion bearing or behind the bearing race. Special tools are available to determine the correct shim thickness, enabling you to assemble the differential correctly on the first attempt. If the tools are not available, the following procedure may be used to determine the correct pinion depth, but may require removal and installation of the differential components several times to obtain the same results.*

9   Temporarily install the pinion into the carrier without the crush sleeve. Carefully tighten the pinion nut until a preload of 15 to 25 in-lbs is required to rotate the pinion shaft.

10   Install the carrier into the housing and temporarily set the backlash (see Step 22). Once backlash a has been set, check the contact pattern to determine where the pinion is located in relation to the ring gear.

11   Clean the gear and apply gear marking compound or white grease **(see illustrations 23.7a and 23.7b)**.

12   Rotate the ring gear several revolutions in one direction and then in the opposite direction. Look at the area where the compound or grease has been rubbed off. This is the wear area or contact point between the ring and pinion **(see illustration 23.8)**.

13   If the pattern is too deep or close to the ring gear flank, the pinion gear needs to be moved away from the ring gear; add or subtract shim thickness as required.

Gear Tooth Patterns

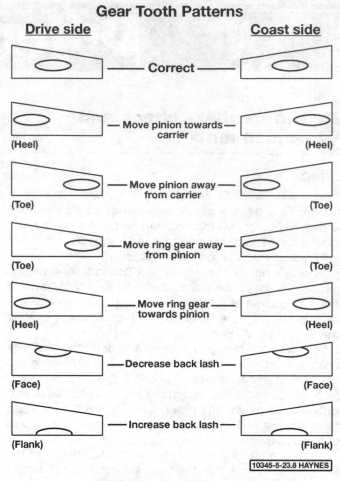

23.8 Typical gear tooth patterns

14   If the pattern is too far or near to the ring gear face, the pinion gear needs to be moved closer to the ring gear; add or subtract shim thickness as required. **Note:** *The proper contact pattern is in the middle of the tooth (between the face and the flank), also called the "pitchline".*

15   Once the correct pinion depth has been found, remove the carrier and pinion gear to set drive pinion bearing preload.

Drive pinion bearing preload

Shim type

16   Install the original shims and slowly tighten the pinion nut until a typical bearing preload of 15 to 25 in-lbs is obtained and the pinion nut is properly torqued. For example, a proper torque reading for a Dana Model 44 axle is 190 ft-lbs and on a Dana Model 60 axle it's 215 ft-lbs. Be very careful and DO NOT back-off the pinion nut once the tightening procedure has begun. **Caution:** *Care should be taken when tightening the pinion nut. It is possible to damage the pinion bearings if the shims are not thick enough.*

17   If the preload is too high, add shim thickness until the preload is correct.

18   If the preload is too low, decrease shim thickness until the preload is correct.

23.25 Attach a dial indicator to the carrier and check the backlash at four different positions on the ring gear

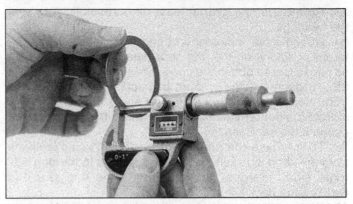

23.32a Measure the shim with a micrometer - this style shim is used between the carrier bearing and the carrier case and is called the inside design

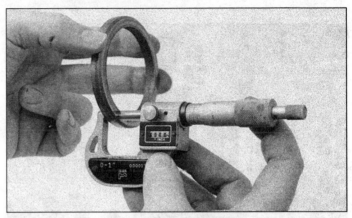

23.32b Measure the shim with a micrometer - this style shim is used between the carrier bearing races and the axle housing and is called the outside design

Crush sleeve type

19 Slide the new crush sleeve onto the pinion shaft and install the pinion shaft into the housing. Install the pinion flange or yoke. Apply thread locking compound on the pinion threads and install a new pinion nut.

20 Using a breaker bar, tighten the nut until the sleeve begins to crush. To properly seat or crush the crush sleeve you will need to hold the pinion flange or yoke to keep it from turning **(see illustration 19.6a)**. This can be very difficult, since it takes approximately 180 ft-lbs to 380 ft-lbs of torque to crush a typical sleeve. Slowly tightening the pinion nut until all endplay has been removed and begin taking rotation torque measurements with the in-lb torque wrench as described in Step 3. Carefully continue tightening the nut until the correct pinion bearing preload is reached. Typically, pinion bearing preload is set at 15 to 25 in-lbs.

21 Extreme care should be taken when setting the pinion bearing preload. Tighten the nut in small increments, while stopping to take preload measurements. It is possible to damage the pinion bearings if the crush sleeve is over-tightened. If the bearing preload is exceeded, you must disassembly the pinion gear, replace the crush sleeve and begin again. **Caution:** *Under no circumstances, DO NOT back off on the pinion nut to decrease pinion bearing preload once the tightening procedure has begun.*

Backlash and carrier bearing preload

Adjusting nut type

Refer to illustration 23.25

22 Tighten the short side (left side) adjuster nut until zero lash is reached.

23 Tighten the long side (right side) adjuster nut until the bearing is seated. Then tighten the adjuster nut two more notches to the bearing preload. At this time don't back off the short side, this will leave two notches of preload on the carrier bearings.

24 Rotate the ring gear at least three revolutions to seat the bearings.

25 Attach a dial indicator and check the backlash at four different positions on the ring gear **(see illustration)**. Adjust

as necessary to obtain the correct backlash.

26 To decrease the amount of backlash, loosen the long side (right side) adjusting nut and tighten the short side (left side) an equal amount of notches. Rotate the ring gear and recheck the backlash.

27 To increase the amount of backlash, loosen the short side (left side) adjusting nut and tighten the long side (right side) an equal amount of notches. Rotate the ring gear and recheck the backlash.

28 Be sure to loosen one side and tighten the other side an equal amount to maintain carrier bearing preload. **Note:** *When making the final adjustment, always tighten the short side (left side) adjuster nut.*

Shim type

Refer to illustrations 23.32a, 23.32b and 23.34

29 Install the carrier and tighten the bearing cap bolts finger tight.

30 Fill the short side (left side) with shims until a snug fit is reached.

31 Fill the long side (right side) with shims until a snug fit is reached.

32 Add, a ten thousandth inch (0.010) thick, shim to each side for carrier bearing preload **(see illustrations)**.

33 Rotate the ring gear at least three revolutions to seat the bearings.

34 Attach a dial indicator and check backlash at four different positions on the ring gear (see illustration). Adjust as necessary to obtain the correct backlash.

35 To decrease the amount of backlash, remove shims from the long side (right side) and add shims to the short side (left side) an equal amount. Rotate the ring gear and recheck the backlash.

36 To increase the amount of backlash, remove shims from the short side (left side) and add shims to the long side (right side) an equal amount. Rotate the ring gear and recheck the backlash.

37 When adjusting backlash (transferring shims from one side to the other) always work with equal amounts (thickness) to maintain carrier bearing preload.

23.34 Attach a dial indicator to the carrier housing and check the backlash at four different positions on the ring gear

24 Carrier and bearings - removal and installation

Removal

1 Raise the rear of the vehicle and support the axle housing securely on jackstands.

2 Remove the wheels and brake drums or discs, and the driveshaft (see Section 7).

3 Before disassembling the carrier, take a pattern check and also measure ring gear runout (see Section 23).

4 Remove the axleshafts (see Section 14).

Integral carrier type

Refer to illustrations 24.6a, 24.6b, 24.7 and 24.8

5 After the axles have been removed it's a good idea to install the pinion shaft and lock bolt to keep the differential gears from rotating out.

6 The differential bearing caps should not be interchanged. They should have a mark or stamp indicating

24.6a Some differential bearing caps have cast marks (arrow) to identify which way the cap faces - the arrow should point to the wheel

24.6b Mark both caps so they can be reinstalled in their original locations

24.7 Slowly loosen the bearing cap bolts 1/4-turn at a time each until they can be removed by hand

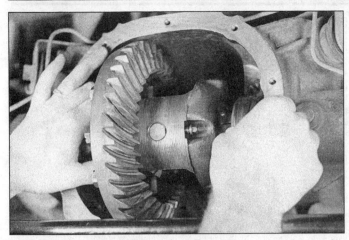

24.8 Pull the carrier to the rear of the housing and remove the bearing cups and shims or bearing adjusting nuts - organize the cups and shims so they can be installed in their original location

24.11 Remove the lock securing bolt and lock

A   Lock retaining bolt   C   Adjuster (backlash)
B   Lock

24.12 Mark both caps so they can be reinstalled in their original locations

24.13 Loosen the bearing cap bolts 1/4-turn at a time each until they can be removed by hand and remove the cap

24.14 Remove the bearing with a puller that grips the bearing cone to prevent damage to the bearing and carrier surface

which side faces out; apply a mark on each one to identify which side it was originally installed **(see illustrations)**.

7   Loosen the bearing cap bolts 1/4-turn at a time each until they can be removed by hand **(see illustration)**. **Caution:** *Once the bearing caps are removed the carrier can fall out of the housing.*

8   Carefully pull the carrier out of the housing **(see illustration)**, once you have pulled it back about an inch remove the differential shims or adjusters and bearing cups. If your not replacing the bearings don't mix up the bearing cups and note which side each of the shims or adjusters were originally installed.

Drop-out type
Refer to illustrations 24.11, 24.12 and 24.13

9   Remove the bolts or nuts securing the differential assembly to the axle housing.

10   Slide the differential assembly out of the differential housing. **Note:** *The drop-out differential assembly is very*

heavy a floor jack or help should be used to remove it from the housing.

11   Remove the carrier adjusting locks and bolts **(see illustration)**.

12   The differential bearing caps should not be interchanged, apply a mark or stamp indicating which side they were originally installed **(see illustrations)**.

13   Loosen the bearing cap bolts 1/4-turn at a time each until they can be removed by hand and remove the carrier **(see illustration)**.

Bearing replacement
Refer to illustrations 24.14, 24.15, 24.16 and 24.17

14   Place the carrier on a bench, using a suitable puller, detach the bearing from the carrier **(see illustration)**. **Caution:** *The puller must grip the bottom of the bearing or cone and not the bearing cage. The cage will not take the pressure required to remove the bearing.*

24.15 Use an adapter(s) to press the bearing onto the carrier

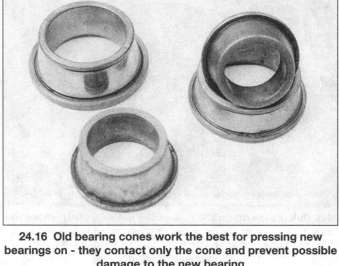

24.16 Old bearing cones work the best for pressing new bearings on - they contact only the cone and prevent possible damage to the new bearing

15   Use a suitable press and adapters to press the bearing onto the carrier **(see illustration)**. The adapter should only touch the cone of the bearing and not the cage.

16   The ideal adapter for this is the old bearing cone with the cage and rollers removed **(see illustration)**.

17   Once the bearing is seated on the carrier, spin the bearing cage and rollers they should spin smoothly and the cage shouldn't be distorted or out of round **(see illustration)**. If the bearing doesn't pass this test replace the bearing again.

Installation

18   Installation is reverse of removal.

19   After the carrier is installed adjust the backlash (see Section 23).

25 Ring and pinion - removal and installation

1   Remove the carrier (see Section 24).

Ring gear

Refer to illustrations 25.2 and 25.3

2   Scribe or paint a mark showing the relationship of the ring gear to the carrier **(see illustration)**.

3   Loosen the ring gear mounting bolts 1/4-turn at a time each until they can be removed by hand **(see illustration)**.

4   Using a plastic or brass hammer tap the back of the ring gear evenly around the gear until the gear is free.

5   Before installation, carefully wipe the carrier and ring gear machined surfaces clean. Use a sanding block to remove any nicks or burrs from the carrier surface.

6   Place the ring gear onto the carrier and align the scribe or paint marks. Apply thread locking compound to the threads and tighten the ring gear bolts only finger tight, working around the ring gear.

7   Once all the bolts are hand tight and the ring gear is seated tighten the ring gear bolts using a criss-cross pattern.

24.17 Once the bearing has been fully seated on the carrier spin the race and rollers if they were nicked or bent the new bearing must be replaced

25.2 If you're going to reuse the same ring gear , but there are no existing marks, mark or scribe the relationship of the ring gear to the carrier

25.3 Loosen each bolt 1/4-turn at a time each , going around the gear

25.11 Remove the pinion and carrier housing from the case - retrieve the shim behind the pinion bearing retainer, it sets the pinion depth on this model (Ford 9-inch shown)

25.12 Hold the pinion flange in place with a suitable tool and loosen the nut

25.13 Tap the pinion out of the bearing retainer using a brass hammer

25.14a Remove the bearing with a collar that grips the bearing cone to prevent damage to the bearing and pinion surface

Pinion gear

Integral carrier type

8   Remove the companion flange (see Section 19).
9   Push the pinion gear and rear bearing to the rear of the housing. If the pinion will not move, use a brass drift and hammer to loosen it from the housing.
10   Remove and discard the crush sleeve. **Caution:** *Never reuse a crush sleeve once it has been used.*

Drop-out type

Refer to illustrations 25.11, 25.12 and 25.13

11   Remove the pinion bearing retainer bolts and remove the retainer and pinion bearings from the carrier housing **(see illustration)**.
12   Remove the companion flange **(see illustration)**.
13   Using a brass or plastic hammer, tap the pinion out of the pinion retainer **(see illustration)**.

Bearing removal

Rear pinion bearing

Refer to illustrations 25.14a, 25.14b and 25.16

14   Place the pinion gear on a bench, using a suitable lock plate and hydraulic press, detach the bearing from the car-

25.14b Use a suitable press to press the rear bearing off of the pinion - once the bearing has started to come off hold the pinion to prevent it from falling and chipping a tooth

rier **(see illustrations)**. **Warning:** *The lock plate must grip the bottom of the bearing or cone and not the bearing cage. The cage will not take the pressure required to remove the bearing.*

25.16 Old bearing cones work the best for pressing new bearings on - they contact only the cone and prevent possible damage to the new bearing

15   Remove any shims that were under the rear bearing and replace as necessary for correct pinion depth (see Section 23).

16   Use a suitable press and adapters to press the bearing onto the pinion **(see illustration)**. The adapter should only touch the cone of the bearing and not the cage.

17   The ideal adapter for this is the old bearing cone with the cage and rollers removed **(see illustration 23.16)**.

18   Once the bearing is seated on the pinion, spin the bearing cage and rollers; they should spin smoothly and the cage shouldn't be distorted or out of round. If the bearing doesn't pass this test, replace it.

Front pinion bearing

Refer to illustrations 25.19, 25.20, 25.21, 25.22, 25.23, 25.24 and 25.25

19   Remove the pinion seal **(see illustration)**, and remove the front bearing.

20   Using a hammer and long brass drift, drive the rear bearing race out of the housing or retainer **(see illustration)**.

21   Using a hammer and long brass drift, drive the front

25.19 Using a pry bar remove the pinion seal

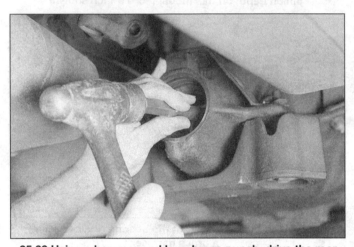

25.20 Using a hammer and long brass punch, drive the rear bearing race out of the housing

bearing race out of the housing or retainer **(see illustration)**.

22   Clean the axle housing or retainer using an aerosol solvent. Wipe any metal or dirt particles out of the housing **(see illustration)**.

25.21 Using a hammer and long brass punch, drive the front bearing race out of the housing

25.22 Be sure to thoroughly clean the inside of the housing, including the front pinion bearing oil drain passage - metal particles tend to accumulate here

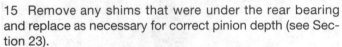

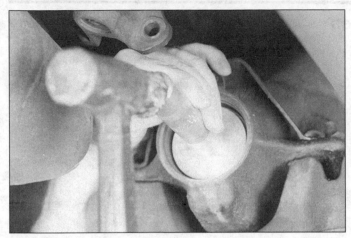

25.23 Install the front bearing race using a race installer

25.24 Install the rear bearing race using a race installer - make sure the race is fully seated into the housing

23 Install the front bearing race using a race installer **(see illustration)**.
24 Install the rear bearing race using a race installer **(see illustration)**.
25 Lubricate the front bearing with gear oil and place the bearing in the housing or retainer. Install the pinion seal **(see illustration)**.
26 Install the pinion gear into the housing. Set the pinion gear depth and pinion bearing preload as described in Section 19
27 The remainder of installation is reverse of removal.
28 After the carrier is installed, adjust the backlash (see Section 23).
29 Install the axleshaft (see Section 14). Install the brake drum and adjust the shoes if necessary.
30 Install the rear wheel, remove the jackstands and lower the vehicle. Tighten the wheel lug nuts.

25.25 Install the front bearing and pinion seal into the housing

Notes

6 Suspension systems

Contents

	Section		Section
Balljoints - check and replacement	8	Solid axle (with leaf springs)	11
Beam type (with coil springs and shocks)	12	Trailing arm type	15
Beam type (with leaf springs)	14	Unequal length A-arm type -	
Beam type (with strut or shock absorber/coil springs)	13	(with coil-over shock absorbers)	3
General information	1	Unequal length A-arm type (with coil springs)	2
MacPherson strut type	7	Unequal length A-arm type -	
Modified strut type (with separate coil springs)	6	(with longitudinal torsion bars)	4
Multi-link type	16	Unequal length A-arm type -	
Rear hub and bearing assembly - removal		(with transverse torsion bars)	5
and installation	17	Wheel stud - replacement	9
Solid axle (with control arms and coil springs)	10		

1 General information

Suspension generally refers to the mechanisms employed to support the frame and body of the vehicle, allowing the wheels to "soak up" irregularities in the road surface. The result is twofold: The wheels tend to stay in contact with the road, and the occupants of the vehicle enjoy a ride that is insulated from road shock. Without suspension, a motor vehicle traveling at today's speeds would not only be uncomfortable, it would be virtually uncontrollable.

Since most of the procedures dealt with in this Chapter involve jacking up the vehicle and working underneath it, a good pair of jackstands will be needed. A hydraulic floor jack is the preferred type of jack to lift the vehicle, and it can also be used to support certain components during various operations.

Warning 1: *Never, under any circumstances, rely on a jack to support the vehicle while working on it. Whenever any of the suspension or steering fasteners are loosened or removed they must be inspected and, if necessary, replaced with new ones of the same part number or of original equipment quality and design. Never attempt to heat or straighten any suspension or steering components. Instead, replace any bent or damaged part with a new one.*

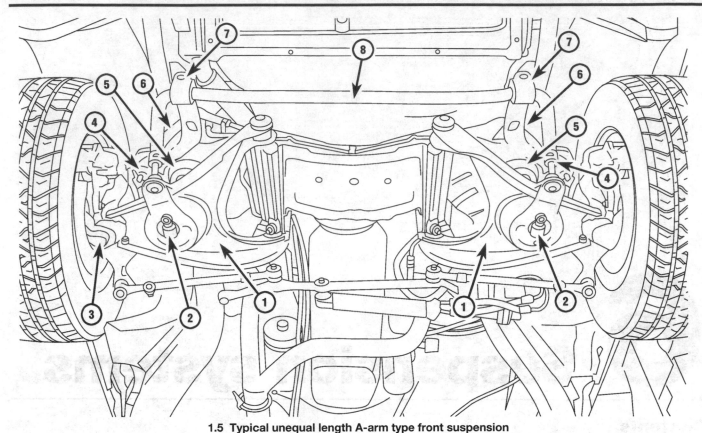

1.5 Typical unequal length A-arm type front suspension

1	Lower control arm	4	Stabilizer bar link	7	Stabilizer bar bushing and
2	Shock absorber	5	Coil spring		mounting clamp
3	Lower balljoint	6	Upper control arm	8	Stabilizer bar

Warning 2: *Some models are equipped with airbags. Always disconnect the negative battery cable, then the positive battery cable and wait two minutes before working in the vicinity of the impact sensors, steering column or instrument panel to avoid the possibility of accidental deployment of the airbag, which could cause personal injury.*

Frequently, when working on the suspension or steering system components, you may come across fasteners which seem impossible to loosen. These fasteners on the underside of the vehicle are continually subjected to water, road grime, mud, etc., and can become rusted or "frozen," making them extremely difficult to remove. In order to unscrew these stubborn fasteners without damaging them (or other components), be sure to use lots of penetrating oil and allow it to soak in for a while. Using a wire brush to clean exposed threads will also ease removal of the nut or bolt and prevent damage to the threads. Sometimes a sharp blow with a hammer and punch is effective in breaking the bond between a nut and bolt threads, but care must be taken to prevent the punch from slipping off the fastener and ruining the threads. Heating the stuck fastener and surrounding area with a torch sometimes helps too, but isn't recommended because of the obvious dangers associated with fire. Long breaker bars and extension, or "cheater," pipes will increase leverage, but never use an extension

pipe on a ratchet - the ratcheting mechanism could be damaged. Sometimes, turning the nut or bolt in the tightening (clockwise) direction first will help to break it loose. Fasteners that require drastic measures to unscrew should always be replaced with new ones.

Front suspension

There are many different designs of front suspension used on today's cars and light trucks. Front suspension systems, with the exception of some four-wheel drive vehicles, are independent, i.e. each wheel can move up and down without appreciably affecting the other. Front suspension can be grouped into two major types: *Unequal length A-arm* (sometimes referred to as "double wishbone" suspension), and *MacPherson strut.* The unequal length A-arm type can be broken down into sub-groups, depending on the type of spring arrangement used: *coil spring* (with the spring positioned between the lower arm and the frame or crossmember), *coil-over shock absorber* (where the coil spring and shock absorber are combined into a single unit, much like a MacPherson strut but without supporting the wheel), and *torsion bar.* The MacPherson strut (which has an integral coil spring) can also be designed as a *modified strut*, where the coil spring is separate from the strut.

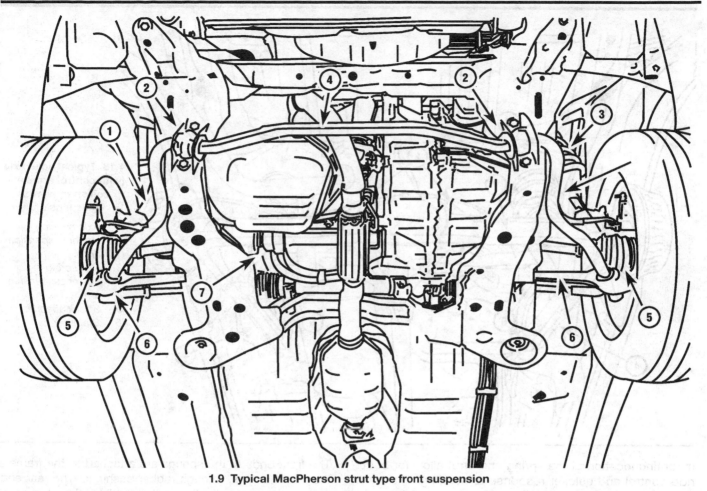

1.9 Typical MacPherson strut type front suspension

1	MacPherson strut assembly	
2	Stabilizer bar bushing and mounting bracket	
3	Coil spring	
4	Stabilizer bar	
5	Outer CV joint	
6	Lower control arm	
7	Inner CV joint	

Unequal length A-arm

The unequal length A-arm suspension systems uses a "short" upper and "long" lower A-shaped control arm **(see illustration)**. The different length arms allow for vertical and side to side control, which is crucial for handling and even tire wear. The design gives the vehicle greater control, driving stability, and better alignment adjustment angles. The front suspension consists of upper and lower control arms, shock absorbers, coil springs or torsion bars, and a stabilizer bar. Some models use a strut bar between the frame and the lower control arms. The designs are similar, except for the mounting of strut bar to the lower control arm. The upper end of each front shock is attached to a bracket on the frame. The lower end of each shock is attached to the lower control arm. On all models, the inner ends of the control arms are attached to the frame; the outer ends are attached to the steering knuckles with balljoints.

On torsion bar models, the longitudinal torsion bars connect either the upper or lower control arms to the frame. The transverse type torsion bars are connected from the frame to the control arms, and are mainly used on RWD Chrysler vehicles.

The balljoints are either bolted or riveted to the control arms or pressed into the control arms. On the riveted type balljoint, the rivets can be drilled out, and a new balljoint bolted to the control arm. On the pressed-in type, the old balljoints need to be pressed out and new balljoints pressed in or pressed into the steering knuckle. **Note:** *Some manufacturers use balljoints that are not removable or serviceable. If the balljoint is damaged or worn, the control arm must be replace as a unit.*

The upper control arms either pivot on a bushing-and-shaft assembly bolted to the frame or on two pivot bolts. The lower control arms pivot on one or two bolts. The shocks and springs are mounted between the lower control arms and the frame. The stabilizer bar is attached to the frame with clamps. The outer ends of the stabilizer bar are attached to the lower control arms with links or to the steering knuckles with links.

MacPherson strut

The MacPherson strut type **(see illustration)**, is a compact unit, incorporating a strut/coil spring design. The upper end of each strut is attached to the vehicle body. The

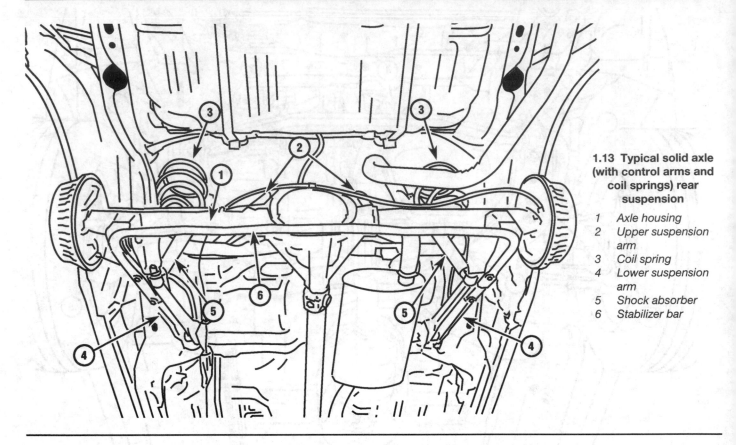

1.13 Typical solid axle (with control arms and coil springs) rear suspension

1   Axle housing
2   Upper suspension arm
3   Coil spring
4   Lower suspension arm
5   Shock absorber
6   Stabilizer bar

mounting location of the spring and strut allow for better ride control and quicker response to steering. The lower end of the strut is connected to the upper end of the steering knuckle. The steering knuckle is attached to a balljoint mounted on the outer end of the control arm. Some types have balljoints riveted or bolted to the control arm while others are not serviceable and need to be replaced as an assembly with the control arm. A stabilizer bar is used on most models. The bar is attached to the frame with a pair of clamps and to the control arms with link rods or braces, and help prevent body roll during turning.

The modified strut type is similar to the MacPherson strut, except for the mounting of the coil spring. The coil spring on the modified strut is mounted separate from the strut between the lower control and the frame.

Rear suspension

There are four basic types of rear suspension systems: *Solid axle*, as used on the majority of rear-wheel drive vehicles, *beam type* (as used on many front-wheel drive vehicles, predominantly with MacPherson struts), *trailing arm*, and *multi-link*. The solid axle suspension can be subdivided into two types: *Control arm with coil springs*, and *leaf springs*.

Solid axle (with leaf springs)

The leaf spring type rear suspension consists of a pair of multi-leaf springs and two shock absorbers. The rear axle assembly is attached to the leaf springs by U-bolts.

The front ends of the springs are attached to the frame at the front hangers, through rubber bushings. The rear ends of the springs are attached to the frame by shackles, which allow the springs to alter their length when the vehicle is in operation. This system is also used on many 4WD drive front suspension systems.

Solid axle (with control arm and coil springs)

The rear suspension consists of shock absorbers, coil springs, and upper and lower suspension arms **(see illustration)**. All models have two lower arms. A track bar is used on some models. The upper end of the track bar is attached to the frame rail and the lower end is attached to a bracket welded onto the axle tube. A stabilizer bar is an option used on many models. The stabilizer bar is attached to the axle housing and to the lower arms. The coil springs are mounted between the underside of the body and the axle housing or the underside of the body and the lower control arms. The shock absorbers are installed between the frame and brackets, welded onto the underside of the axle assembly on all models.

Beam type

The strut/coil spring type rear suspension system uses strut/coil springs, a pair of lateral suspension arms and a trailing arm, or radius rod, at each corner. The upper ends of the struts are attached to the vehicle body and their lower ends are attached to the upper ends of the rear knuckles. The lower ends of the knuckles are attached to the outer ends of the lateral arms. The strut/knuckle/hub

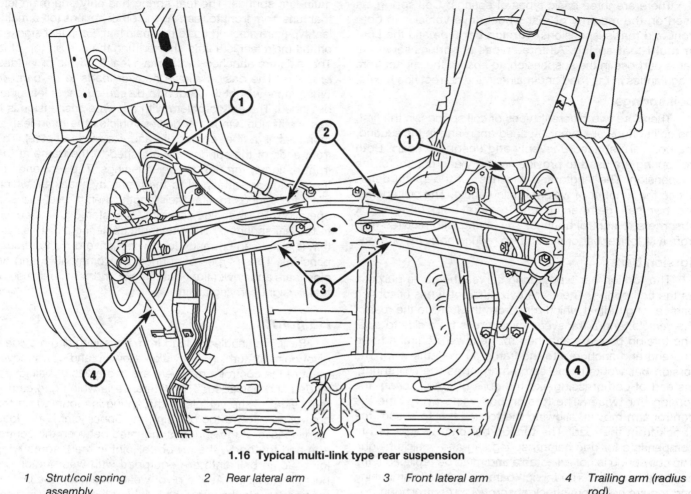

1.16 Typical multi-link type rear suspension

| 1 | Strut/coil spring assembly | 2 | Rear lateral arm | 3 | Front lateral arm | 4 | Trailing arm (radius rod) |

assemblies are positioned laterally by the trailing arms. A stabilizer bar is attached to the vehicle by a pair of brackets and to the struts by link rods.

Trailing arm

The trailing arm rear suspension system also uses strut/coil springs, a pair of lateral suspension arms, and a trailing arm, or radius rod, at each corner. The upper ends of the struts are attached to the vehicle body and their lower ends are attached to the upper ends of the rear knuckles. The lower ends of the knuckles are attached to the outer ends of the lateral arms. The strut/knuckle/hub assemblies are positioned laterally by the trailing arms. A stabilizer bar is attached to the vehicle by a pair of brackets and to the struts by link rods.

Multi-link

The multi-link rear suspension consists of independent knuckle assemblies that are located by trailing arms and parallel lateral link rods, two shock/strut assemblies and an optional stabilizer bar **(see illustration)**.

Shock absorbers/struts

The shock absorber is an oil or gas filled chamber or cartridge. The shock absorbers main function is to reduce or stop vehicle bounce not to, as the name would imply, absorb shock. The shock absorber is mounted between the frame or body and the lower control arm or leaf spring. The shock absorber uses a special piston that is pushed or pulled through oil or gas to regulate or dampen the up and down movement of the suspension. Over time the oil and gas can break down and, in the case of oil filled shocks, leave a visible oil residue on the outside of the shock.

The strut is basically a shock absorber mounted with a coil spring between the body and the steering knuckle and the mainstay of the suspension. The strut is a compact unit which works well for today's vehicles that require more room for different engine and transaxle configurations. There are two basic strut types: sealed and non-sealed. The sealed type can't be serviced and the non sealed has a replaceable cartridge or "shock absorber".

Springs

Most of the front and rear suspensions used on today's vehicles incorporate many of the same type of components and work in a similar manner. The real differences are the way the components are arranged on the vehicle.

Haynes suspension, steering and driveline manual

There are three basic types of springs: **Coil spring**, as used on the majority of rear wheel drive vehicles in both front and rear suspensions, **Torsion bar** type and the **Leaf or multi-leaf spring**. All three types of springs serve the same purpose in the suspension, to absorb the impact from irregularities in the road and maintain the correct ride height.

Coil springs

There are two different types of coil springs. in the first, the coil springs are evenly spaced apart and in the second, the coils are spaced unevenly and unequally apart. Both coil spring are used to provide flexibility and control for the suspension. The length and thickness of the spring determines the amount of weight it can handle. The coil springs are not self dampening and require help from shock absorbers to control bounce. The spring also requires help from a stabilizer bar or strut bar to reduce body roll or sway.

Torsion bars

The torsion bar type spring serves the same purpose as the coil spring. There are two different types of torsion bars, a longitudinal which runs from the front to the rear of the vehicle, and a transverse which runs from side to side. The torsion bars are attached to the lower control arm on one end and anchored to the frame on the other end. The torsion bar works in the same manner as the coil spring. Instead of compressing like the coils of a coil spring, the torsion bar twists. The torsion bar twists every time the control arm moves, allowing the torsion bar to absorb the shock from the road. The torsion bar can be adjusted to compensate for ride height, spring sag and vehicle height. The coil spring is not adjustable and must be replaced if the height is incorrect. The torsion bars are not self dampening and require help from shock absorbers to control bounce.

Leaf springs

The leaf spring type consists of a pair of single-leaf or multi-leaf springs. The leaf spring has only one main leaf that runs from front to rear. All the other springs (on a multi-leaf type) are stacked on to the main leaf. Each leaf stacked on the main leaf is a little shorter than the one on top of it. The leafs are attached to the main leaf by rivets or welded brackets. The design allows heavier loads to be carried. When more weight is needed to be carried, more leafs can be added. The only problem with this type suspension is it has a stiff ride. On many newer vehicles the main leaf or monoleaf, is a fiber composite and not made of steel. The front ends of the springs are attached to the frame at the front hangers, through rubber bushings. The rear ends of the springs are attached to the frame by shackles, which allow the springs to alter their length when the vehicle is in operation. The rear axle assembly is centered and attached to the leaf springs by U-bolts and a centering pin. This system is also used on many four wheel drive drive front suspensions. These types of springs are commonly found on older cars and newer light duty trucks and vans, because of there weight carrying ability.

Balljoints

Balljoints handle the load from the spring and provide a pivot point for turning. They also allow up and down movement of the control arm. There are two types of ball joints, loaded and unloaded. The difference is which one carries the weight of the vehicle. If the spring is mounted on the control arm above the balljoint, the upper joint is the load carrying joint. If the spring is mounted between the control arm and the body, the lower balljoint is the load carrying joint. Some balljoints are equipped with wear indicators built into the joint. As the pivot wears internally, the indicator recedes into the joint. The height of the indicator is measured, once the measurement is bellow the limit the balljoint must be replaced.

Front suspension

2 Unequal length A-arm type (with coil springs)

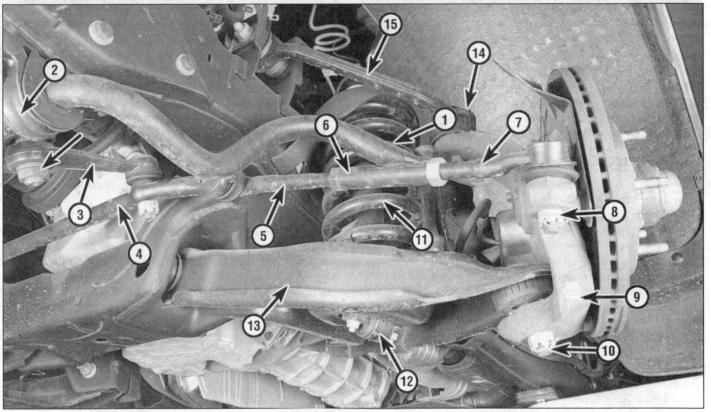

A close up view of a typical A-arm type suspension (with coil springs)

1	Stabilizer bar	6	Tie-rod adjuster tube	11	Coil spring
2	Stabilizer bar bushing and mounting clamp	7	Outer tie-rod end	12	Shock absorber
3	Pitman arm	8	Tie-rod ballstud/nut	13	Lower control arm
4	Center link	9	Steering knuckle	14	Upper control arm balljoint
5	Inner tie-rod end	10	Lower balljoint ballstud/nut	15	Upper control arm

The unequal length A-Arm design uses a "short" upper and "long" lower A-shaped control arm **(see illustration)**. The different length arms allow for vertical and side to side control, which is crucial for handling and even tire wear. The design gives the vehicle greater control, driving stability and better alignment adjustment angles. The front suspension consists of upper and lower control arms, shock absorbers, coil springs and a stabilizer bar. The upper end of each front shock is attached to a bracket on the frame. The lower end of each shock is attached to the lower control arm. The inner ends of the control arms are attached to the frame; the outer ends are attached to the steering knuckles with balljoints.

The balljoints are either bolted or riveted to the control arms or pressed into the control arms. On the riveted type

balljoint, the rivets can be drilled out, and new ones bolted to the control arm. On the press type, the old balljoints need to be pressed out and new balljoints pressed in or pressed into the steering knuckle. **Note:** *Some manufactures use upper and lower control arm balljoints that are not removable or serviceable. If the balljoint is damaged or worn, the control arm must be replace as a unit.*

The upper control arms pivot on a bushing-and-shaft assembly bolted to the frame. The lower control arms pivot on one or two bolts. The shocks and springs are mounted between the lower control arms and the frame. The stabilizer bar is attached to the frame with clamps. The outer ends of the stabilizer bar are attached to the lower control arms with links or to the steering knuckles with links.

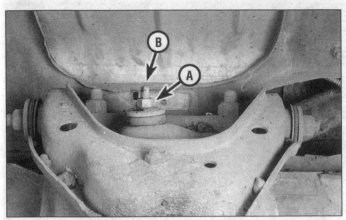

2.1 When removing the shock absorber upper mounting nut (A), hold the flats on the shock rod (B) to prevent the rod from turning (some models use two nuts in place of a lock nut)

2.4 To detach the lower end of the shock absorber from the lower control arm on a "bar mounted" shock absorber, remove these two bolts (arrows), then pull the shock down through the hole in the arm

Shock absorber

Removal

Refer to illustrations 2.1 and, 2.4

1   Using a backup wrench on the stem, remove the upper mounting nut **(see illustration)**.
2   Remove the retainer and grommet.
3   Raise the vehicle and place it securely on jackstands.
4   Working underneath the vehicle, remove the nut and through bolt, nut or bolts which attaches the lower end of the shock absorber to the lower control arm **(see illustration)**. Pull the shock out from below.
5   Remove the lower grommet and retainer from the stem.
6   Fully extend the shock absorber prior to installation. The remaining installation steps are the reverse of removal. Be sure to tighten the upper mounting nut and the lower mounting bolts securely.

Installation

13   Guide the shock absorber assembly up into the fenderwell and insert the three upper mounting studs through the holes in the body. Once the studs protrude from the holes, install the nuts so the assembly won't fall back through, but

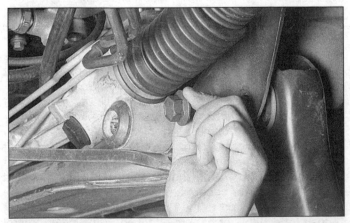

2.22 If the steering gear obstructs access to the control arm bolts, unbolt and raise it up to provide sufficient clearance to remove the front pivot bolt for the lower control arm

don't tighten the nuts completely yet. The shock absorber is heavy and awkward, so get an assistant to help you, if possible.
14   Insert the lower end of the shock absorber into the damper fork. Make sure the aligning tab on the back of the shock body enters the slot in the damper fork.
15   Connect the damper fork to the lower control arm, tightening the self-locking nut securely. Now tighten the damper fork pinch bolt securely
16   Attach the brake hose to its bracket and tighten the bolt securely.
17   Install the wheel and lug nuts, lower the vehicle and tighten the lug nuts securely
18   Tighten the upper mounting nuts securely.

Coil spring

Removal

Refer to illustrations 2.22, 2.23, 2.25a, 2.25b and 2.25c
Warning: *The following procedure is potentially dangerous if the proper safety precautions are not taken. You should use a coil spring compressor to safely perform this procedure.*
Note: *This type of coil spring is found on vehicles using a strut assembly and a lower control arm. The coil spring is mounted between the lower control arm and the front crossmember. This procedure describes removing the lower control arm pivot bolts and detaching the control arm from the frame. On some models, it may be easier to detach the balljoint from the steering knuckle instead.*
19   Loosen the wheel lug nuts on the side to be disassembled. Raise the vehicle, support it securely on jackstands and remove the wheel.
20   Disconnect the tie-rod end from the steering knuckle (see Chapter 7).
21   Disconnect the stabilizer bar link from the lower control arm.
22   On vehicles with rack and pinion steering, you may have to remove the steering gear bolts (see Chapter 7) and raise it slightly to remove the front pivot bolt from either

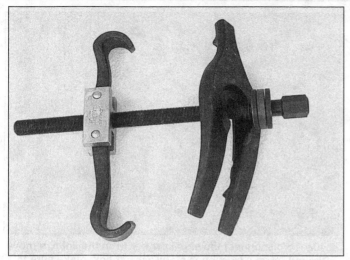

2.25a To detach the lower control arm from the front suspension crossmember, remove these nuts (arrows) and bolts . . .

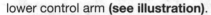

2.23 A typical aftermarket internal spring compressor tool: The hooked arms grip the upper coils of the spring, the plate is inserted below the lower coils, and when the nut on the threaded rod is turned, the spring is compressed

lower control arm **(see illustration)**.

23 Install a spring compressor tool, with a plate between the coils near the top of the spring **(see illustration)**.

24 Guide the compression rod up through the hole in the lower control arm and the coil spring, then insert the end of the compression rod into the upper plate. Install the lower plate, ball nut, thrust washer and bearing and the forcing nut on the compression rod. Tighten the forcing nut until the coil spring can be wiggled, indicating that all spring pressure has been taken up.

25 Remove the lower control arm-to-crossmember nuts and pivot bolts **(see illustrations)**.

26 Remove the spring compressor compression rod then maneuver the coil spring out from between the lower control arm and crossmember.

Installation

27 Install the coil spring insulator on the top of the spring.

28 Install the spring in between the lower control arm and

the spring upper pocket in the suspension crossmember.

29 Position the bottom of the spring so that the pigtail covers only one of the drain holes, but leaves the other one open.

30 Locate the spring in the upper seat in the crossmember. Install the spring compressor tool as described in Steps 3 and 4, then tighten the forcing nut until the lower control arm bushing holes align with the pivot bolt holes in the crossmember.

31 Install the lower control arm pivot bolts and nuts with the bolt heads facing out (away from each other). Don't tighten them completely at this time.

32 Remove the spring compressor tool. Position a floor jack under the outer end of the lower control arm and raise it to simulate a normal ride position. Now tighten the pivot bolt nuts securely.

33 Reattach the steering gear (see Chapter 7) and tighten the steering gear bolts, if necessary.

34 Reconnect the stabilizer bar link to the lower control arm.

35 Reattach the tie-rod end to the steering knuckle (see Chapter 7).

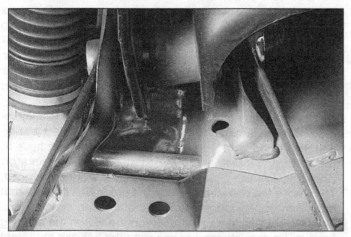

2.25b . . . using two wrenches

2.25c Remove the lower control arm pivot bolt

2.39 To detach the stabilizer bar from the frame, remove the nuts (arrows) that attach the brackets to the frame

2.40a To disconnect the stabilizer bar from the links, remove the nut (arrow) from each stabilizer bar link; make sure to note the order in which the bushings and washers are installed - they must be installed in the same sequence when the stabilizer is installed

36 Install the front wheel, remove the jackstands and lower the vehicle. Tighten the wheel lug nuts securely.
37 Have the alignment checked by a dealer service department or an alignment shop.

Stabilizer bar

Removal

38 Apply the parking brake. Loosen the front wheel lug nuts, raise the front of the vehicle and support it securely on jackstands. Remove the wheels.

Link type

Refer to illustrations 2.39, 2.40a and 2.40b

39 Remove the bolts which attach the stabilizer bar brackets to the underside of the vehicle **(see illustration)**.
40 Detach the stabilizer bar link bolts from the lower control arms **(see illustrations)**. Note the order in which the spacers, washers and bushings are arranged on the link bolt.
41 Remove the bar from under the vehicle.

Mount/bracket type

42 Apply penetrating oil to the stabilizer mounting bolts.
43 Remove the bolts that secure the stabilizer brackets to the control arm and remove the brackets.
44 Remove the bolts that secure the stabilizer mounting plates and remove both plates.
45 Remove the stabilizer bar from the vehicle, complete with bushings and brackets. The bushings need not be removed from the stabilizer bar unless either the bushings or the bar are being replaced.

Installation

46 Inspect the bushings to be sure they are not hardened, cracked or excessively worn, and replace if necessary.
47 Installation is the reverse of the removal procedure.

Upper control arm

Removal

Refer to illustrations 2.51, 2.53a and 2.53b
48 Loosen the wheel lug nuts, raise the front of the vehicle

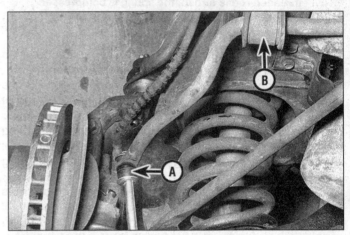

2.40b Typical stabilizer bar setup with the links attached to the lower control arms

A   Link   B   Stabilizer bar bracket

and support it securely on jackstands. Remove the wheel.
49 Position a floor jack, with a wood block on the jack head (to act as a cushion), under the lower control arm as close to the balljoint as possible. Raise the jack slightly to take the spring pressure off the upper control arm. **Warning:** *The jack must remain in this position throughout the entire procedure.*
50 Remove the cotter pin from the upper balljoint stud and loosen the stud nut one or two turns.
51 Rap on the steering knuckle near the upper balljoint stud with a hammer to loosen the stud in the steering knuckle. If the stud won't come loose, you might have to resort to a "picklefork" type of balljoint stud separator. Some models use a pinch bolt to retain the upper ball joint. To remove the ball joint; remove the pinch bolt and nut that retain the upper balljoint stud to the steering knuckle **(see illustration)**. Using a prybar, spread the slot in the steering knuckle far enough apart to separate the balljoint stud from the knuckle. **Caution:** *The use of a picklefork balljoint sepa-*

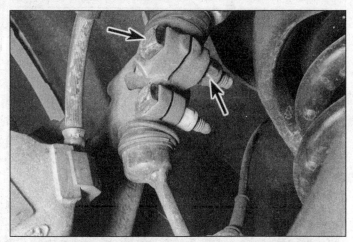

2.51 Detach the upper control arm balljoint stud from the steering knuckle, remove the nut and pinch bolt (arrows) if equipped - on models without pinch bolts, the upper arm is attached to the steering knuckle with a balljoint stud nut

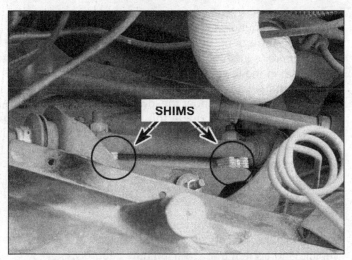

SHIMS

2.53b On models with shims, note the position of the alignment shims and return them to their original position

2.53a To disconnect the upper control arm from the frame, remove these two bolts (arrows)

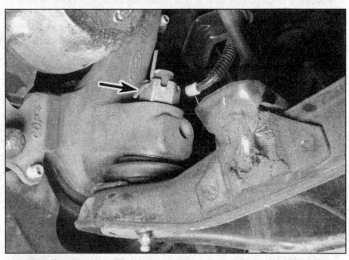

2.60a To separate the lower control arm from the steering knuckle, remove the cotter pin (arrow), loosen the castle nut one or two turns and give the steering knuckle a few sharp raps on the boss area adjacent to the balljoint stud to loosen the stud

rator will usually result in balljoint boot damage.

52 Remove the nut from the upper balljoint stud.

53 Remove the upper control arm retaining bolts **(see illustrations)** and remove the upper control arm assembly.

Inspection

54 Inspect the control arm bushings for cracks and tears. If they're damaged or worn, take the control arm to an automotive machine shop to have new bushings installed. This procedure requires a number of specialized tools, so it's not worth tackling at home.

Installation

55 Position the upper control arm pivot shaft on the frame and install the two attaching bolts and washers. Tighten the bolts to the securely.

56 Insert the upper balljoint stud into the steering knuckle and install the ball stud nut or pinch bolt and nut. Tighten the nut securely. On models without pinch bolts, continue

to tighten the nut, if necessary, until the cotter pin hole in the stud is in line with the nut slots, then install a new cotter pin.

Lower control arm

Removal

Refer to illustrations 2.60a, 2.60b and 2.61

57 Loosen the wheel lug nuts. Raise the front of the vehicle and support it securely on jackstands. Remove the front wheel.

58 Remove the shock absorber and disconnect the stabilizer bar or link if equipped.

59 Remove the front coil spring if equipped (see coil spring removal in this Section).

60 Remove the cotter pin and loosen the balljoint stud nut **(see illustration).** Break the balljoint loose from the steering knuckle by rapping the steering knuckle boss sharply with a

2.60b Apply penetrating oil to the balljoint stud/steering knuckle area, then strike the steering knuckle boss sharply with a large hammer until the balljoint stud is freed (be sure the nut is loosened a couple of turns, but not removed) - this will probably take a few tries, because of the extremely tight fit between the two parts

2.61 The lower control arm pivot bolts/nuts (arrows) can be removed AFTER the coil spring has been restrained with the special tool

hammer **(see illustration)**. **Note:** *A picklefork type balljoint separator may damage the balljoint seals. Separate the arm from the steering knuckle.*

61 Remove the lower control arm-to-crossmember nuts and pivot bolts **(see illustration)**. It may be necessary to remove the steering gear mounting bolts and reposition the gear to allow bolt removal. **Note:** *On some models it is necessary to remove the lower control arm pivot bolts to remove the coil spring prior to this step.*

Inspection

62 Inspect the bushings for cracks and tears. If they're damaged or worn, take the control arm to an automotive machine shop to have new bushings installed. This procedure requires a number of specialized tools, so it's not worth tackling at home.

Installation

63 Insert the balljoint stud into the steering knuckle, install the castle nut and tighten it securely. Continue to tighten the nut, if necessary, until the hole in the stud is in line with

the slot in the nut. Install a new cotter pin.

64 Install the spring compressor in accordance with the manufacturer's instructions and place the spring in position on its seat in the lower control arm. Make sure the spring is properly seated.

65 Place a floor jack under the lower control arm, raise the lower control arm and guide the upper end of the spring into position. Make sure the insulator is properly installed on top of the spring.

66 Install the lower control arm pivot bolts and nuts. Don't fully tighten the nuts at this time. Remove the floor jack.

67 Install the shock absorber and stabilizer bar or link if equipped.

68 Install the strut rod/radius rod if equipped.

69 Install the wheel, remove the jackstands, lower the vehicle and tighten the wheel lug nuts securely.

70 With the vehicle on the ground, tighten the lower control arm pivot bolt nuts securely.

Balljoints

Note: *See Section 8 for balljoint check and replacement procedures.*

3 Unequal length A-arm type (with coil-over shock absorbers)

A close up view of the coil-over shock absorber type suspension

1	Stabilizer bar	3	Shock absorber/coil spring assembly	5	Lower control arm
2	Upper control arm	4	Lower ball joint	6	Stabilizer link
				7	Upper ball joint

The fully-independent unequal length A-arm type (with coil-over shock absorbers) front suspension allows each wheel to compensate for road surface irregularities without any appreciable effect on the other wheel. The suspension at each front wheel consists of a shock absorber/coil spring assembly situated between the upper control arm support and the lower control arm **(see illustration)**. A steering knuckle is located between the upper and lower arms by a pair of balljoints, one in each control arm. A stabilizer bar controls vehicle roll during cornering. The stabilizer bar is attached to the frame by a pair of steel clamps and to the lower control arms by links.

Shock absorber/coil spring assembly

Removal

Refer to illustrations 3.2, 3.4, 3.5a and 3.5b

1   Loosen the front wheel lug nuts, raise the vehicle, place it securely on safety stands and remove the front wheel.

2   Mark the spring with paint so the assembly can be reinstalled with the correct orientation **(see illustration)**.

3.2 Mark the outside of the spring with paint so the shock/coil spring assembly can be installed in the correct orientation

3.4 Remove the bolt and nut from the lower end of the shock/coil spring assembly

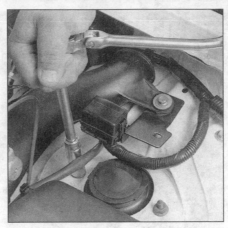

3.5a Remove the upper mounting nuts - DO NOT REMOVE THE CENTER NUT UNDER THE CAP!

3.5b Pry the lower control arm downward and remove the shock/coil spring

3   Loosen the lower control arm pivot bolts so the control arm can drop lower than its normal installed position. Support the control arm so it doesn't strain the brake hose.

4   Remove the shock absorber's lower bolt and nut **(see illustration)**.

5   Working in the engine compartment, remove the shock absorber's upper mounting nuts **(see illustration)**. Pry down on the lower control arm, lower the shock free of the fender and remove it from the vehicle **(see illustration)**.

Inspection

6   Check the shock absorber body for leaking fluid, dents, cracks and other obvious damage which would warrant replacement.

7   Check the coil spring for chips or cracks in the spring coating (this will cause premature spring failure due to corrosion). Inspect the spring seat for cuts and general deterioration.

8   If any undesirable conditions exist, proceed to the shock absorber disassembly procedure (below). **Note:** *When disposing of a used damper, return it to your automotive parts store or dealer - some struts are gas charged and require special disposal procedures.*

Installation

9   Installation is the reverse of removal. Be sure to tighten all fasteners securely. Tighten the lower control arm pivot bolts and shock absorber lower bolt loosely at first, then tighten them securely after the vehicle's weight is resting on the wheels.

Shock absorber/coil spring

10   If the shock absorbers or coil springs exhibit the telltale signs of wear (leaking fluid, loss of damping capability, chipped, sagging or cracked coil springs) explore all options before beginning any work. The shock absorbers are not serviceable and must be replaced if a problem develops. However, assemblies complete with springs may be available on an exchange basis, which eliminates much time and work. Whichever repair/replacement method you choose, check on the cost and availability of parts before

disassembling your vehicle. **Warning:** *Disassembling the shock absorber/coil spring is potentially dangerous and utmost attention must be directed to the job, or serious injury may result. Use only a high-quality spring compressor and carefully follow the manufacturer's instructions furnished with the tool. After removing the coil spring from the assembly, set it aside in a safe, isolated area.*

Disassembly

Refer to illustrations 3.13 and 3.15

11   Remove the shock absorber/coil spring assembly following the procedure described previously. Mount the assembly in a vise. Line the vise jaws with wood or rags to prevent damage to the unit and do not tighten the vise excessively.

12   Pry the cap from the top center of the shock/coil spring assembly. Loosen the damper shaft nut that's beneath the cap, but DO NOT remove it yet!

13   Following the tool manufacturer's instructions, install the spring compressor (which can be obtained at most auto parts stores or equipment yards on a daily rental basis) on the spring and compress it sufficiently to relieve all pressure from the upper spring seat **(see illustration)**. This can be verified by wiggling the spring.

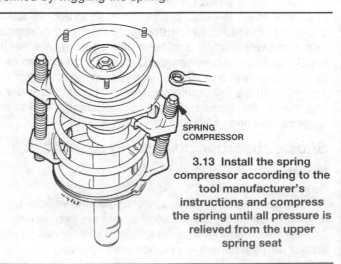

SPRING COMPRESSOR

3.13 Install the spring compressor according to the tool manufacturer's instructions and compress the spring until all pressure is relieved from the upper spring seat

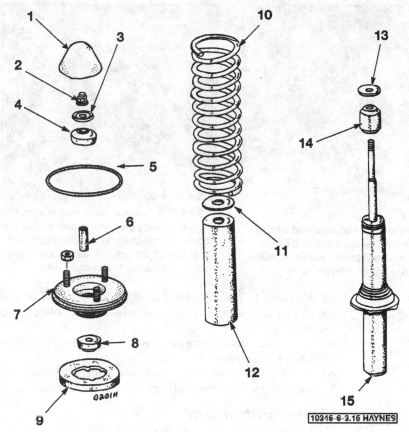

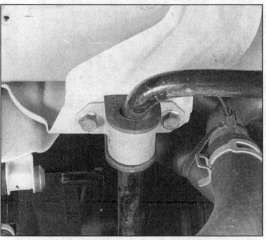

3.27 To detach the front stabilizer bar from the body, unbolt the brackets

3.28 To detach it from the suspension, remove the nuts and bolts from the links (arrows)

3.15 Exploded view of a typical front shock absorber/coil spring assembly

1	Damper cap	9	Spring mounting rubber
2	Self-locking nut	10	Spring
3	Damper mounting washer	11	Dust cover plate (if
4	Upper mounting rubber		equipped)
5	Seal	12	Dust cover
6	Damper mounting collar	13	Bump stop plate
7	Damper mounting base	14	Bump stop
8	Lower mounting rubber	15	Damper unit

14 Remove the damper shaft nut. It may be necessary to hold the shaft from turning while loosening the nut.

15 Remove the nut and washer. Mark the outer side of the spacer plate and mounting plate so they can be reinstalled in the same orientation to the spring. Remove the spacer plate and mounting plate **(see illustration)**. Check the mounting plate for cracking and general deterioration. If there is any doubt about its condition, replace it.

16 Lift the rebound stopper (the accordion-like rubber piece) from the damper shaft. Check the rubber for cracking and hardness, replacing it if necessary.

17 Carefully lift the compressed spring from the assembly and set it in a safe place. **Warning:** *Keep the ends of the spring pointed away from your body.*

Reassembly

18 Place the coil spring onto the lower insulator, with the end of the spring resting in the step (lowest part of the insulator) **(see illustration 3.2)**.

19 Extend the damper rod to its full length and install the rebound stopper.

20 Install the mounting plate and spacer plate so their paint marks will face outward when the shock/coil spring assembly is installed.

21 Install the nut and partially tighten it.

22 Remove the spring compressor tool.

23 Tighten the dampener shaft nut securely.

24 Install the shock/coil spring assembly following the procedure outlined previously.

Stabilizer bar

Removal

Refer to illustrations 3.27 and 3.28

25 Raise the vehicle and support it securely on jackstands.

26 Remove the under cover from under the front.

27 Remove the bracket bolts **(see illustration)**. Lower the brackets away from the frame.

28 Unbolt the link at each end of the stabilizer bar **(see illustration)**.

3.37 To detach the front end of the radius rod, remove the plug in the plastic splash shield, then remove this nut from the rod

3.38 To detach the rear end of the radius rod, remove the two bolts (arrows) which attach the radius rod to the lower control arm

3.44 Remove the pivot bolt and nut (upper arrows) from the outer end of the control arm - also remove the balljoint bolt (lower arrow); it's accessible from the top of the control arm

29 Take the stabilizer bar out form under the vehicle and remove the rubber bushings

Inspection

30 Check the rubber bushings and link grommets for cracks and tears. Replace all damaged bushings; replace the links if the grommets are damaged.

Installation

31 Align the rubber bushings with the marks on the stabilizer, then install the brackets and bolt them to the frame.

32 Install the links.

33 Tighten the bracket bolts and link bolts.

34 Install the undercover, then lower the vehicle.

Radius rod

Refer to illustrations 3.37 and 3.38

35 Loosen the wheel lug nuts, raise the front of the vehicle and place it securely on jackstands. Remove the wheel.

36 Remove the plug from the plastic splash shield.

37 Remove the nut from the front end of the radius rod in the front crossmember **(see illustration)**.

38 Remove the bolts which attach the rear end of the radius rod to the lower control arm **(see illustration)** and remove the rod.

39 Installation is the reverse of removal. Be sure to tighten all fasteners securely.

40 Drive the vehicle to an alignment shop and have the front end alignment checked and, if necessary, adjusted.

Lower control arm

Refer to illustrations 3.44, 3.45a and 3.45b

41 Loosen the front wheel lug nuts, raise the vehicle, place it securely on safety stands and remove the front wheel.

42 Disconnect the stabilizer bar link bolt from the lower control arm **(see illustration 2.4)**.

43 Remove the shock absorber lower bolt.

44 Remove the balljoint through-bolt and the upper bolt that secure it to the lower control arm **(see illustration)**.

45 Remove the inner pivot bolts that secure the lower control arm to the frame **(see illustrations)**. Lower the inner end of the arm away from the frame, pull it away from the balljoint and remove it from under the vehicle.

3.45a Mark the position of the adjusting cams (arrows) . . .

3.45b . . . then remove the pivot bolts and nuts from the inner end of the control arm

3.52a The upper and lower balljoint stud nuts are secured by cotter pins (arrows)

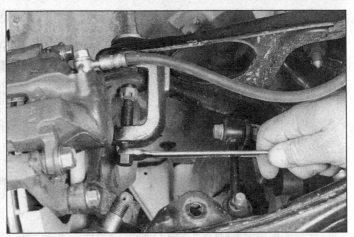

3.52b Remove the cotter pin and undo the nut partway, then use a small puller to separate the balljoint stud from the knuckle

46 Inspect the front and rear bushings in the lower control arm. If either bushing is torn or cracked, take the control arm to an automotive machine shop and have the old bushings pressed out and new ones pressed in.

47 Installation is the reverse of the removal procedure. Tighten the balljoint bolts securely while the vehicle is still raised. Tighten the remaining fasteners (lower arm pivots, stabilizer link, shock absorber lower bolt) loosely at first, then install the wheels and lower the vehicle. Tighten the fasteners securely with the vehicle's weight resting on the wheels.

48 Have front end alignment checked by a dealer or front end shop.

Upper control arm

Removal

Refer to illustrations 3.52a, 3.52b and 3.53

49 Loosen the front wheel lug nuts, raise the vehicle, place it securely on safety stands and remove the front wheel. If you're working on an ABS model, detach the retaining band that secures the wheel speed sensor harness.

50 Remove the under cover from the front of the vehicle.

51 Remove the shock absorber lower mounting bolt. Support the lower control arm from below.

52 Remove the cotter pin and nut from the upper balljoint, then separate the balljoint from the upper control arm with small puller **(see illustrations)**.

53 Hold the pivot bolt with a wrench and undo the nut (one long bolt passes through the front and rear pivot points of the upper control arm) **(see illustration)**. Remove the nut and take the upper control arm off the vehicle.

Inspection

54 Inspect the front and rear bushings in the lower control arm. If either bushing is torn or cracked, take the control arm to an automotive machine shop and have the old bushings pressed out and new ones pressed in.

55 Check the balljoint dust boot for cracks or deterioration. If there are any problems, carefully tap the dust boot off of the control arm with a hammer and chisel. Be sure not

3.53 Hold the bolt head (right arrow) with a wrench and undo the nut (left arrow)

to hit the balljoint stud with the chisel.

56 Pack the inside of a new dust boot with grease, then press the dust boot onto the balljoint stud with an appropriate size socket. Leave a small gap (no more than 3/64 inch) between the dust boot and the control arm. Don't push the dust boot on too far or the retaining ring will be damaged.

Installation

57 Installation is the reverse of the removal procedure. Tighten the balljoint nut securely while the vehicle is still raised, then install a new cotter pin. Tighten the remaining fasteners (upper control arm pivot bolt, shock absorber lower bolt) loosely at first, then install the wheels and lower the vehicle. Tighten the fasteners securely with the vehicle's weight resting on the wheels.

58 Have front end alignment checked by a dealer or front end shop.

Balljoints

Note: *See Section 8 for balljoint check and replacement procedures.*

4 Unequal length A-arm type (with longitudinal torsion bars)

Typical view of a longitudinal torsion bar type suspension

1	Center link	6	Outer tie-rod (tie-rod end)	10	Strut rod
2	Idler arm	7	Steering arm	11	Shock absorber
3	Pitman arm	8	Steering knuckle	12	Lower control arm
4	Inner tie-rod (tie-rod end)	9	Upper control arm	13	Torsion bar
5	Tie-rod adjuster sleeve				

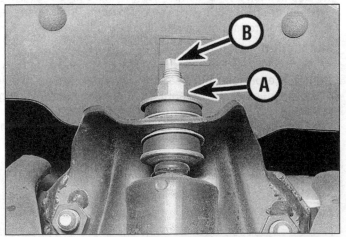

4.1 When removing the shock absorber upper mounting nut (A), hold the flats on the shock rod (B) to prevent the rod from turning (some models use two nuts in place of a locknut)

The unequal length A-arm type (with longitudinal torsion bars) front suspension **(see illustration)** is fully independent. Each wheel is connected to the frame by a steering knuckle, upper and lower balljoints and upper and lower control arms. All models use shock absorbers and torsion bars. The upper end of each front shock is attached to a bracket on the frame. The lower end of each shock is attached to the lower control arm. Longitudinal torsion bars connect the lower control arms to the frame.

Shock absorber
Refer to illustrations 4.1, 4.4a, 4.4b, 4.4c, 4.6a and 4.6b

1   Using a backup wrench on the stem, remove the upper mounting nut **(see illustration)**.
2   Remove the retainer and grommet.
3   Raise the vehicle and place it securely on jackstands.
4   Working underneath the vehicle, remove the nut and through bolt or the nut which attaches the lower end of the

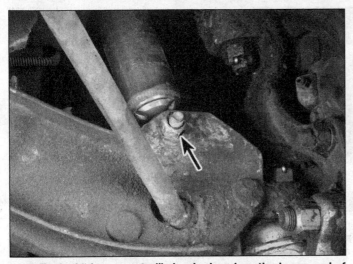

4.4a Typical "ring mounted" shock absorber, the lower end of the shock absorber is attached to the lower arm with a nut (arrow) and horizontal through-bolt - be sure to note the direction in which the bolt head and the nut are facing, so you can install them facing in the same direction during reassembly

shock absorber to the lower control arm **(see illustrations).** Pull the shock out from below.

5   Remove the lower grommet and retainer from the stem.

6   Fully extend the shock absorber prior to installation. The remaining installation steps are the reverse of removal **(see illustrations).** Be sure to tighten the upper mounting nut and the lower mounting bolts securely.

4.4b Typical "bayonet mounted" shock absorber, the lower end of the shock absorber is attached to the lower control arm with a nut - again, note the order in which the retainer washer and rubber bushing are installed, with the washer between the nut and the bushing

4.4c Typical "bar mounted" shock absorber, the lower end of the shock absorber is attached to the lower control arm by two bolts

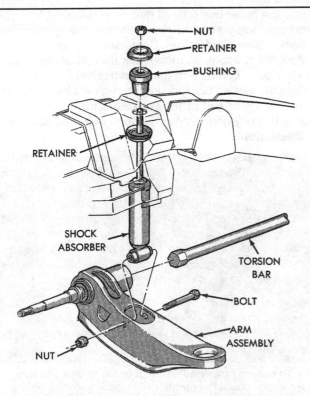

4.6a An exploded view of a typical front shock absorber installation (lower shock eye and through-bolt)

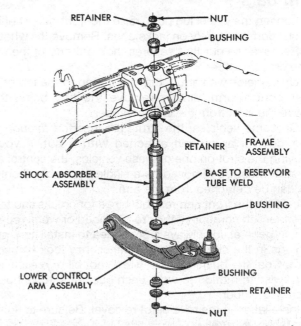

4.6b An exploded view of a typical front shock absorber installation (threaded lower extension with nut)

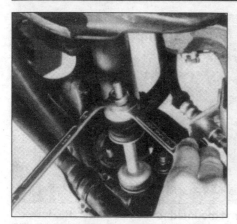

4.8 Remove the nuts from the stabilizer bar link bolts and remove the bolts, washers, spacers and bushings

4.9 A bracket. bushing and two bolts connect the stabilizer bar to each side of the frame

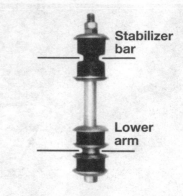

4.11 When properly assembled, the stabilizer bar links should be arranged like this

Stabilizer bar

Refer to illustrations 4.8, 4.9 and 4.11

7   Jack up the front of the vehicle and place it securely on jackstands.

8   Remove the nuts, bolts and cushions from both ends of the stabilizer bar at the lower suspension arms and disconnect the stabilizer bar **(see illustration)**.

9   Remove the stabilizer bar bushings and brackets from the frame and remove the stabilizer bar **(see illustration)**.

10   Inspect all parts for wear or damage. Replace parts as necessary with new ones.

11   Installation is the reverse of the removal procedures. Make sure the bolts, cushions and nuts are assembled in their proper order **(see illustration)** and tighten all bolts and nuts securely.

Strut bars

12   Loosen the wheel lug nuts, raise the front of the vehicle and support it securely on jackstands. Remove the wheel.

13   Remove the nut that secures the front end of the strut to the crossmember.

14   On vehicles with struts that are bolted to the top of the lower control arm, simply remove the two bolts, then remove the strut from the crossmember.

15   On some vehicles, the struts are inserted *through* the lower control arm, then attached with a nut. If you're removing the strut on one of these vehicles, the control arm and strut must be removed as a single assembly, then the strut can be detached from the arm.

16   Inspect the front and rear bushings for cracks and tears and other deterioration. If they're damaged or worn, replace them. In general, it's always a good idea to install new. strut bushings in the crossmember whether they look damaged or not. Coat the bushings with water (not oil or grease) and use a twisting motion to work each bushing into its hole in the crossmember.

17   Installation is the reverse of removal. Be sure to tighten all fasteners securely

Torsion bars

GM type torsion bar

Refer to illustrations 4.19, 4.20, 4.21, 4.22, 4.23a, 4.23b, 4.24, 4.25 and 4.26

18   Loosen the front wheel lugs nuts, raise the vehicle and place it securely on jackstands. Remove the wheel.

19   Count the number of threads showing on the torsion bar adjuster bolt and mark the relationship of the bolt to the torsion bar adjuster nut **(see illustration)**.

20   In the torsion bar adjuster arm, there's a small dimple. Install a small puller with its bolt centered on this dimple **(see illustration)**.

21   Turn the puller bolt until all tension is removed from the torsion bar adjuster arm, then remove the torsion bar adjuster nut **(see illustration)**. Remove the puller.

22   Mark the relationship of the forward end of the torsion bar to the lower control arm **(see illustration)**.

23   Push the torsion bar forward, through the lower control arm, until the rear end of the bar clears the crossmember **(see illustration)** and remove the torsion bar adjuster arm **(see illustration)**.

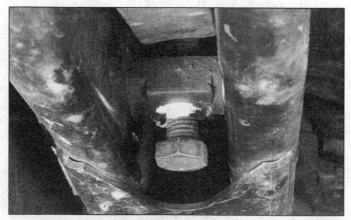

4.19 To ensure proper adjustment of the torsion bar upon reassembly, count the number of threads showing on the torsion bar adjuster bolt and mark the relationship of the bolt to the torsion bar adjuster nut as insurance

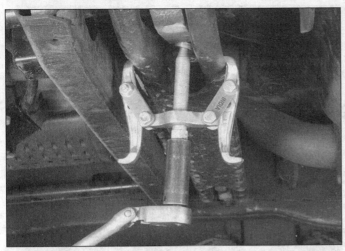

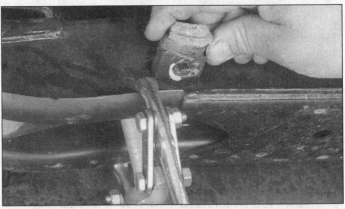

4.21 With tension removed from the adjuster nut, remove the adjuster nut

4.20 Install a small puller as shown, with the fingers hooked around the flange running along each side of the crossmember; make sure the puller bolt is centered on the dimple in the torsion bar adjuster arm; tighten the puller bolt until all tension is removed from the adjuster nut

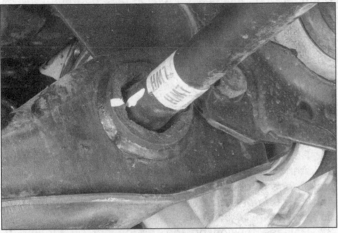

4.23a Slide the torsion bar forward through the lower control arm far enough to pull the rear end of the bar out of the crossmember . . .

24 Pull the torsion bar down and to the rear as far as it will go. If the front end of the bar hangs up in the lower control arm, drive it out of the control arm with a brass drift (see illustration).

25 Installation is the reverse of removal. Be sure to clean out the hexagonal hole in the lower control arm and lube it

4.22 Mark the relationship of the torsion bar to the lower control arm as shown

4.23b . . . and remove the torsion bar adjuster arm. Hold your hand under the arm as you slide out the torsion bar to prevent the arm from falling

4.24 If the torsion bar hangs up in the lower control arm, knock it out with a brass drift

4.25 Clean up any corrosion inside the hex hole in the control arm and lube it with multi-purpose grease before installing the torsion bar

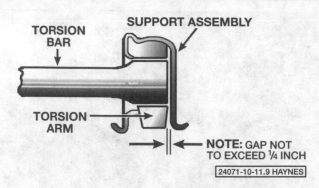

4.26 The clearance measurement indicated here should be checked after torsion bar installation; it should not exceed 1/4-inch

4.32 To release tension on the torsion bar, turn this adjusting bolt (arrow) counterclockwise

with multi-purpose grease **(see illustration)** before inserting the torsion bar into the arm. Also apply some grease to the hex ends of the torsion bar, to the top of the adjuster arm and to the adjuster bolt. Make sure that the alignment marks you made on the torsion bar and the control arm are aligned. Make sure that the torsion bar adjuster bolt is tightened until the same number of threads are showing and the marks you made on the adjuster bolt and nut are aligned.

26 Measure the torsion arm-to-crossmember clearance after installation **(see illustration)** and adjust the torsion bar fore and aft as necessary to bring it within this clearance.

27 Install the wheel, remove the jackstands and lower the vehicle.

28 Tighten the wheel lug nuts securely.

29 Measure the vehicle's ride height on each side, from equal points on the frame to the ground. If the side that has been worked on is higher or lower than the other side, turn the torsion bar adjusting screw accordingly until the vehicle sits level. This may take a few tries, and it's important to roll the vehicle back and forth and jounce the front end between adjustments, to settle the suspension and get an accurate reading.

Chrysler type (longitudinal) torsion bar

Caution: *Don't interchange torsion bars between vehicles. They're identified for use by length and thickness (depending on carline, body, engine, etc.) Don't even interchange torsion bars from one side to the other. They're marked either "L" or "R" on the ends to indicate the side they must be used on.*

Removal

Refer to illustrations 4.32, 4.33a and 4.33b

30 Remove the upper control arm rebound bumper if equipped.

31 Raise the front of the vehicle and support it securely on jackstands positioned under the frame. **Note:** *Don't place jackstands under the control arms - put them under the frame. The front suspension must be in full rebound (under no load) for this procedure.*

32 Release tension on the torsion bar by turning the adjusting bolt counterclockwise **(see illustration)**. **Warning:** *On some models, the load on both torsion bars must*

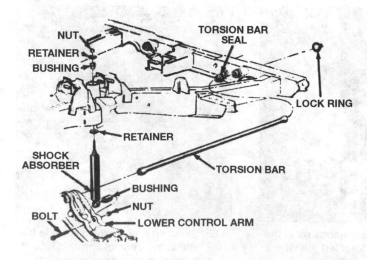

4.33a An exploded view of a typical torsion bar assembly

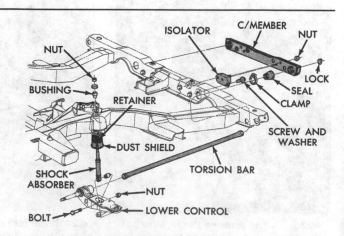

4.33b An exploded view of a typical torsion bar assembly used on some early model Chrysler Vehicles

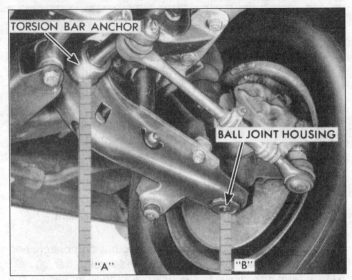

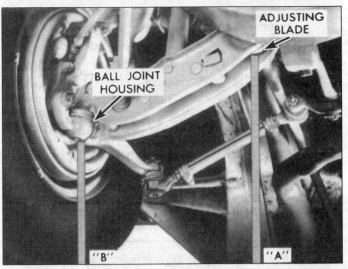

4.44a On 1971 through 1973 Chryslers and on 1974 B and C-body vehicles, measure the distance from the torsion bar anchor to the floor (measurement "A") and from the balljoint housing to the floor (measurement "B") - the difference is the ride height; on 1975 and later vehicles, measure the distance from the torsion bar anchor to the floor (measurement "A")

4.44b On all other 1971 through 1974 vehicles, measure the distance from the lowest point of one adjusting blade to the floor (measurement "A") and from the lowest point of the steering knuckle arm, at the centerline, on the same side, to the floor (measurement "B") - the difference is the ride height

be released before proceeding. Otherwise, the rubber isolator in the rear crossmember remains under a load - and could cause severe damage or personal injury.

33 Remove the lock-ring from the torsion bar rear crossmember **(see illustrations)**.

34 Wrap the torsion bar with a rag to protect the bar from damage, then clamp up a pair of vise-grips to the protected part of the bar and tap on the vise-grips with a hammer to disengage the front end of the torsion bar from the control arm. Remove the vise-grips and rag from the torsion bar, slide the rear dust boot or seal off the front end of the torsion bar, then pull the torsion bar through its hole in the crossmember.

Installation

35 Clean and inspect the seal. If the seal is cracked or torn, replace it.

36 Put a little chassis grease in the hole in the crossmember and slide the torsion bar through the hole.

37 Coat the inside walls of the seal bore with a little grease, install the seal onto the front end of the torsion bar and slide it toward the rear end of the torsion bar (you can't place the seal in its final position until the torsion bar is installed).

38 Wrap the torsion bar with a rag again, clamp it up with a pair of vise-grips and drive the front end of the bar into place in the hex opening of the control arm by tapping against the vise-grips. Remove the vise-grips and shop rag from the torsion bar. Slide the new seal all the way back so that its lip is fully engaged with the groove in the crossmember hole.

39 Install a new lock-ring on the rear of the torsion bar. Make sure it's fully seated.

40 After installation, the vehicle ride height must be set.

Adjustment

Refer to illustrations 4.44a, 4.44b and 4.44c

Note: *Ride height adjustment must always be performed on a level surface.*

41 Check and adjust the tire pressure.

42 Place the vehicle on level, solid pavement. It must be unloaded with a full tank of fuel.

43 Jounce the front bumper up and down several times at least four inches to settle the suspension.

44 The procedure for measuring the ride height - and where to measure it - differs slightly on various vehicles:

a) *On 1971 through 1973 Chryslers and on 1974 B and C-body vehicles* **(see illustration)**, *measure the distance from the lowest point of the front torsion bar anchor at the rear of the lower control arm flange to the floor (measurement "A") and from the lowest point of the balljoint housing on the same side to the floor (measurement "B").*

b) *On all other 1971 through 1974 vehicles* **(see illustration)**, *measure the distance from the lowest point of one adjusting blade to the floor (measurement "A") and from the lowest point of the steering knuckle arm, at the centerline, on the same side, to the floor (measurement "B"). Now subtract measurement "B" from measurement "A." In other words, the ride height dimension is actually the difference between the higher point, (measurement "A") and the lower point (measurement "B").*

c) *On 1975 through 1980 vehicles, measure the distance from the lowest point of the lower control arm torsion bar anchor at a point one inch forward of the rear face of the anchor, to the floor* **(see illustration 11.27a)**.

d) *On 1981 vehicles, measure the distance from the bottom of the front frame rail, between the radiator yoke*

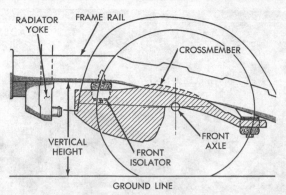

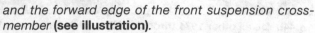

4.44c On 1981 vehicles, measure the distance from the bottom of the front frame rail, between the radiator yoke and the forward edge of the front suspension crossmember

4.48 Apply alignment marks to the torsion bar, anchor arm and torque arm to ensure proper realignment

and the forward edge of the front suspension cross-member **(see illustration)**.

45 Make adjustments by turning the adjusting bolt. After each adjustment, jounce the front bumper again. The final adjustment must be in the up direction. **Note:** *Measure both sides even if only one side was replaced.*

46 Front vehicle height should not vary more than 1/4-inch from the dimension originally measured, and should be with- in 1/8 to 1/4-inch side-to-side (i.e. there should be no more than 1/8-inch difference in vehicle height at the left and right sides).

Toyota type torsion bar

Removal

Refer to illustrations 4.48, 4.49, 4.50 and 4.51

47 Jack up the front of the vehicle and place it securely on jackstands.

48 Slide back the rubber boots and place alignment marks on the torsion bar, anchor arm and torque arm to facilitate proper alignment upon installation **(see illustration)**.

49 Remove the locknut securing the anchor arm and tor-sion bar and measure the protruding bolt end A **(see illus-tration)**. Record the measurement.

50 Unbolt the torque arm from the lower suspension arm

(see illustration). Loosen the adjusting nut completely and remove the anchor arm, torque arm and torsion bar.

51 Inspect the parts for wear or damage, including the splines on the torque arm **(see illustration)**. **Note:** *There are left and right identification marks on the rear end of the torsion bars* **(see illustration)**. *Be careful not to interchange them. Replace any damaged or worn parts with new ones.*

Installation and adjustment

52 Apply a light coat of multi-purpose grease to the tor-sion bar splines.

53 Set the alignment marks and attach the torsion bar to the torque arm, tightening the nuts securely.

54 Set the alignment marks and attach the anchor arm to the torsion bar.

55 Tighten the adjusting nut until the bolt protrusion is equal to that recorded before removal.

56 Measure the vehicle's ride height on each side, from equal points on the frame to the ground. If the side that has been worked on is higher or lower than the other side, turn the torsion bar adjusting nut accordingly until the vehicle sits level. This may take a few tries, and it is important to roll the vehicle back-and-forth and jounce the front end between adjustments, to settle the suspension and get an accurate reading.

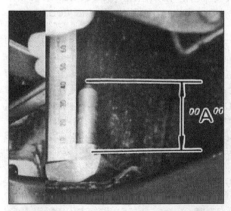

4.49 Remove the locknut while holding the adjusting nut with another wrench (left), then measure the anchor arm bolt protrusion (right)

4.50 The torque arm is held to the lower suspension arm by two bolts and nuts. Don't loosen these nuts unless the anchor arm bolt adjusting nut has been completely loosened

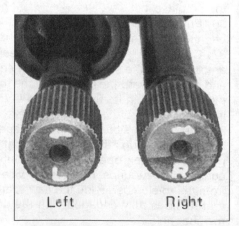

4.51 The torsion bars are marked on the rear ends - don't mix them up

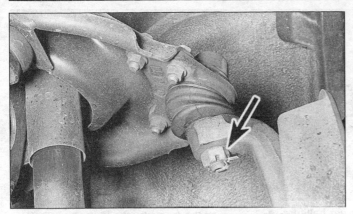

4.64 Use a balljoint separator to pop the balljoint out of the steering knuckle - make sure the lower suspension arm is supported by a floor jack before doing this!

4.66a Record the positions and number of adjusting (alignment) shims as they are removed

4.63 Remove the upper balljoint- to-steering knuckle nut

57 Apply multi-purpose grease to the lips of the boots and install them on the torque arm and anchor arm.

58 Tighten the locknut securely, using another wrench to prevent the adjusting nut from turning.

59 When installation is correct, the front wheel alignment and ride height should be checked by a front end alignment and repair shop.

Upper control arm

Removal

Refer to illustrations 4.63, 4.64, 4.66a, 4.66b, and 4.71

60 Jack up the front of the vehicle and place it securely on jackstands.

61 Remove the front wheels. Support the lower suspension arm with a floor jack and raise it slightly.

62 Remove the brake caliper and hang it out of the way with a piece of wire.

63 Remove the cotter pin and castle nut retaining the upper balljoint to the steering knuckle **(see illustration)**.

64 Using a balljoint separator, remove the upper balljoint from the steering knuckle **(see illustration)**.

65 Remove the bolts and nuts retaining the upper balljoint to the upper suspension arm, noting how it is installed.

66 Remove the upper arm mounting bolts and the adjusting shims (if equipped) **(see illustration)**. **Note:** *Do not lose the shims. Record the position and the thickness of the shims so that they can be reinstalled in their original locations. If the upper control arm is equipped with an eccentric (cam-type) pivot bolt, make matchmarks on the cams and the frame to facilitate reassembly. Remove the upper control arm nut, washer, cams (if equipped), bushings and pivot bolt* **(see illustration)**.

67 Inspect the upper arm for damage. Replace it with a new one if damage is found.

68 Inspect the bushings for wear or damage. If wear or damage is found, replace them. A hydraulic press is needed to perform this job, so it is advisable to take the arm to a repair shop or an auto parts store equipped for this type of work.

Installation

69 If the bushings have been replaced, don't tighten the upper arm shaft bolts yet (make sure they are loose enough

for the shaft to turn).

70 To complete the installation, reverse the remaining removal steps, tightening all bolts, except the upper arm shaft bolts, securely.

71 Raise the lower suspension arm to simulate normal ride height, then tighten the upper arm shaft bolts securely **(see illustration)**.

72 Have the front end alignment checked and, if necessary, adjusted.

Lower control arm

Removal

Refer to illustrations 4.77, 4.78, 4.82 and 4.83

Note: *On some Chrysler models, which use strut bars that*

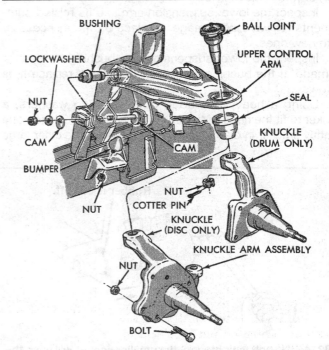

4.66b An exploded view of a typical early upper control arm with eccentric cam type adjustment

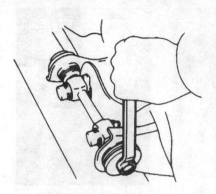

4.71 Tighten the upper arm shaft bolts only when the vehicle is at normal ride height

4.77 Remove the lower balljoint-to-lower suspension arm retaining bolts

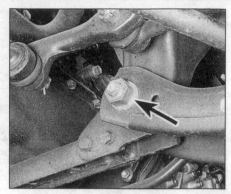

4.78 The lower suspension arm is held to the frame by the lower arm shaft, which is splined to the torque arm on the other end

are inserted through bushings in the lower control arms, you'll have to remove the arm and strut as a single assembly, then disassemble them off the vehicle.

73 Remove the torsion bar.

74 Disconnect the stabilizer bar from the lower suspension arm.

75 Disconnect the strut bar from the lower suspension arm.

76 Remove the front shock absorber.

77 Disconnect the lower balljoint from the lower suspension arm by removing the retaining bolts **(see illustration)**.

78 Remove the lower suspension arm shaft nut **(see illustration)**.

79 Remove the torque arm and lower arm shaft from the lower suspension arm, then pull down on the lower suspension arm to remove it.

80 Inspect the lower suspension arm and its related components for wear or damage. Replace parts as necessary with new ones.

81 Inspect the lower suspension arm bushing for wear or damage. If the bushing is worn or damaged, replace it as follows.

82 Using a heavy, threaded bolt, nut, two washers, a socket to fit the head of the bolt and two large sockets - one slightly smaller in diameter than the bushing, the other large

enough to fit over the bushing rubber - remove the bushing from the crossmember. **Note:** *As the bushing is removed, the rubber on the rear side will be cut off* **(see illustration)**.

83 Apply soapy water to the front rubber part of the new bushing, reverse the position of the removal apparatus and install the new bushing **(see illustration)**.

Installation

84 To install the lower suspension arm, place the arm in position and insert the lower shaft into the arm.

85 Finger tighten, but do not torque the mounting nut at this time.

86 Install the torque arm to the lower suspension arm and tighten the torque arm nuts and bolts securely.

87 To complete the installation, reverse the remaining removal steps, except for the tightening of the lower suspension arm shaft nut.

88 Remove the jackstands and lower the vehicle.

89 Bounce the vehicle several times to stabilize the lower suspension arm bushing.

90 Tighten the lower suspension arm shaft securely, then adjust the vehicle ride height.

Balljoints

Note: *See Section 8 for balljoint check and replacement procedures.*

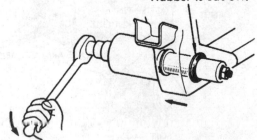

4.82 As the bolt is tightened, the smaller socket will push the bushing out of the lower arm and into the larger socket

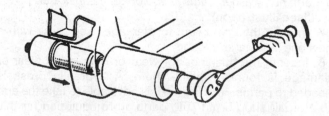

4.83 The new bushing is installed in the lower arm by reversing the positions of the sockets - be sure to lubricate the bushing with soapy water (don't use oil or grease)

5 Unequal length A-arm type (with transverse torsion bars)

Typical view of a transverse torsion bar type suspension

1	Right torsion bar	5	Left torsion bar anchor	9	Pitman arm
2	Right torsion bar anchor	6	Left torsion bar pivot cushion	10	Idler arm
3	Right torsion bar pivot cush-		bushing	11	Center link
	ion bushing	7	Stabilizer bar	12	Tie-rod
4	Left torsion bar	8	Lower control arm		

The Unequal length A-arm type (with transverse torsion bars) front suspension **(see illustrations)** is fully independent. Each wheel is connected to the frame by a steering knuckle, upper and lower balljoints and upper and lower control arms. All models use shock absorbers and torsion bars. The upper end of each front shock is attached to a bracket on the frame. The lower end of each shock is attached to the lower control arm. Transverse torsion bars connect the frame to the control arms.

Shock absorber

1 Using a backup wrench on the stem, remove the upper mounting nut **(see illustration 4.1)**.
2 Remove the retainer and grommet.
3 Raise the vehicle and place it securely on jackstands.
4 Working underneath the vehicle, remove the nut and through bolt or the nut which attaches the lower end of the

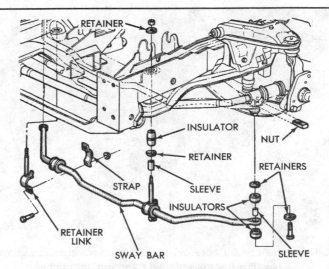

5.2a An exploded view of stabilizer bar assembly

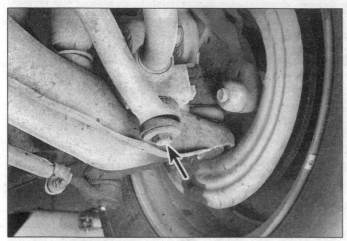

5.2b To disconnect the ends of the stabilizer bar from the control arms, remove these link bolts (arrow)

5.9 To detach the stabilizer bar assembly from the frame or crossmember, remove these bracket bolts (arrow)

shock absorber to the lower control arm **(see illustrations 4.4a, 4.4b and 4.4c)**. Pull the shock out from below.

5   Remove the lower grommet and retainer from the stem.

6   Fully extend the shock absorber prior to installation. The remaining installation steps are the reverse of removal **(see illustrations 4.6a and 4.6b)**. Be sure to tighten the upper mounting nut and the lower mounting bolts securely.

Stabilizer bar

Refer to illustrations 5.2a, 5.2b and 5.9

7   Raise the vehicle and place it securely on jackstands.

8   Remove the nuts from the link bolts and remove the link bolts **(see illustrations)**.

9   Unscrew the bracket bolts **(see illustration)** from the frame or crossmember and remove the stabilizer bar.

10   Remove the rubber bushings.

11   Inspect all parts for wear and damage.

12   Installation is the reverse of removal. Be sure to tighten all fasteners securely.

Torsion bar

Removal

Refer to illustrations 5.14a, 5.14b, 5.15, 5.17, 5.18 and 5.19

Caution: *Don't interchange torsion bars between vehicles. They're marked either "L" or "R" on the ends to indicate the side on which they must be used.*

13   Raise the front of the vehicle and support it securely on jackstands positioned under the frame. **Note:** *Don't place jackstands under the control arms - put them under the frame. The front suspension must be in full rebound (under no load) for this procedure.*

14   Release tension on the torsion bar by turning the anchor adjusting bolts in the frame crossmember counter-clockwise **(see illustrations)**. Remove the anchor adjusting bolt on the torsion bar to be removed.

15   Before disconnecting a torsion bar, you must align the stabilizer bar and lower control arm attaching points to facilitate disassembly, component realignment and attach-ment during reassembly. To do so, raise the lower control

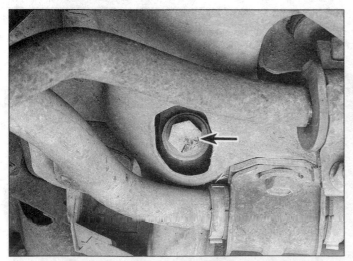

5.14a The left (driver's side) torsion bar anchor adjusting bolt . . .

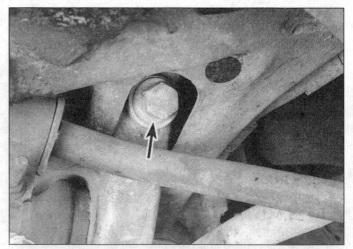

5.14b . . . and the right (passenger's side) torsion bar anchor adjusting bolt; both adjusting bolts, which are hidden in the crossmember, face toward the rear of the vehicle

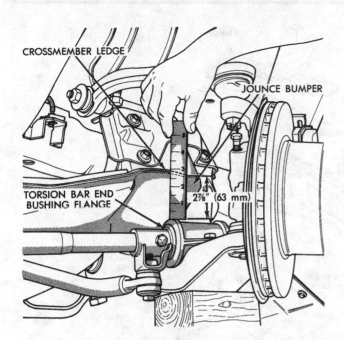

5.15 Before disconnecting a torsion bar, align the stabilizer bar and lower control arm attaching points by raising the lower control arms until the clearance between the crossmember ledge (at the jounce bumper) and the torsion bar end-bushing is 2-7/8 inches - support the control arms at this "design height" until the torsion bar has been reinstalled

arms until the clearance between the crossmember ledge (at the jounce bumper) and the torsion bar end-bushing is 2-7/8 inches (see illustration). Support the lower control arms at this "design height".

16   Remove the stabilizer bar-to-control arm attaching bolt and retainers.

17   Remove the two bolts that connect the torsion bar end-bushing to the lower control arm (see illustration).

18   Remove the two bolts that attach the torsion bar pivot cushion bushing to the crossmember (see illustration) and

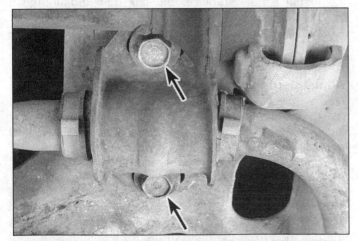

5.18 Remove the two bolts (arrows) that attach the torsion bar pivot cushion bushing to the crossmember and remove the torsion bar and anchor assembly from the crossmember

5.17 Remove these two nuts (arrows) and the two bolts that connect the torsion bar end-bushing to the lower control arm

remove the torsion bar and anchor assembly from the crossmember.

19   Separate the anchor from the torsion bar (see illustration).

Inspection

20   Inspect the boots for damage - cracks, tears and deterioration - and replace as necessary.

21   Inspect the lower control arm and pivot cushion bushings for cuts, tears, deterioration and other damage and replace as necessary.

22   Make sure that the seals at either end of the pivot cushion bushings are also in good shape; if they're damaged, moisture can get into the cushion. If the seals are damaged, inspect the torsion bar area under the bushing very carefully. If corrosion is evident in this area, replace the torsion bar.

23   Inspect the torsion bar paint for chips. If the paint is damaged, touch it up. Don't allow a torsion bar to remain uncoated or it will rust.

24   Inspect the torsion bar adjusting bolt and swivel for signs of corrosion or other damage. Replace as necessary.

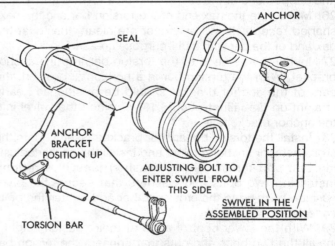

5.19 An exploded view of the hex end of the torsion bar and the anchor bracket and swivel assembly

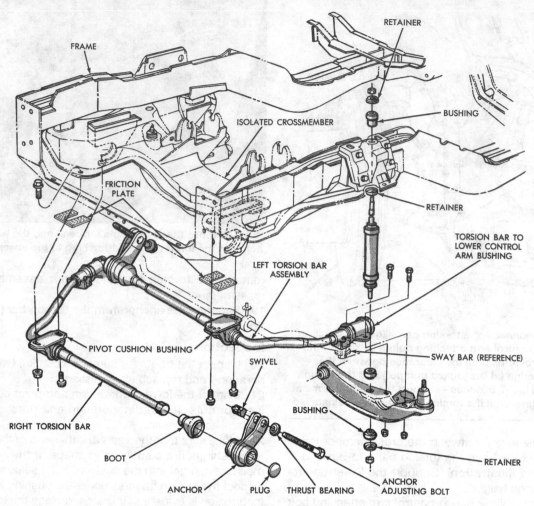

5.28 An exploded view of the transverse torsion bar assembly

Installation

Refer to illustration 5.28

25 Carefully slide the seal over the end of the torsion bar (the cupped end of the seal should face toward the hex).

26 Make sure the hex end of the torsion bar and the hex-shaped receptacle in the anchor are clean, then coat the hex end of the bar with multi-purpose grease.

27 Install the hex end of the torsion bar into the anchor bracket. With the torsion bar in a horizontal position, the ears of the anchor bracket should be positioned nearly straight up **(see illustration 5.19)**. Position the swivel into the anchor bracket ears.

28 Install the torsion bar anchor bracket assembly into the crossmember anchor retainer and install the anchor adjusting bolt and thrust bearing (washer) **(see illustration)**. Install the two bolts and washers that secure the pivot cushion bushing to the crossmember and tighten the bolts finger-tight.

29 With the lower control arms at their "design height," install the two bolts and nuts that connect the torsion bar bushing to the lower control arm. Tighten them securely.

30 Make sure the anchor bracket is fully seated in the crossmember, then tighten the pivot cushion retainer securely.

31 Position the seal over the anchor bracket.

32 Install a new stabilizer bar link bolt and tighten it securely.

33 Load the torsion bar by turning the adjusting bolt clockwise.

34 Lower the vehicle and adjust the vehicle ride height.

Adjustment

Refer to illustration 5.38

35 Check and adjust tire pressure.

36 Place the vehicle on a level, solid surface. The vehicle must be unloaded, but with a full tank of fuel.

37 Jounce the front bumper up and down several times at least four inches to settle the suspension.

38 On 1976 through 1980 vehicles, measure the distance from the lowest point of the lower control arm inner pivot bushing to the floor. On 1981 and later vehicles, measure the distance from the head of the front suspension front crossmember isolator bolt to the ground **(see illustration)**.

39 Compare your measurements with each side and adjust as necessary. To increase ride height, turn the

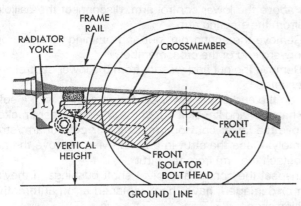

5.38 Measure the distance from the head of the front suspension front crossmember isolator bolt to the ground

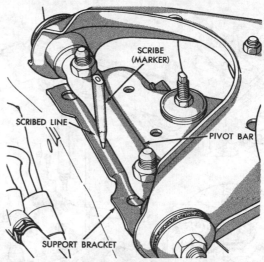

5.43a To insure basic suspension alignment after reassembly, scribe a line on the support bracket along the inner edge of the pivot bar

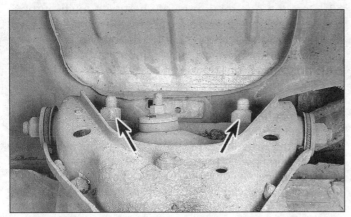

5.43b To disconnect the inner end of the upper control arm from a later vehicle, remove these two nuts (arrows) - or, on some models, bolts - from the pivot bar

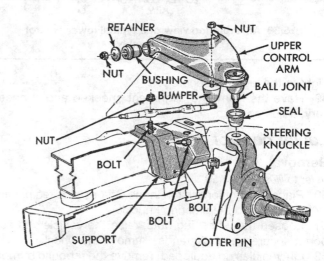

5.43c An exploded view of the upper control arm

adjusting bolts clockwise; to decrease height, turn the bolts counterclockwise. After each adjustment, jounce the vehicle as described in Step 25. Both sides must be measured even if only one side has been adjusted or is to be adjusted. Ride heights shouldn't vary more than 1/4-inch from the dimensions measured, and they should be within 1/4-inch from side-to-side.

Upper control arm

Removal

Refer to illustrations 5.43a, 5.43b and 5.43c

40   Loosen the wheel lug nuts, raise the front of the vehicle and support it securely on jackstands. Remove the wheel.

41   Position a floor jack, with a wood block on the jack head (to act as a cushion), under the lower control arm as close to the balljoint as possible. Raise the jack slightly to take the spring pressure off the upper control arm. **Warning:** *The jack must remain in this position throughout the entire procedure.*

42   Disconnect the upper balljoint stud from the steering knuckle. Using a piece of wire or rope tied to the frame rail or some other component, support the top of the steering knuckle so it doesn't fall and stretch the brake hose.

43   Remove the splash shield, scribe a line on the support

bracket along the inner edge of the pivot bar to insure that the suspension is still basically aligned after reassembly **(see illustration)**, remove the pivot bolts or nuts **(see illustrations)** and remove the upper control arm.

44   Inspect the control arm bushings for wear. Replace them if necessary. A hydraulic press may be required to remove and install the bushings, in which case you may have to take them to a dealer service department or other repair shop to have this done.

Installation

45   Position the arm in the frame brackets and install the bolt, washers, cams (if equipped) and nut.

46   Line up the previously applied matchmarks (if applicable). Tighten the mounting bolts and nuts securely.

47   Attach the balljoint to the steering knuckle. Tighten the balljoint nut securely.

48   Install the wheel and lug nuts, then lower the vehicle.

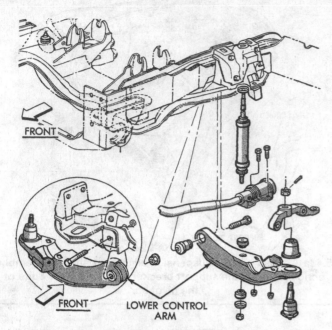

5.56 An exploded view of a typical lower control arm assembly

Tighten the lug nuts securely.

49   Have the front end alignment checked and, if necessary, adjusted.

Lower control arm

Removal

Refer to illustration 5.56

50   Place the ignition switch in the Off or Unlocked position (so you can turn the steering knuckle).

51   Loosen the wheel lug nuts, raise the vehicle and support it securely on jackstands. Remove the wheel.

52   On vehicles so equipped, remove the rebound bumper. If necessary, raise the lower control arm with a floor jack to provide clearance.

53   On vehicles with front drum brakes, remove the drum, the shoe assembly and the brake backing plate. On vehicles with front disc brakes, remove the brake caliper and hang it out of the way, then remove the hub/disc assembly (see Chapter 4, *Front wheel bearing check, repack and adjustment*).

54   Remove the disc splash shield, if equipped, then disconnect the shock absorber from the lower control arm.

55   Disconnect the stabilizer bar from the lower control arm.

56   Disconnect the torsion bar from the lower control arm **(see illustration)**.

57   Disconnect the tie-rod end from the steering arm. Remove the steering knuckle arm-to-brake support bolts and remove the steering knuckle arm.

58   On vehicles with integral balljoint/steering arm assemblies located *below* the control arm, disconnect the balljoint stud from the lower control arm. On vehicles with steering

arms *above* the lower control arm, disconnect the balljoint stud from the steering arm.

59   Remove the spring pin, nut and bushing retainer from the forward end of the crossmember.

60   Remove the nut and washer from the lower control arm pivot shaft.

61   Tap the end of the lower control arm shaft with a soft-blow hammer and knock it out of the crossmember bracket. Remove the lower control arm, or lower control arm/strut assembly. Place the strut in a bench vise, remove the nut and detach the arm from the strut.

62   Inspect the control arm pivot shaft bushings. If they're worn or damaged, have them replaced at an automotive machine shop.

Installation

63   Coat the control arm pivot shaft with multi-purpose grease, install the lower control arm and install the pivot shaft washer and nut. Don't tighten the nut yet.

64   On drum-brake models, position the brake backing plate on the steering knuckle and install the two upper bolts and nuts finger tight. Position the steering arm on the steering knuckle, install the two bolts and nuts and tighten them finger tight (the upper bolts go through the steering knuckle and through the brake backing plate; the lower bolts go through the steering arm, the steering knuckle and the brake backing plate). Tighten the upper, then the lower, bolt nuts securely.

65   On vehicles with integral balljoint/steering arm assemblies, insert the balljoint stud into the lower control arm; on vehicles with the balljoint pressed or screwed into the lower arm, insert the balljoint stud into the steering arm. Tighten the balljoint stud nut securely. Install a new cotter pin.

66   Inspect the tie-rod end seal and replace it if damaged. Connect the tie-rod end to the steering knuckle arm and tighten the nut securely.

67   Connect the shock absorber to the control arm and tighten it finger tight.

68   Install the torsion bar.

69   On drum-brake models, install the brake shoe assembly and the drum; on disc-brake models, install the brake disc (see Chapter 4, *Front wheel bearing check, repack and adjustment*), the caliper support and the caliper.

70   Raise the lower control arm with a floor jack to simulate normal ride height. Tighten the strut nut at the crossmember securely, and install the strut pin. Tighten the lower control arm shaft nut bushings securely.

71   Install the wheel and tighten the wheel lug nuts hand tight.

72   Lower the vehicle and tighten the wheel lug nuts securely.

73   Adjust the vehicle ride height.

74   Drive the vehicle to an alignment shop and have the alignment checked, and if necessary, adjusted.

Balljoints

Note: *See Section 8 for balljoint check and replacement procedures.*

6 Modified strut type (with separate coil springs)

A close-up view of a modified strut type front suspension

1	Stabilizer bar	7	Coil spring	13	Steering gear dust boot outer clamp
2	Stabilizer bar clamp	8	Strut assembly	14	Steering gear dust boot
3	Stabilizer bar-to-lower control arm link	9	Steering knuckle	15	Steering gear dust boot inner clamp
4	Lower control arm	10	Balljoint	16	Steering gear
5	Control arm pivot bolt/nut	11	Tie-rod end		
6	Control arm pivot bolt/nut	12	Tie-rod end jam nut		

Warning: *Whenever any of the suspension or steering fasteners are loosened or removed they must be inspected and if necessary, replaced with new ones of the same part number or of original equipment quality and design. Never attempt to heat, straighten or weld any suspension or steering component. Instead, replace any bent or damaged part with a new one.*

Note: *These vehicles use a combination of standard and metric fasteners on the various suspension and steering components, so it would be a good idea to have both types of tools available when beginning work.*

Each wheel is connected to the frame by a steering knuckle, a balljoint, a lower control arm and a strut assembly positioned vertically between the steering knuckle and the vehicle body **(see illustration)**. A coil spring is mounted between the lower control arm and the front suspension crossmember. Body side roll is controlled by a stabilizer bar attached to the frame and to the lower control arms.

Strut

Removal

Refer to illustrations 6.3, 6.4 and 6.5

1   Loosen the front wheel lug nuts, raise the vehicle and support it securely on jackstands. Remove the wheels.
2   Place a floor jack under the lower control arm and raise it slightly. **Caution:** *The lower control arm must be supported throughout the entire procedure.*
3   If the vehicle is equipped with ABS, remove the nut

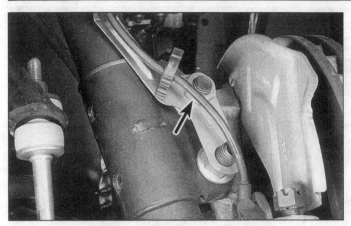

6.3 If the vehicle is equipped with ABS, remove this nut (arrow) and detach the bracket for the ABS lead

6.4 To detach the lower end of the strut from the steering knuckle, remove these two bolts (arrows); note the alignment marks painted on the strut, along the edge of the strut bracket, to ensure that correct camber is restored when the new strut is installed

(see illustration) which attaches the bracket for the ABS lead to the strut.

4   Remove the strut-to-steering knuckle nuts and bolts (see illustration).

5   Unscrew the three upper mount-to-body retaining bolt and nuts (see illustration).

6   Separate the strut assembly from the steering knuckle and remove it from the vehicle. Don't allow the steering knuckle to fall outward, as the brake hose could be damaged.

7   Installation is the reverse of removal. Be sure to tighten all fasteners securely.

8   Remove the jack from under the lower control arm, install the wheels, lower the vehicle and tighten the lug nuts securely.

Stabilizer bar

Removal

Refer to illustrations 6.10a, 6.10b and 6.11

9   Raise the vehicle and support it securely on jackstands. Apply the parking brake.

6.5 To detach the upper end of the strut from the vehicle, remove this bolt and these nuts (arrows)

6.10a To disconnect the stabilizer bar from the links, remove the nut (arrow) from each stabilizer bar link; make sure to note the order in which the bushings and washers are installed - they must be installed in the same sequence when the stabilizer is installed

6.10b To disconnect either link from a lower control arm, remove this nut; be sure to note the order in which the washers and bushings are installed

6.11 To detach the stabilizer bar from the frame, remove these nuts (arrows) that attach the bushing clamps to the frame (left bushing clamp shown)

10 Remove the stabilizer bar-to-link nuts **(see illustration)**, noting how the washers and bushings are positioned. Clamp a pair of locking pliers to the stabilizer bar link to prevent it from turning. Inspect the bushings for wear. If they're cracked or torn, replace them. To detach the lower end of either link from the lower control arm, remove the lower nut **(see illustration)** located underneath the control arm.

11 Remove the stabilizer bar bracket nuts **(see Illustration)** and detach the bar from the vehicle.

12 Pull the brackets off the stabilizer bar and inspect the bushings for cracks, hardening and other signs of deterioration. If the bushings are damaged, cut them off the bar and discard them.

Installation

13 Lubricate the new stabilizer bar bushings with a rubber lubricant (such as a silicone spray) and slide the bushings onto the bar. The bushings should be installed with the seam facing forward.

14 Push the brackets over the bushings and raise the bar up to the frame. Install the bracket nuts but don't tighten them completely at this time.

15 Install the stabilizer bar link washers and rubber bushings and tighten the nuts.

16 Tighten the bracket nuts and the stabilizer link nuts securely.

Coil spring

Removal

Refer to illustrations 6.21, 6.23a, 6.23b, 6.23c and 6.23d

Warning: *The following procedure is potentially dangerous if the proper safety precautions are not taken. You should use a coil spring compressor to safely perform this procedure.*

17 Loosen the wheel lug nuts on the side to be disassembled. Raise the vehicle, support it securely on jackstands and remove the wheel.

18 Disconnect the tie-rod end from the steering knuckle.

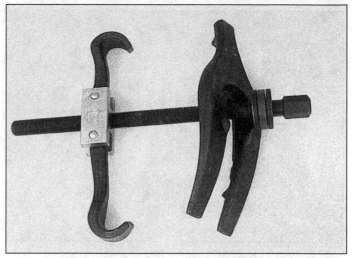

6.21 A typical aftermarket internal spring compressor tool: The hooked arms grip the upper coils of the spring, the plate is inserted below the lower coils, and when the nut on the threaded rod is turned, the spring is compressed

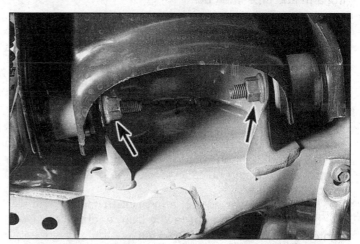

6.23a To detach the lower control arm from the front suspension crossmember, remove these nuts (arrows) and bolts . . .

19 Disconnect the stabilizer bar link from the lower control arm.

20 On vehicles with rack and pinion steering, it may be necessary to slightly raise the steering gear in order to remove the front pivot bolt from either lower control arm. If necessary, remove the steering gear mounting bolts.

21 Install a spring compressor tool, with a plate between the coils near the top of the spring **(see illustration)**.

22 Guide the compression rod up through the hole in the lower control arm and the coil spring, then insert the end of the compression rod into the upper plate. Install the lower plate, ball nut, thrust washer and bearing and the forcing nut on the compression rod. Tighten the forcing nut until the coil spring can be wiggled, indicating that all spring pressure has been taken up.

23 Remove the lower control arm-to-crossmember nuts and pivot bolts **(see illustrations)**.

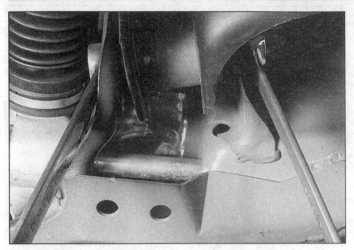

6.23b . . . using two wrenches

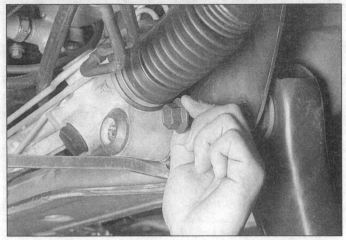

6.23c Note that the steering gear has been unbolted and raised up to provide sufficient clearance to remove the front pivot bolt for the lower control arm

24 Remove the spring compressor compression rod then maneuver the coil spring out from between the lower control arm and crossmember.

Installation

25 Install the coil spring insulator on the top of the spring.

26 Install the spring in between the lower control arm and the spring upper pocket in the suspension crossmember.

27 Position the bottom of the spring so that the pigtail covers only one of the drain holes, but leaves the other one open.

28 Locate the spring in the upper seat in the crossmember. Install the spring compressor tool as described in Steps 21 and 22, then tighten the forcing nut until the lower control arm bushing holes align with the pivot bolt holes in the crossmember.

29 Install the lower control arm pivot bolts and nuts with the bolt heads facing out (away from each other). Don't tighten them completely at this time.

30 Remove the spring compressor tool. Position a floor jack under the outer end of the lower control arm and raise it to simulate a normal ride position. Now tighten the pivot bolt nuts securely.

31 Reattach the steering gear and tighten the steering gear bolts securely.

32 Reconnect the stabilizer bar link to the lower control arm.

33 Reattach the tie-rod end to the steering knuckle.

34 Install the front wheel, remove the jackstands and lower the vehicle. Tighten the wheel lug nuts securely.

35 Have the alignment checked by a dealer service department or an alignment shop.

Lower control arm

36 Loosen the wheel lug nuts, raise the vehicle and support it securely on jackstands. Remove the wheel.

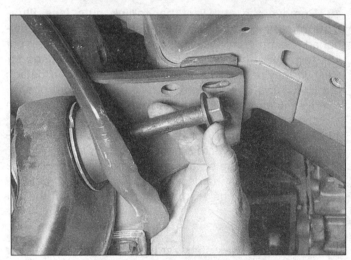

6.23d Remove the lower control arm pivot bolt

37 Remove the coil spring.

38 Unscrew the balljoint nut, strike the steering knuckle sharply with a hammer to break the ballstud loose and remove the control arm from the vehicle. A "pickle fork" type balljoint separator can be used, but it will damage the balljoint seal, so avoid a pickle fork unless you're planning to replace the balljoint.

39 Installation is the reverse of the removal procedure. Be sure to tighten all of the fasteners and lug nuts securely.

40 Have the alignment checked by a dealer service department or an alignment shop.

Balljoints

Note: *See Section 8 for balljoint check and replacement procedures.*

7 MacPherson strut type

The MacPherson Strut design uses a combination strut and shock absorber assembly which is mounted directly to the steering knuckle. A control arm, which pivots on the engine cradle, is also attached to the steering knuckle by way of a balljoint (see illustration).

Strut/coil spring assembly

Note: *On 1988 and later GM W body models the strut cartridge can be replaced with the strut and spring assembly installed in the vehicle. Because it isn't necessary to remove and disassemble the strut and spring assembly for strut cartridge replacement, removal of the assembly should only be required to repair damage to the spring, seats and strut/knuckle components.*

Removal

Refer to illustration 7.2, 7.4 and 7.6

1 Loosen the front wheel lug nuts, raise the front of the vehicle and support it securely on jackstands. Remove the wheels.

2 Unclip the brake hose from the strut bracket (see illustration) and detach it from the bracket. If the vehicle is equipped with ABS, detach the speed sensor wiring harness from the strut by removing the clamp bracket bolt (see illustration 7.4).

3 Remove the disc brake caliper and hang it out of the way with a piece of wire.

4 Remove the strut-to-knuckle nuts (see illustration) and knock the bolts out with a hammer and punch on.

5 Separate the strut from the steering knuckle. Be careful not to overextend the inner CV joint and don't let the knuckle fall outward, as this could damage the brake hose.

6 Support the strut and spring assembly with one hand and remove the three strut-to-shock tower nuts (see illustration). Remove the assembly out from the fenderwell.

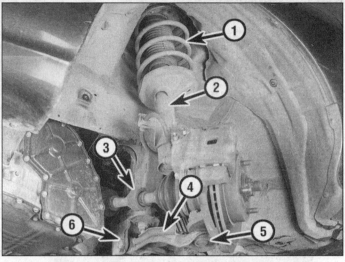

A close-up view of a MacPherson strut type front suspension

1	Coil spring	4	Lower control arm
2	Strut assembly	5	Balljoint
3	CV joint axle	6	Control arm pivot bolt/nut

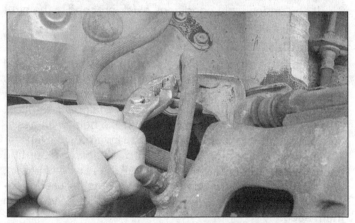

7.2 Remove the retaining clip with a pair of pliers and detach the brake hose from the strut

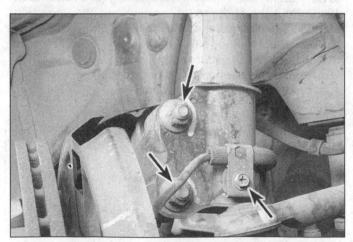

7.4 To detach the strut assembly from the steering knuckle, remove the ABS line bracket bolt (right arrow), remove the two nuts (left arrows), then drive out the strut-to-knuckle bolts with a hammer and punch

7.6 To detach the upper end of the strut assembly from the body, remove the upper mounting nuts (arrows) only

7.17 Mark the position of the strut mount cover before removing the nuts

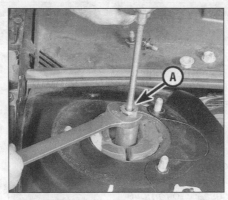

7.18 Remove the strut shaft nut using this tool (arrow) with a large box or open wrench on it, and an extension with a Torx bit through the top to hold the strut shaft from turning

7.19a Remove the strut mount bushing (arrow) . .

Inspection

7   Check the strut body for leaking fluid, dents, cracks and other obvious damage which would warrant repair or replacement.8   Check the coil spring for chips or cracks in the spring coating (this will cause premature spring failure due to corrosion). Inspect the spring seat for cuts, hardness and general deterioration.

9   If any undesirable conditions exist, proceed to the strut disassembly procedure.

Installation

10   Guide the strut assembly up into the fenderwell and insert the upper mounting studs through the holes in the shock tower. Once the studs protrude from the shock tower, install the nuts so the strut won't fall back through. This is most easily accomplished with the help of an assistant, as the strut is quite heavy and awkward.

11   Insert the steering knuckle into the lower mounting flange of the strut assembly and install the two bolts. Again, it is recommended that new bolts and nuts be used (genuine OEM parts or equivalent). Install the nuts and tighten them securely.

12   Guide the brake hose through its bracket in the strut

and install the retaining clip.

13   Remove the jack from under the control arm on (if used) models and install the disc brake caliper.

14   Install the wheel and lug nuts, then lower the vehicle and tighten the lug nuts securely.

15   Tighten the upper mounting nuts securely.

16   Drive the vehicle to an alignment shop to have the front end alignment checked, and if necessary, adjusted.

Strut cartridge - 1988 and later GM "W" body models only

Refer to illustrations 7.17, 7.18, 7.19a, 7.19b, 7.20a, 7.20b and 7.21

17   Mark the position of the strut mount cover **(see illustration)** and remove the nuts and the cover. **Note:** *The vehicle should be on the ground with full weight on the suspension.*

18   Remove the strut shaft nut, using a tool available at auto parts stores that grips the nut, while allowing a Torx bit to be used to keep the shaft from turning **(see illustration)**.

19   Pry out the strut mount bushing and the upper strut bumper **(see illustrations)**. You may have to compress the strut shaft down into the cartridge using a length of pipe that just fits over the strut shaft.

20   Remove the strut cartridge nut using a tool available at auto parts stores **(see illustrations)**.

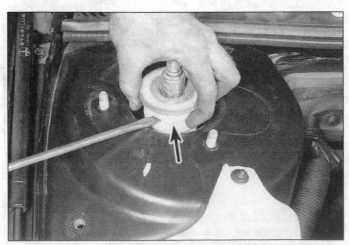

7.19b . . . and the upper strut bumper (arrow)

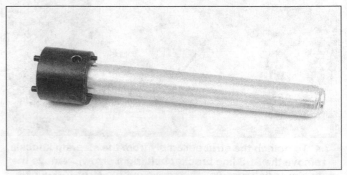

7.20a Strut cartridge nut removal tool for GM struts - the lugs at the large end fit into the slots in the nut

7.20b In use, the tool fits down over the strut shaft and a breaker bar can be used to turn it

7.21 Lift the cartridge out

7.30 Install the spring compressor according to the tool manufacturer's instructions and compress the spring until all pressure is relieved from the upper spring seat

21 Grasp the strut cartridge and lift it out **(see illustration)**.

22 If the old cartridge had been leaking oil, use a suction pump to remove the damper fluid from the strut body and pour it into an approved oil container.

23 Insert the replacement strut cartridge, which is a self-contained unit, into position and install the nut. Tighten the cartridge nut securely

24 Install the bumper.

25 Temporarily install the strut shaft nut enough to grip it with locking pliers to raise the shaft If It doesn't come up by itself, then remove the nut and install the bushing.

26 Install the strut shaft nut and tighten the nut securely.

27 Install the strut mount cover and nuts and tighten the nuts.

Strut or coil spring assembly replacement

28 If the struts or coil springs exhibit the telltale signs of wear (leaking fluid, loss of damping capability, chipped, sagging or cracked coil springs) explore all options before beginning any work. The strut/shock absorber assemblies are not serviceable and must be replaced if a problem develops. However, strut assemblies complete with springs may be available on an exchange basis, which eliminates

much time and work. Whichever route you choose to take, check on the cost and availability of parts before disassembling your vehicle. **Warning:** *Disassembling a strut is a potentially dangerous undertaking and utmost attention must be directed to the job, or serious injury may result. Use only a high-quality spring compressor and carefully follow the manufacturer's instructions furnished with the tool. After removing the coil spring from the strut assembly, set it aside in a safe, isolated area.*

Disassembly

Refer to illustrations 7.30, 7.31, 7.32, 7.33 and 7.34

29 Remove the strut and spring assembly following the procedure described in the previous Section. Mount the strut assembly in a vise. Line the vise jaws with wood or rags to prevent damage to the unit and don't tighten the vise excessively.

30 Following the tool manufacturer's instructions, install the spring compressor (which can be obtained at most auto parts stores or equipment yards on a daily rental basis) on the spring and compress it sufficiently to relieve all pressure from the upper spring seat **(see illustration)**. This can be verified by wiggling the spring.

31 Remove the damper shaft nut **(see illustration)**.

32 Remove the upper suspension support **(see illustra-

7.31 Remove the damper shaft nut

7.32 Lift the suspension support off the damper shaft

7.33 Remove the spring seat from the damper shaft

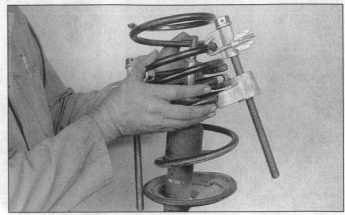

7.34 Remove the compressed spring assembly - keep the ends of the spring pointed away from your body

tion). Inspect the bearing in the suspension support for smooth operation. If it does not turn smoothly, replace the suspension support. Check the rubber portion of the suspension support for cracking and general deterioration. If there is any separation of the rubber, replace it.

33 Lift the spring seat and upper insulator from the damper shaft **(see illustration).** Check the rubber spring seat for cracking and hardness, replacing it if necessary.

34 Carefully lift the compressed spring from the assembly **(see illustration)** and set it in a safe place. **Warning:** *Carry the spring carefully and never place any part of your body near the end of the spring!*

35 Slide the dust boot off the damper shaft.

36 Check the lower insulator (if equipped) for wear, cracking and hardness and replace it if necessary.

Reassembly

Refer to illustration 7.38, 7.39a and 7.39b

37 If the lower insulator is being replaced, set it into position with the dropped portion seated in the lowest part of the seat. Extend the damper rod to its full length and install the dust boot.

38 Carefully place the coil spring onto the lower insulator, with the end of the spring resting in the lowest part of the

insulator **(see illustration).**

39 Install the upper insulator and the spring seat. Make sure the cutout on the spring seat is facing out (away from the vehicle), in line with the strut-to-knuckle attachment points **(see illustrations).**

40 Install the dust seal and suspension support to the damper shaft.

41 Install the nut and tighten it securely.

42 Install the strut/ coil spring assembly.

Stabilizer bar

Removal

43 Apply the parking brake. Loosen the front wheel lug nuts, raise the front of the vehicle and support it securely on jackstands. Remove the wheels.

Link type

Refer to illustrations 7.44 and 7.45

44 Remove the bolts which attach the stabilizer bar brackets to the underside of the vehicle **(see illustration).**

45 Detach the stabilizer bar link bolts from the lower control arms **(see illustration).** Note the order in which the spacers, washers and bushings are arranged on the link bolt.

46 Remove the bar from under the vehicle.

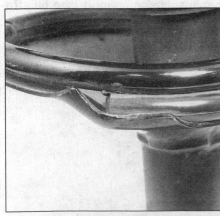

7.38 When installing the spring, make sure the end fits into the recessed portion of the lower seat (arrow)

7.39a Make sure this cutout in the upper seat . . .

7.39b . . . is facing out (toward the strut-to-knuckle flanges)

7.44 To detach the stabilizer bar from the frame, remove the nuts (arrows) that attach the brackets to the frame

7.45 To prevent the link from turning while you're breaking loose the nut on top that attaches it to the stabilizer bar, hold the link with a pair of locking pliers

7.48 Unbolt the stabilizer bar bracket from the control arm

Mount/bracket type

Refer to illustrations 7.47, 7.48 and 7.50

47   On some models it will be necessary to remove the front part of the exhaust.

48   Remove the bolts that secure the stabilizer brackets to the control arm and remove the brackets **(see illustration)**.

49   Remove the bolts that secure the stabilizer mounting plates to the engine cradle and remove both plates **(see illustration)**.

50   Remove the stabilizer bar from its recesses in the engine cradle, complete with bushings and brackets. The bushings need not be removed from the stabilizer bar unless either the bushings or the bar are being replaced **(see illustration)**.

Installation

51   Inspect the bushings to be sure they are not hardened, cracked or excessively worn, and replace if necessary.

52   Installation is the reverse of the removal procedure.

Lower control arm

Removal

Refer to illustrations 7.55a, 7.55b, 7.55c, 7.55d, 7.56 and 7.57

53   Loosen the wheel lug nuts on the side to be disman-

7.49 Remove the stabilizer bar mounting plate from the engine cradle

tled, raise the front of the vehicle, support it securely on jackstands and remove the wheel.

54   Disconnect the stabilizer link from the control arm.

55   Remove the cotter pin and loosen the balljoint stud nut **(see illustrations)**. Separate the balljoint stud from the steering knuckle with a "picklefork"-type balljoint separator

7.50 Remove the stabilizer bar complete with bushings and brackets

7.55a Remove the cotter pin . . .

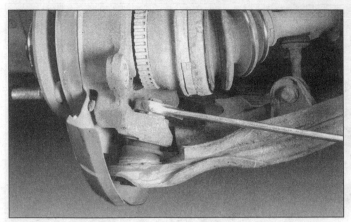

7.55b ... loosen the balljoint stud nut and back it off as far as it will go (without actually removing it) ...

7.55c ... then pop the balljoint stud loose with a "picklefork" (be sure to grease the pickle fork to protect the balljoint dust boot)

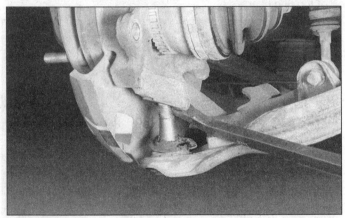

7.55d Separate the control arm from the steering knuckle by prying it down with a prybar or large screwdriver

7.56 Remove the front pivot stud nut (arrow)

(see illustration). Be sure to grease the picklefork to protect the balljoint dust boot. Separate the arm from the steering knuckle (see illustration).

56 Remove the front pivot bolt nut and/or bolt (see illustration).

57 Remove the rear bushing clamp bolts (see illustration). Remove the control arm.

Installation

58 Installation is the reverse of removal. Tighten all of the fasteners securely. Be sure to install a new cotter pin through the balljoint stud.

59 Install the wheel and lug nuts, lower the vehicle and tighten the lug nuts securely.

60 It's a good idea to have the front wheel alignment checked and, if necessary, adjusted after this job has been performed.

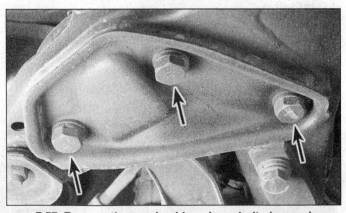

7.57 Remove the rear bushing clamp bolts (arrows)

Balljoints

Note: *See Section 8 for balljoint check and replacement procedures.*

8 Balljoints - check and replacement

Check

Refer to illustrations 8.2, 8.4 and 8.5
Note: *This check assumes that the wheel bearings are properly adjusted.*

Upper balljoint

1   Raise the vehicle and place floor jacks under the lower arms.

2   Have an assistant grasp the lower edge of the tire and move the wheel in and out. As the wheel is being moved in and out, watch the upper end of the steering knuckle and the upper arm **(see illustration)**. If there's any movement between the upper end of the steering knuckle and the upper arm, replace the balljoint.

Lower balljoint - without wear indicator

Unequal length A-arm and modified strut type

3   Raise the front of the vehicle and support it securely on jackstands placed under the lower arms. Place a large pry-bar between the lower arm and the steering knuckle and attempt to pry the components apart. If any movement is noted, replace the balljoint.

MacPherson strut type

4   Raise the vehicle and support it securely on jackstands placed under the frame rails. Place a large prybar under the balljoint and resting on the wheel. Pry up on the balljoint, checking for movement between the balljoint and steering knuckle **(see illustration)**. If any movement is noted, replace the balljoint.

Lower balljoint - with wear indicator

5   To do a quick visual inspection of the lower balljoints, wipe the grease fitting and inspection surface so they're free of dirt and grease. The inspection area **(see illustration)** is the round boss into which the grease fitting is threaded. The inspection area should project outside the cover. If the inspection area is inside the cover, replace the balljoint.

8.2 To check a balljoint, raise the front of the vehicle and place it securely on jackstands, support the lower control arm with a floor jack, then try to rock the wheel in and out; if there's any play in the balljoint, replace it

Replacement

Note: *Some manufacturers use upper and lower control arm balljoints that are not removable or serviceable. If the balljoint is damaged or worn, replace the control arm.*

6   Loosen the wheel lug nuts, raise the vehicle and support it securely on jackstands. Apply the parking brake. Remove the wheel.

7   Place a floor jack under the lower control arm. Raise the jack just far enough to unload the torsion bar. **Caution:** *The jack must remain under the control arm during removal and installation of the balljoint to hold the control arm in position.*

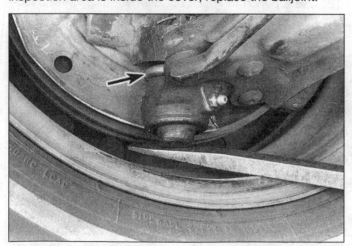

8.4 Check for movement between the balljoint and steering knuckle when prying up

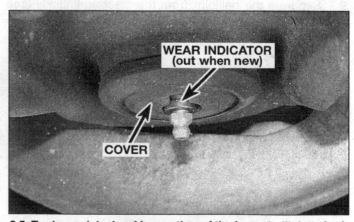

8.5 To do a quick visual inspection of the lower balljoints, look at the inspection area (the round boss into which the grease fitting is threaded) and note whether the inspection area projects outside the cover; if it doesn't, replace the balljoint

8.9 Use a "picklefork" -type balljoint tool to separate the upper balljoint from the steering knuckle

8.10 Use progressively larger diameter drill bits to remove the balljoint rivet heads,: start with a 1/8-inch bit, then proceed to a 1/4-inch, then use a 1/2-inch bit to drill off the heads

8.12a Once the rivet heads are removed, use a punch to knock out the rivet shanks

8.12b If any rivet material remains, the balljoint may not want to come off; use a chisel to knock it loose

Riveted type ball joints

Upper balljoint

Refer to illustrations 8.9, 8.10, 8.12a, 8.12b, 8.13a, 8.13b, 8.14a, 8.14b and 8.14c

8   Remove the cotter pin from the upper balljoint stud and back off the nut two turns.

9   Separate the balljoint from the steering knuckle **(see**

illustration).

10   Using a 1/8-inch drill bit, drill a 1/4-inch deep hole in the center of each rivet head **(see illustration)**.

11   Using a 1/2-inch diameter drill bit, drill off the rivet heads, but don't drill through the balljoint retainer and into the control arm.

12   Use a punch to knock out the rivet shanks **(see illus-**

8.13a To assemble the new balljoint, install the grease fitting . . .

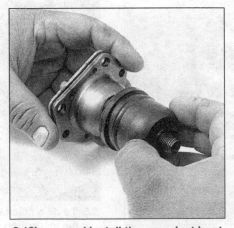

8.13b . . . and install the new dust boot

8.14a Install the balljoint in the upper control arm as shown . . .

8.14b . . . install the metal dust boot shield (if equipped) . . .

8.14c . . . and install the bolts and nuts

8.24a A grinder is the fastest way to remove the big rivets from a lower control arm balljoint on older models

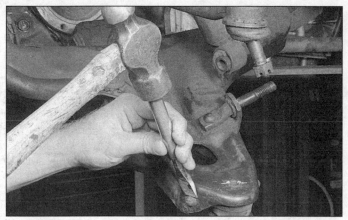

8.24b If you don't have a grinder, a big sharp chisel works well too

8.24c Once the rivet heads are completely removed, use a drill bit to remove the rivet material inside the holes in the control arm (be careful not to enlarge the holes)

tration), then use a chisel to knock the old balljoint loose (see illustration). Remove it from the control arm.

13   Assemble the new balljoint (see illustrations).

14   Position the new balljoint on the control arm and install the bolts and nuts supplied in the kit (see illustrations). Be sure to tighten the nuts to the torque specified on the balljoint kit instruction sheet. If no torque is provided, tighten the nuts securely.

15   Insert the balljoint stud into the steering knuckle, install the nut and tighten it securely. Remove the floor jack from under the lower control arm.

16   Install a new cotter pin, tightening the nut slightly if necessary to align a slot in the nut with the hole in the balljoint stud. Pack the balljoint with grease.

17   Install the wheel and hand tighten the wheel lug nuts. Remove the jackstands and lower the vehicle. Tighten the wheel lugs nuts securely.

18   Have the alignment check and, if necessary, adjusted by an alignment shop.

Lower balljoint

Refer to illustrations 8.24a, 8.24b, 8.24c, 8.24d and 8.25

19   There are a few types of lower balljoints used: the first uses one large rivet in the top and two rivets on the side, one in front and one in back. The following procedure

depicts the replacement of this type balljoint. The other uses a riveted balljoint that's replaced the same way as the upper balljoint shown above.

20   Loosen the wheel lug nuts, raise the vehicle and support it securely on jackstands. Apply the parking brake. Remove the wheel.

21   Place a floor jack under the lower control arm. Raise the jack just far enough to unload the torsion bar. **Caution:** *The jack must remain under the control arm during removal and installation of the balljoint to hold the control arm in position.*

22   Remove the cotter pin from the lower balljoint stud and back off the nut two turns.

23   Separate the balljoint from the steering knuckle (see illustration 8.9).

24   Again, using progressively larger drill bits - 1/8-inch, then 1/4-inch, then 1/2-inch - will eventually remove the rivets. But these models use bigger rivets, so to speed things up, use a grinder (see illustration) once you've drilled out the rivet heads. If you don't have a grinder, use a large sharp chisel to knock off the heads (see illustration). Once the heads are completely removed, use a big drill bit to drill out the material in the hole (see illustration). Finally, use that big sharp chisel again to knock the balljoint loose (see illustration).

25   Position the new balljoint on the control arm and install

8.24d Use the big chisel to knock the balljoint out of the lower control arm; the two side rivets can be a problem because until ALL the rivet material has been removed from them, the balljoint won't budge

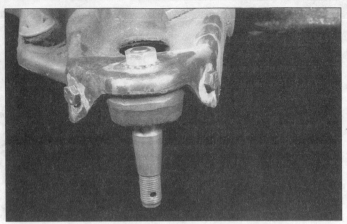

8.25 Install the new balljoint with the nuts and bolts supplied with the kit

the bolts and nuts supplied in the kit **(see illustration)**. Be sure to tighten the nuts to the torque specified on the balljoint kit instruction sheet.

26 Insert the balljoint stud into the steering knuckle, install the nut and tighten it securely. Remove the floor jack from under the lower control arm.

27 Install a new cotter pin, tightening the nut slightly if necessary to align a slot in the nut with the hole in the balljoint stud.

28 Pack the balljoint with grease.

29 Install the wheel and hand tighten the wheel lug nuts. Remove the jackstands and lower the vehicle. Tighten the wheel lugs nuts securely.

30 Have the alignment check and, if necessary, adjusted by an alignment shop.

Press or screw type

31 Loosen the wheel lug nuts, raise the front of the vehicle and support it securely on jackstands. Remove the wheel.

Upper balljoint
Refer to illustrations 8.35a, 8.35b, 8.35c and 8.35d

32 Remove the brake caliper and tie it out of the way.

33 Remove the cotter pin from the upper balljoint stud and back off the nut several turns.

34 Place a jack or jackstand under the outer end of the lower control arm. **Warning:** *The jack or jackstand must remain under the control arm during removal and installation of the balljoint to hold the spring and control arm in position.*

35 Separate the balljoint from the steering knuckle as follows:

a) *Remove the cotter pin and loosen the nut on the upper balljoint stud a few turns, but don't remove it.*

b) *Install the special tool as* **(see illustrations)** *shown. Position the balljoint stud remover over the lower balljoint stud, allowing the tool to rest on the knuckle arm. Set the tool nut securely against the upper stud. Tighten the tool to apply pressure to the upper stud and*

strike the knuckle sharply with a hammer to loosen the stud. Don't try to force out the stud with the tool alone.

c) *If you you're not worried about damaging the balljoint boot, i.e. you're planning to replace the balljoint anyway, it's okay to use a picklefork type balljoint separator* **(see illustration)**.

d) *You can also loosen the balljoint stud nut and give the knuckle or steering arm a few sharp raps with a hammer* **(see illustration)**, *but we don't recommend this method unless you're working on a vehicle which makes the use of any other means difficult because of the shape of the knuckle, its close proximity to the ball stud nut, etc.*

36 If the upper balljoint has a hex on top, unscrew it from the control arm using a large socket of the correct size will work. After you remove the balljoint, clean the balljoint bore in the control arm. If the balljoint is pressed in, i.e. there is no hex on top with which to unscrew it, remove the upper control arm and have the old balljoint pressed out and a

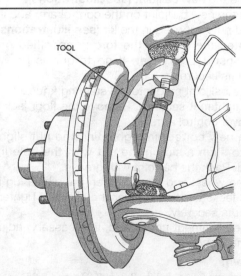

TOOL

8.35a The best way to separate an upper balljoint stud from the steering knuckle (or to split a lower balljoint stud from the steering knuckle arm or the lower control arm, depending on the design) is with a special balljoint separator tool

8.35b You can make your own version of the factory tool with a large bolt and nut (same diameter and thread pitch as bolt), a large washer and a big socket - this photo shows our homemade tool set up for removing an upper balljoint stud

8.35c On some models, you won't be able to use the factory tool or the homemade setup because the balljoint studs don't align - on these models, you'll have to split the upper balljoint stud from the knuckle with a "picklefork" tool (but remember, it will damage the balljoint boot)

8.35d If you don't have a picklefork tool, you can also rap the knuckle sharply with a hammer to loosen the upper balljoint stud

8.39a This is the same homemade setup we showed you in illustration 8.35b, except that the whole assembly has been turned around so that it now presses on the lower balljoint stud (this particular vehicle has a balljoint in the lower control arm; the balljoint stud is attached to the steering knuckle arm)

new unit pressed in at an automotive machine shop. **Note:** *If you can rent or borrow a balljoint removal/installation press, the control arm doesn't have to be removed from the vehicle. Follow the tool manufacturer's instructions.*

Lower balljoint

Refer to illustrations 8.39a, 8.39b and 8.39c

Note: *Lower balljoints on the vehicles covered by this manual can be in the steering knuckle arm, with the stud facing down (lower control arm below); or in the steering knuckle arm, with the stud facing up (lower control arm above); or in the lower control arm, with the stud facing up (steering knuckle arm above).*

37   If the balljoint is an integral part of the steering knuckle arm, disconnect the tie-rod end from the steering knuckle arm (see Chapter 7).

38   Remove the cotter pin from the lower balljoint stud and back off the nut several turns.

39   Separate the lower balljoint stud from the steering knuckle arm **(see illustration)** or from the lower control arm **(see illustrations)**.

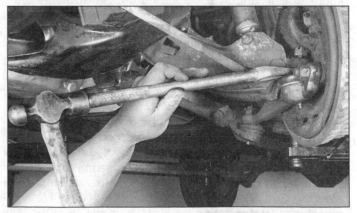

8.39b The procedure shown in illustration 8.39a for splitting a lower balljoint just won't work, because the upper and lower balljoint studs don't align; instead, separate the balljoint from the lower control arm with a picklefork (note the balljoint-in-the-steering arm, with the stud facing up, through the lower control arm)

8.39c Or, if you don't have a picklefork, use a hammer - this is very tricky on some early models like this one because there's not a lot of knuckle surface to hit! Turn the brake and knuckle assembly all the way to the left to give yourself a clear target

8.50a Pull the cotter pin out of the stud nut . . .

8.50b . . . and loosen, but do not remove, the nut

8.51 Lift the steering knuckle up and off the balljoint stud

40 If the lower balljoint is an integral part of the steering knuckle arm, unbolt the steering arm from the steering knuckle **(see illustration 8.9)**, install a new steering arm and reattach it to the lower control arm.

41 If the balljoint is an integral part of the lower control arm, remove the lower control arm, then take the arm to a dealer device department or other repair shop to have the old balljoint pressed out and the new one installed. **Note:** *If you can rent or borrow a balljoint removal/installation press, the control arm doesn't have to be removed from the vehicle. Follow the tool manufacturer's instructions.*

42 While the balljoint stud is disconnected from the lower control arm, steering knuckle and/or steering arm, inspect the tapered holes in the control arm, steering knuckle or steering arm. Remove any accumulated dirt. If out-of-roundness, deformation or other damage is noted, the lower control arm, steering knuckle and/or steering arm must be replaced with a new one.

43 Inspect the balljoint boot(s) for cracks and tears. If a boot is damaged, replace it. You can install a new balljoint boot and retainer using a large socket and a hammer.

44 Installation is the reverse of removal. After the suspension has been reassembled, raise the lower control arm with a floor jack to simulate normal ride height, then tighten all fasteners securely.

45 If a cotter pin doesn't line up with an opening in the castle nut, tighten (never loosen) the nut just enough to allow installation of the cotter pin.

46 Install the grease fittings and lubricate the new balljoints.

47 Install the wheels and lower the vehicle. Tighten the lug nuts securely.

48 The front end alignment should be checked, and if necessary, adjusted, by a dealer service department or alignment shop.

Bolt-on type balljoints

Refer to illustrations 8.50a, 8.50b, 8.51 and 8.52

49 Loosen the wheel lug nuts, raise the vehicle and support it securely on jackstands. Remove the wheel. Support the lower control arm with a jack. **Caution:** *The lower control arm must remain supported during this entire procedure.*

50 Remove the cotter pin from the balljoint stud nut **(see illustration)**. Loosen the nut, but don't remove it from the

8.52 Remove the balljoint through-bolt and upper bolt to detach the balljoint from the lower control arm

stud **(see illustration)**.

51 Separate the balljoint from the steering knuckle with a small puller **(see illustration 3.61b)**. Lift the steering knuckle off the balljoint stud **(see illustration)**.

52 Remove the balljoint mounting bolts and nut **(see illustration)**.

53 To install the balljoint, insert the balljoint threaded stud through the hole in the lower control arm and install the nut, but don't tighten the nut yet.

54 Install the balljoint mounting bolts and nut and tighten them securely.

55 Tighten the balljoint stud nut securely, then secure it with a new cotter pin.

56 Install the wheel and lug nuts. Lower the vehicle and tighten the lug nuts securely.

9 Wheel stud - replacement

Refer to illustration 9.3

1 Raise the front of the vehicle and support it on jack-stands and remove the wheel.

2 Remove the brake caliper, brake disc and splash shield (if equipped).

3 Position the stud to be replaced, so that the stud can be maneuvered out from the hub once it has been pressed out. Install a lug nut onto the end of the stud and, using a small C-clamp tool or equivalent, press the stud from the hub flange **(see illustration)**.

4 Remove the lug nut, then the stud.

5 Position the stud hole, so that the stud can be maneuvered out from the hub once it has been pressed out. Insert the new stud in the hole, making sure the serration's are aligned with those made by the original bolt.

6 Place four flat washers over the outside end of the stud, then thread a lug nut onto the stud.

7 Tighten the lug nut until the stud head seats against the rear of the hub. Remove the lug nut and washers.

8 Reinstall the splash shield, brake disc and caliper.

9 Install the wheel and lower the vehicle to the ground. Tighten the lug nuts securely.

9.3 Press out a wheel stud using a small C-clamp tool

Rear suspension

10 Solid axle (with control arms and coil springs)

Typical solid axle (with control arms and coil springs) rear suspension

1 Stabilizer bar
2 Lower suspension arm pivot bolt/nut
3 Lower suspension arm-to-rear axle bolt/nut
4 Upper suspension arm pivot bolt/nut
5 Shock absorber
6 Shock absorber-to-rear axle bolt/nut
7 Coil spring

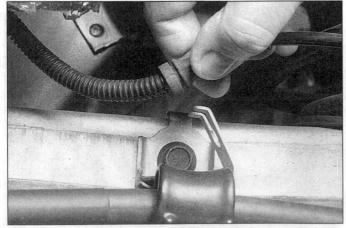

10.2 If the vehicle is equipped with ABS, disengage the ABS lead from its bracket on the lower suspension arm

The solid axle (with control arms and coil springs) rear suspension **(see illustration)** consists of the axle, two coil springs, a pair of shock absorbers, and four - two lower and two upper - suspension arms. Body roll is controlled by a stabilizer bar attached to the two lower suspension arms.

Stabilizer bar

Refer to illustrations 10.2 and 10.3
1 Raise the rear of the vehicle and support it securely on jackstands. Place the jackstands under the frame, not under the axle housing. The axle housing must be free to hang down so that the shock absorbers are fully extended.
2 Detach the ABS lead, if equipped **(see illustration)**, from the lower suspension arm.
3 Remove the four bolts (two on each arm) that attach the stabilizer bar to the lower suspension arms **(see illustration)** and remove the bar.

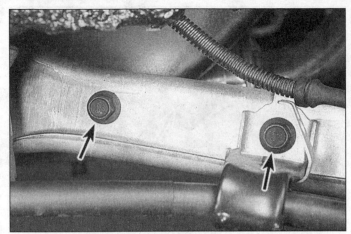

10.3 To detach the rear stabilizer bar from the lower suspension arms, remove these two bolts (arrows) from each arm (left lower suspension arm shown)

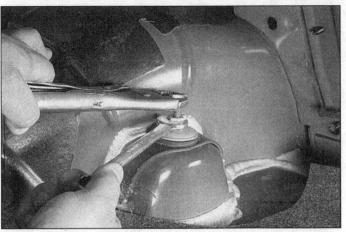

10.8 Locate the upper retaining nut for the shock absorber inside the trunk; to prevent the piston rod from turning, hold it with a wrench or locking pliers while you break loose the nut

4   Installation is the reverse of the removal procedure. Be sure to tighten all fasteners securely.
5   Remove the jackstands and lower the vehicle.

Shock absorber

Refer to illustrations 10.8 and 10.9
6   Raise the rear of the vehicle and support it securely on jackstands placed under the frame, not under the axle housing.
7   Support the rear axle with a floor jack to prevent it from dropping when the shock absorber is disconnected.
8   Locate the upper retaining nut for the shock absorber inside the trunk **(see illustration)**. To prevent the piston rod from turning, hold it with a wrench or locking pliers while you break loose the nut. Remove the nut, washer and upper insulator. Discard the nut.
9   Remove the lower retaining nut and bolt **(see illustration)** and remove the shock absorber from the vehicle.
10   Installation is the reverse of removal. Be sure to tighten the upper and lower shock absorber fasteners securely.
11   Remove the jackstands and lower the vehicle.

Coil spring

Removal

Note: *Rear coil springs should always be replaced in pairs.*
12   Loosen the wheel lug nuts, raise the rear of the vehicle and support it securely on jackstands placed under the frame. Remove the wheel and block the front wheels.
13   Remove the rear stabilizer bar.
14   Support the rear axle assembly with a floor jack placed under the axle housing.
15   Pass a length of chain up through the suspension arm and coil spring, then bolt the chain together. This will prevent the spring from flying out before it's fully extended. Be sure to leave enough slack in the chain to allow the coil spring to extend fully.
16   Detach the parking brake cable clips from the lower suspension arm.
17   Place a floor jack under the lower suspension arm pivot

10.9 Remove the lower retaining nut and bolt (arrows) and disengage the shock absorber from the axle bracket

bolt and remove the pivot bolt and nut.
18   Slowly lower the jack under the suspension arm until all tension is removed from the coil spring. Remove the chain, coil spring and insulators from between the suspension arm and the spring upper seat.

Installation

19   Set the upper insulator on top of the spring, using tape to hold it in place, if necessary.
20   Place the spring between its seat on the axle tube and the frame seat, so that the pigtail (the lower end of the spring) is at the rear and pointing toward the left side of the vehicle. Install the safety chain.
21   Raise the lower suspension arm with the floor jack and install the arm pivot bolt and nut.
22   Repeat Steps 12 through 15 for the other spring.
23   Reattach the parking brake cables to the lower suspension arms.
24   Install the rear stabilizer bar.
25   Install the wheel and lug nuts. Lower the vehicle and tighten the lug nuts securely.

10.28 To disconnect an upper rear suspension arm from the axle, remove this bolt (arrow) and nut; note that the bolt head faces out

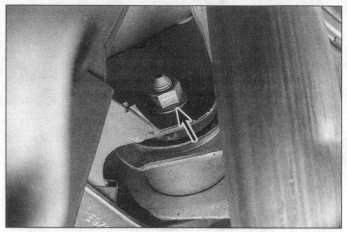

10.29 To disconnect an upper rear suspension arm from the vehicle, remove this nut (arrow) and bolt; note that the nut faces out

Suspension arms

Warning: *If you're going to remove both the upper and the lower arms at the same time, then remove the coil springs first.*

Note: *If one upper arm requires replacement, replace the other upper arm as well. The manufacturer recommends installing new fasteners when reassembling the rear suspension components.*

26 Loosen the rear wheel lug nuts. Raise the rear of the vehicle and support it securely on jackstands placed beneath the frame rails. Block the front wheels. Remove the rear wheels.

27 Position a jack under the differential housing to support the axle.

Upper arm

Refer to illustrations 10.28 and 10.29

28 Remove the upper arm-to-rear axle pivot bolt and nut **(see illustration)**.

29 Remove the upper arm-to-frame pivot bolt and nut **(see illustration)** and remove the arm from the vehicle.

30 Inspect the bushings at both ends of the arm. If either

bushing is damaged or worn, have it replaced by an automotive machine shop.

31 Position the leading end of the new suspension arm in the frame bracket. Install a new pivot bolt and nut with the bolt head facing forward. Don't fully tighten the nut at this time.

32 Attach the other end of the arm to the axle housing with a new pivot bolt and nut. The bolt head should be facing to the rear. It may be necessary to jack up the rear axle to align the holes. Don't fully tighten the nut yet.

33 Repeat Steps 26 through 31 and replace the other upper suspension arm, then proceed to Step 32.

Lower arm

Refer to illustrations 10.37 and 10.39

Note: *If one lower arm requires replacement, replace the other lower arm as well. Also, the manufacturer recommends installing new fasteners when reassembling the rear suspension components.*

34 Detach the ABS lead **(see illustration 10.2)**, if equipped, from the lower suspension arm.

35 Remove the stabilizer bar.

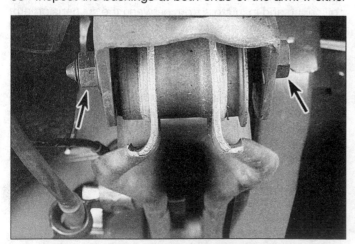

10.37 To disconnect a lower rear suspension arm from the axle bracket, remove this nut and bolt (arrows)

10.39 To disconnect a lower rear suspension arm from the vehicle, remove the nut and pivot bolt

10.48 To disconnect the axle damper from the axle bracket, remove this nut (arrow)

10.49 To disconnect the axle damper from the vehicle bracket, remove this nut (arrow)

36 Chain the coil spring to the lower suspension arm.

37 Place a jack under the lower suspension arm-to-axle pivot bolt and remove the pivot bolt and nut **(see illustration)**.

38 Carefully lower the jack until all tension is removed from the spring.

39 Remove the lower arm-to-frame pivot bolt and nut **(see illustration)**, then remove the arm from the vehicle.

40 Inspect the bushings at both ends of the arm. If either bushing is damaged or worn, have it replaced by an automotive machine shop.

41 Position the new lower arm in the frame mounting bracket and install a new pivot bolt and nut, with the nut facing out. Do not tighten the nut completely at this time.

42 Position the other end of the lower arm in the axle bracket and install a new bolt and nut (again, nut facing out). Do not fully tighten the nut at this time.

43 Repeat Steps 34 through 40 and replace the other lower suspension arm.

44 Remove both jacks (from underneath the differential and which-ever lower suspension arm you just replaced) and place the two jacks underneath either end of the axle tube, then raise the axle until the rear of the vehicle is supported by the axle (this is the normal ride height). Tighten all pivot bolt nuts securely.

45 Install the stabilizer bar.

Upper or lower arm

46 Install the wheel and lug nuts. Remove the jackstands and floor jack, lower the vehicle and tighten the lug nuts securely.

Axle damper

Removal

Refer to illustrations 10.48 and 10.49

47 Loosen the wheel lug nuts, raise the vehicle and support it on jackstands. Remove the wheel and place a floor jack under the rear axle to support it, just in case it shifts when the axle damper is removed.

48 Remove the axle damper front attaching nut and bolt from the axle bracket **(see illustration)**.

49 Remove the rear attaching nut **(see illustration)**. Remove the bolts that hold the rear mounting bracket to the frame sidemember and remove the damper from the vehicle.

Installation

50 Place the rear bracket onto the axle damper and install the nut (don't tighten it yet).

51 Position the rear bracket on the frame sidemember and install the bolts, tightening them securely.

52 Swing the axle damper into the mount on the rear axle and install the pivot bolt and attaching nut, tightening them securely.

53 Tighten the rear retaining nut securely.

54 Install the wheel and lug nuts and lower the vehicle. Tighten the lug nuts securely.

11 Solid axle (with leaf springs)

Typical leaf spring type suspension

1	Shock absorbers
2	Lower shock absorber mount

3	Multi-leaf spring assemblies
4	Leaf spring anchor plates

The solid axle (with leaf springs) rear suspension **(see illustration)** consists of a pair of multi-leaf springs and two shock absorbers. The rear axle assembly is attached to the leaf springs by U-bolts. The front ends of the springs are attached to the frame at the front hangers, through rubber bushings. The rear ends of the springs are attached to the frame by shackles. Some models also use a stabilizer bar, bolted to the frame and the axle, to reduce vehicle roll during cornering.

Stabilizer bar

Refer to illustrations 11.2 and 11.3

1   Raise the rear of the vehicle and support it securely on jackstands. Block the front wheels to keep the vehicle from rolling off the stands.

2   Remove the lower nuts, washers and bolts from the stabilizer bar-to-frame bracket links **(see illustration)**.

3   Remove the nuts from the U-bolts and remove the stabilizer bar clamps **(see illustration)**.

4   Remove the stabilizer bar assembly.

5   Inspect the stabilizer bar and link bushings for cracks, tears and other deterioration. Replace as necessary.

6   Installation is the reverse of removal. Be sure to tighten all fasteners securely.

11.2 To detach the stabilizer bar from the lower end of the link, remove this nut and bolt

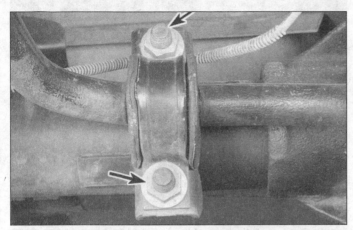

11.3 To detach the stabilizer bar from the rear axle, remove these nuts (arrows) from both pairs of U-bolts

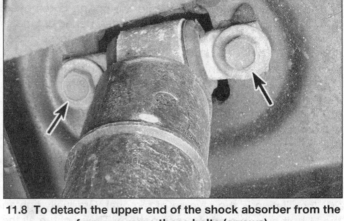

11.8 To detach the upper end of the shock absorber from the frame, remove these bolts (arrows)

11.9 To detach the lower end of the shock absorber from the anchor plate, remove this nut, washer and bolt (arrows)

11.13 To detach the anchor plate, remove these four nuts from the U-bolts

Shock absorber

Refer to illustrations 11.8 and 11.9

7 Raise the rear of the vehicle and support securely on jackstands. Block the front wheels so the vehicle doesn't roll off the stands.

8 Remove the shock absorber upper mounting bolts from the frame **(see illustration)**.

9 Remove the lower mounting nut, washer and bolt from the anchor plate bracket **(see illustration)**.

10 Remove the shock absorber.

11 Installation is the reverse of removal. Make sure you install the nuts and bolts facing in the proper direction. Tighten all fasteners securely.

Leaf spring/shackle

Refer to illustrations 11.13, 11.16 and 11.18

12 Raise the rear of the vehicle and support it securely on jackstands. Block the front wheels to keep the vehicle from rolling off the stands. Support the axle with a floor jack and raise it slightly to relieve the tension on the leaf springs.

13 Remove the four U-bolt nuts and washers **(see illustration)**.

14 Remove the anchor plate.

15 Remove the U-bolts and spacer.

16 Remove the nut from the front spring-mount bolt **(see illustration)**.

11.16 Remove the nut from the front leaf spring mounting bolt (arrow), but don't remove the bolt until the rear of the leaf spring has been detached and the leaf spring is supported

17 Raise the axle slightly off the springs with the floor jack. Be sure the axle is stable. The axle must remain supported by the jack during the entire time the spring is removed from the vehicle. If this will be an extended period of time, it would be a good idea to support the axle with jackstands at this point.

18 While an assistant supports the rear of the spring, remove the shackle-to-frame bracket nut, washers and bolt **(see illustration)**. Lower the rear of the spring to the ground. **Note:** *On some pick-ups and four-door models, it will be necessary to remove the fuel tank for access to the bolt.*

19 While having an assistant support the front of the spring, remove the front spring-mount bolt, then remove the spring assembly from the vehicle.

20 Installation is the reverse of removal. Gradually tighten the U-bolt nuts in a criss-cross pattern. Then tighten all the fasteners securely.

21 If the bushings at the ends of the spring are worn or deteriorated, an automotive machine shop or dealer service department can press the old ones out and press new ones in.

Track bar

22 Raise the rear of the vehicle and support it securely on jackstands. Block the front wheels to keep the vehicle from rolling off the stands.

23 Support the rear axle at its normal ride height position with a floor jack.

11.18 To detach the rear of the leaf spring, have an assistant support the spring, then remove this nut (arrow) and withdraw the shackle bolt

24 Remove the nut, washer and bolt and detach the upper end of the track bar from the right frame rail bracket.

25 Remove the nut, washer and bolt and detach the lower end of the track bar from the left axle bracket.

26 Remove the track bar.

27 Installation is the reverse of removal. Be sure to tighten both nuts securely.

Rear axle assembly

28 For removal procedures, see Chapter 5.

12 Beam type (with coil springs and shock absorbers)

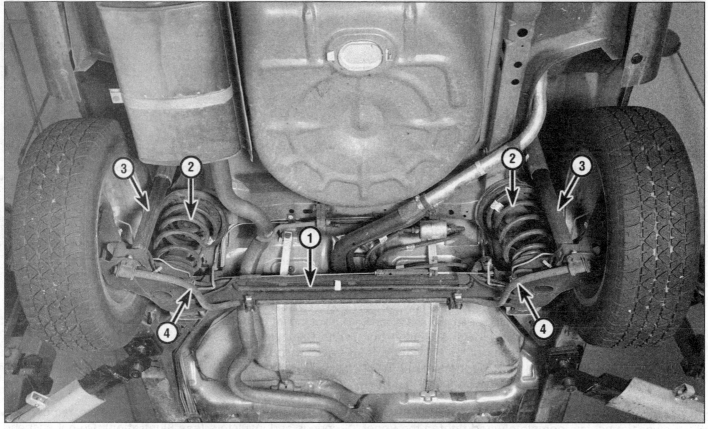

Typical beam type suspension (with coil springs and shock absorbers)

1	Rear axle assembly	2	Coil spring	3	Shock absorber	4	Stabilizer bar

The beam type (with coil springs and shock absorbers) rear suspension consists of a rear axle assembly, two coil springs, two shock absorbers and a track bar. The rear axle has two control arms welded to it, which are used to mount the axle assembly to the body. These control arms, together with the track bar and shock absorbers, maintain the proper geometric relationship of the axle assembly to the body under the forces created by accelerating, braking and cornering. On some models, a non-serviceable stabilizer bar is welded to the inside of the axle housing.

Shock absorbers

Refer to illustrations 12.1 and 12.3

1   Remove the upper mounting nut **(see illustration)**. The damper rod must be kept from turning while the nut is loosened.

2   Raise the rear of the vehicle enough to take the weight off of the suspension, but do not lift the tires off the ground. Support the vehicle with jackstands placed at suitable locations under the car's frame. Do not place them under the rear axle. If more clearance is needed under the vehicle, the

vehicle can be raised higher, but then the rear axle must also be supported with jackstands.

12.1 Locate the upper retaining nut (arrow) for the shock absorber on top of the frame. It may be necessary to prevent the piston rod from turning by holding it with a wrench or locking pliers while you break loose the nut

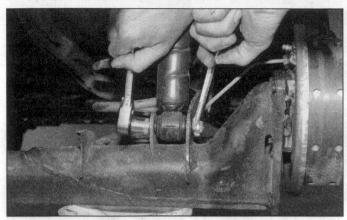

12.3 Remove the shock absorber lower mounting bolt

12.10 Remove both mounting nuts and bolts on the track bar

3   Remove the shock absorber lower attaching bolt and nut and remove the shock **(see illustration)**. It may be necessary to use a screwdriver to pry the lower end of the shock out of its mounting bracket.

4   The shock should be compressed and then extended its full length a few times to check for any free movement of the shaft, noise, or fluid leakage. If any of these conditions are found the shocks should be replaced with a new set.

5   To install, extend the shock to its full length and place it in its lower mount. Then feed the lower attaching bolt through the mount and shock and install the nut loosely.

6   Lower the vehicle enough to guide the shock's upper stud through the body opening and loosely install the upper attaching nut.

7   Tighten the lower attaching nut.

8   Lower the vehicle completely and tighten the upper attaching nut. Again the damper rod must be kept from turning while tightening the upper nut.

Track bar

Refer to illustration 12.10

9   Raise the rear of the vehicle and support it securely on jackstands. Support the rear axle with a jack or jackstand.

10   Remove the nuts and bolts securing the track bar at both ends **(see illustration)**.

11   Remove the track bar.

12   Inspect the bushings for hardening, cracking or excessive wear. If they exhibit any of these conditions, the track bar must be replaced.

13   To install, place the left end of the track bar in the body mount and loosely install the bolt and nut. The open side of the bar must face to the rear.

14   Place the other end of the bar in the axle mount and loosely install the bolt and nut. Both nuts must face the rear of the vehicle.

15   Raise the rear axle to simulate normal ride height.

16   Tighten both nuts securely, then lower the vehicle to the ground.

Coil springs and insulators

Refer to illustration 12.25

17   Raise the rear of the vehicle and support it with jack-

stands placed under the frame.

18   Support the rear axle with a floor jack.

19   Remove the wheels and brake drums.

20   Disconnect the parking brake cable by loosening the adjustment nut and prying forward on the parking brake equalizer lever to disconnect the forward cable from the equalizer lever.

21   Twist the equalizer lever and disengage it from the pivot mount on the body.

22   Remove the bolts attaching the brake line brackets to the chassis on both the left and right sides.

23   Remove the track bar.

24   Remove the shock absorber lower attaching nuts and bolts from both shock absorbers.

25   Slowly lower the rear axle enough to remove the springs and insulators **(see illustration)**. Do not suspend the rear axle by the brake hoses, as this will damage the hoses.

26   If the insulators are worn, cracked or damaged, they should be replaced.

27   Inspect the springs for cracks or other damage. If they exhibit any of these conditions, or if the vehicle has been sagging in the rear, the springs should be replaced. The rear springs should always be replaced as a pair.

28   Installation is the reverse of the removal procedure.

12.25 After lowering the rear axle, remove the coil springs

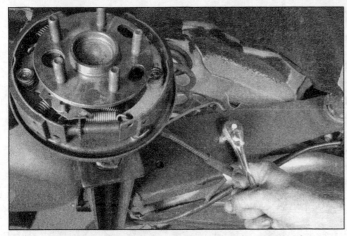

12.31 Disconnect the brake line brackets from the rear axle control arms

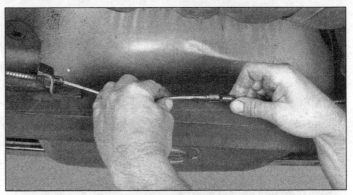

12.33 Disconnect the right parking brake cable from the left one

Note: *When installing the springs, be sure they are in the proper position.*

29   Adjust the parking brake.

Rear axle assembly

Removal

Refer to illustrations 12.31, 12.33, 12.35, 12.38, 12.41a and 12.41b

30   Remove the coil springs.

31   Disconnect the brake lines from the control arms **(see illustration)**.

32   Disconnect and cap the rigid brake lines from both rear brake cylinders.

33   Disconnect the right rear parking brake cable from the left rear parking brake cable **(see illustration)**.

34   Disconnect the right rear and left rear parking brake cables from their brackets on the rear axle.

35   Remove the bolts securing the rear parking brake cable guide to the underbelly and allow the entire parking brake assembly to hang from the right rear backing plate **(see illustration)**.

36   Remove the bolts securing the hub and bearing assembly to the rear axle and remove the assemblies along with the brake backing plates.

37   While an assistant steadies the rear axle on the jack, remove the bolts securing the control arm brackets to the body.

38   Lower the rear axle and remove it from under the vehicle **(see illustration)**.

39   If the rear axle is being replaced, remove the control arm brackets from the control arms and install them on the new axle.

40   Inspect the control arm bushings for cracking, hardening or other damage and replace if necessary.

Installation

41   Installation of the rear axle is the reverse of the removal procedure. Raise the rear axle to simulate normal ride height before tightening the track bar and shock absorber fasteners securely.

42   Bleed the brake system and adjust the parking brake.

12.35 Allow the parking brake assembly to hang from the right backing plate

12.38 Carefully remove the rear axle

13 Beam type
(with strut or shock absorber/coil springs)

Typical beam type axle suspension

1 Shock absorber and coil spring assembly	*2 Axle beam*

1   The beam type (with strut or shock absorber/coil springs) rear suspension **(see illustration)** utilizes a strut or integrated shock absorber/coil spring assemblies. The upper end of each shock is attached to the vehicle body. The lower end of each shock is attached to the axle beam. The axle beam is attached to the vehicle by a pair of arms welded to each end of the beam.

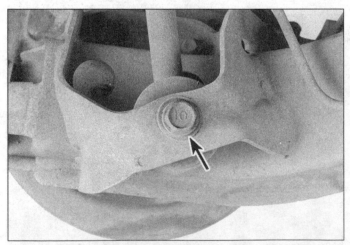

13.3 To detach the lower end of a rear shock from the axle beam, remove this bolt (arrow)

Shock absorber/coil spring

Removal

Refer to illustrations 13.3 and 13.5

1   Loosen the rear wheel lug nuts, raise the rear of the vehicle and support it securely on jackstands. Remove the wheels.

2   Support the axle beam near the lower shock mount with a floor jack.

3   Remove the shock-to-axle beam bolt **(see illustration)**.

4   To access the upper mounting nut, remove the luggage compartment side cover.

5   Remove the upper mounting nut **(see illustration)** while an assistant supports the shock so it doesn't fall. Guide the shock out of the fenderwell.

Inspection

6   Check the strut body for leaking fluid, dents, cracks and other obvious damage which would warrant repair or replacement.

7   Check the coil spring for chips or cracks in the spring coating (this will cause premature spring failure due to corrosion). Inspect the spring seat for cuts and general deterioration.

8   If any undesirable conditions exist, proceed to the strut disassembly procedure (see Section 7, Steps 28 through 42).

13.5 To detach the upper end of a rear shock from the body, remove this nut (arrow)

13.19 To detach the axle beam from the vehicle, remove the pivot nut (arrow) and bolt from the bracket for each trailing arm

Installation

9   Maneuver the shock and coil spring assembly up into the fenderwell and insert the mounting studs through the holes in the body. Install the nut, but don't tighten it yet.

10   Position the lower end of the shock between the mounting bracket and the axle beam, install the bolt and tighten it securely.

11   Install the wheel and lug nuts, lower the vehicle and tighten the lug nuts securely.

12   Tighten the shock upper mounting nut securely.

Axle beam

Warning: Dust created by the brake system may contain asbestos, which is harmful to your health. Never blow it out with compressed air and don't inhale any of it. Do not, under any circumstances, use petroleum-based solvents to clean brake parts. Use brake cleaner or denatured alcohol only.

Removal

Refer to illustration 13.19

13   Loosen the wheel lug nuts, raise the rear of the vehicle and place it securely on jackstands. Block the front wheels and remove the rear wheels.

14   Remove the rear brake drums and the brake assemblies.

15   Disconnect the brake lines from the wheel cylinders and disconnect the brake hoses from the brake lines. Plug the brake lines to prevent moisture and contamination from entering the brake system.

16   Detach the brake backing plates from the axle beam and suspend them from the coil springs with pieces of wire. It isn't necessary to remove the parking brake cable from the backing plate.

17   Support the axle beam with a floor jack.

18   Disconnect the lower ends of the shock absorbers from the axle beam.

19   Remove the pivot bolts from the forward ends of the axle beam trailing arms **(see illustration)**.

20   Remove the axle beam assembly.

Inspection

21   Inspect the trailing arm bushings for cracks, deformation and wear. If they're damaged or worn out, take the axle beam assembly to a dealer service department or an automotive machine shop to have the old ones pressed out and new ones pressed in.

Installation

22   Installation is the reverse of removal. Be sure to tighten all fasteners securely.

23   Lower the vehicle and tighten the lug nuts securely.

24   Bleed the brakes.

14 Beam type (with leaf springs)

Typical beam type suspension (with leaf springs)

| 1 | Leaf spring | 2 | Shock absorber | 3 | Axle |

The beam type (with leaf springs) rear suspension features a tubular axle with leaf springs **(see illustration)**. Damping is handled by vertically-mounted shock absorbers located between the axle and the chassis. Some models are equipped with a rear stabilizer bar.

Rear shock absorbers

Refer to illustrations 14.2, 14.3a and 14.3b

Removal

1   Loosen the wheel lug nuts, raise the rear of the vehicle

14.2 When removing the shock absorbers, the axle must be raised slightly to remove tension from the shocks - this is easily done with a jack

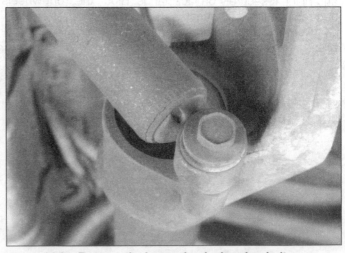

14.3a Remove the lower shock absorber bolt . . .

14.3b ... followed by the upper bolt

and support it securely on jackstands.

2   Support the axle with jack and remove the rear wheels **(see illustration)**.

3   Remove the lower and upper shock mounting bolts **(see illustrations)** and detach the shock absorber.

Installation

4   Hold the new shock absorber in position and install the bolts. Tighten the upper bolt securely and the lower bolt snugly. Remove the jack, lower the vehicle and tighten the lower bolt securely.

Rear sway bar

Refer to illustration 14.6

Removal

5   Raise the rear of the vehicle and support it securely on jackstands.

6   Remove the two lower sway bar link arm bolts **(see illustration)**.

7   Loosen the four sway bar bushing retainer bolts.

8   Hold the sway bar in place, remove the bushing retainer bolts, then lower the sway bar from the vehicle.

Installation

9   Connect the link arms with the bolts finger tight.

10   Place the sway bar in position on the axle with the slits in the bushings facing up and install the bolts finger tight.

11   Lower the vehicle weight onto the suspension and tighten the bolts securely.

Leaf spring

Removal

Refer to illustrations 14.14, 14.16, 14.18 and 14.19

12   Loosen the rear wheel lug nuts, raise the rear of the vehicle and support it securely on jackstands. Remove the rear wheels.

13   Jack up the rear axle slightly, just until the weight is off

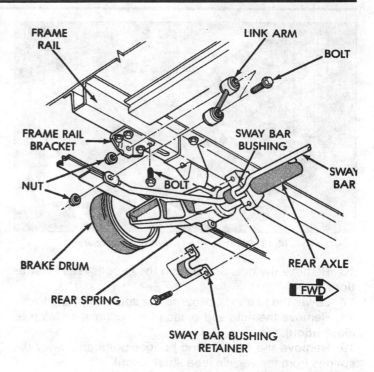

14.6 To remove the sway bar, remove the lower bolt from each link arm, loosen the two bolts on each bushing retainer, support the sway bar and remove the retainer bolts

the rear springs, then support the axle at this height with jackstands.

14   If the springs are being removed, disconnect the brake proportioning valve link from the left side spring **(see illustration)** if equipped. If you're removing the rear axle, disconnect the parking brake cables and brake hoses from the axle.

15   Remove the bolts from the lower ends of the shock absorbers.

14.14 Mark the nut location (arrow), remove the nut and bolt, then disconnect the proportioning valve link (if equipped)

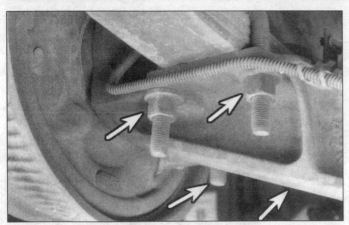

14.16 Remove the U-bolt nuts (arrows)

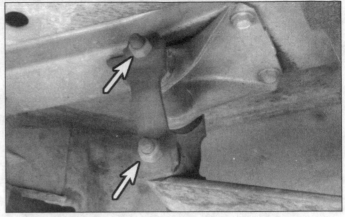

14.18 Remove the spring shackle nuts (arrows)

16 Remove the nuts and detach the U-bolts (see illustration).

17 Lower the rear axle, allowing the springs to hang free.

18 Remove the nuts and detach the spring shackles (see illustration).

19 Remove the front spring hanger bolts and lower the springs from the vehicle (see illustration).

Installation

20 Raise the front ends of the springs into position and install the spring hanger bolts and tighten the bolts.

21 Raise the rear end of the spring into place and connect the shackles, with the nuts finger tight.

22 Raise the axle with the jack until it is centered under the center bolt, install the U-bolts, plate and nuts. Tighten the nuts securely.

23 Connect the lower ends of the shock absorbers and install the bolts finger tight. Connect the brake hoses and cables, if disconnected. Install the wheels.

24 Lower the vehicle weight onto the suspension and tighten the front pivot bolt (if loosened), shock absorber bolts and shackle nuts.

25 If the brake hoses were disconnected, bleed the brakes.

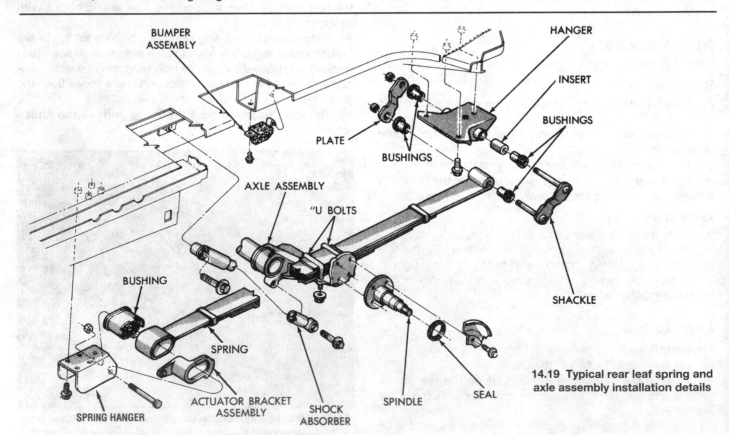

14.19 Typical rear leaf spring and axle assembly installation details

15 Trailing arm type

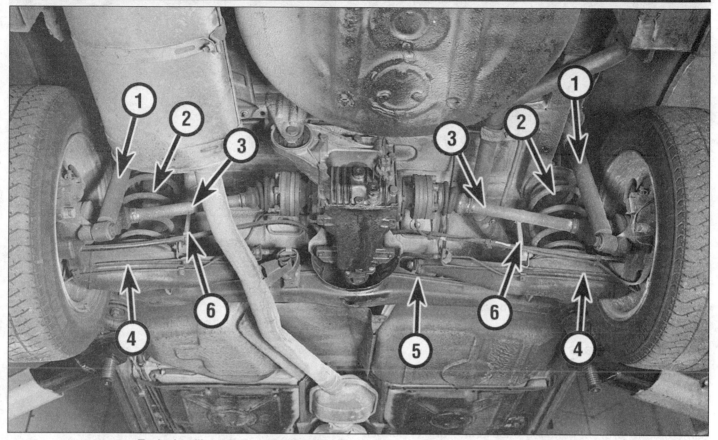

Typical trailing arm type rear suspension (with shock absorbers and coil springs)

1	*Shock absorber*	3	*Driveaxle*	5	*Rear axle carrier*
2	*Coil spring*	4	*Trailing arm*	6	*Stabilizer bar link*

The trailing arm type rear suspension system **(see illustration)** uses coil springs and telescopic shock absorbers or coil-over shock absorbers. On shock absorber models the upper ends of the shocks are attached to the body; the lower ends are connected to trailing arms. A stabilizer bar is attached to the trailing arms via links and to the body with clamps. On coil-over spring models the upper ends are attached to the body; the lower ends are connected to the trailing arms. The rear suspension is otherwise similar.

Rear stabilizer bar

Refer to illustrations 15.2, 15.3a and 15.3b
Note: *The rear stabilizer bar is mounted basically the same way on all models. Follow these general removal and installation procedures keeping in mind any variations.*
1   Raise the rear of the vehicle and support it securely on jackstands. Block the front wheels to keep the vehicle from rolling.
2   Remove the stabilizer bar bracket bolts or nuts **(see illustration)**.

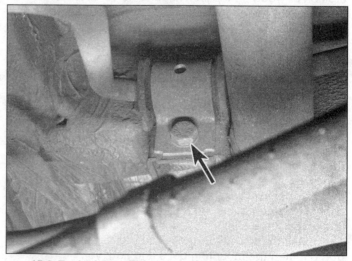

15.2 Typical rear stabilizer bar bracket bolts (arrows)

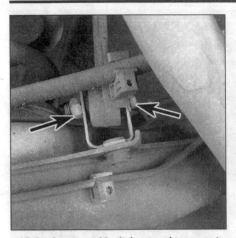

15.3a A nut and bolt (arrows) connect each rear stabilizer bar link to the rear trailing arms

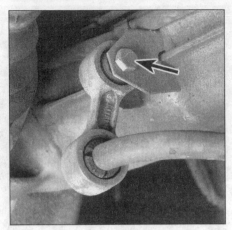

15.3b Typical bolt (arrow) connecting rear stabilizer bar link to trailing arm (coil-over spring type)

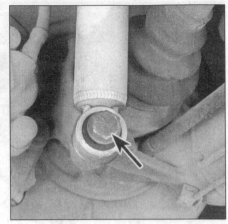

15.8 Remove the shock absorber lower mounting bolt (arrow)

3   Disconnect the stabilizer bar from the link at each end of the bar **(see illustrations)** and detach the stabilizer bar.
4   Inspect and, if necessary, replace any worn or defective bolts, washers, bushings or links.
5   Installation is the reverse of removal. Tighten all fasteners securely.

Shock absorbers

Removal

Refer to illustrations 15.8 and 15.9
Note: *Although shock absorbers don't always wear out simultaneously, replace both left and right shocks at the same time to prevent handling peculiarities and abnormal ride quality.*
6   If a shock absorber is to be replaced with a new one, it is recommended that both shocks on the rear of the vehicle be replaced at the same time.
7   Raise the rear of the vehicle and support it securely on jackstands. Support the trailing arm with a floor jack. Place a block of wood on the jack head to serve as a cushion.
8   Remove the shock absorber lower mounting bolt **(see illustration)**.
9   On some models, working inside the trunk, you can remove the trim to access the upper mounting nuts; on some models, you'll have to remove the rear seatback to get at the upper mounting nuts. As you remove the mounting nuts **(see illustration)**, have an assistant support the shock from below so it doesn't fall out.
10   Look for oil leaking past the seal in the top of the shock body. Inspect the rubber bushings in the shock eye. If they're cracked, dried or torn, replace them. To test the shock, grasp the shock body firmly with one hand and push the damper rod in and out with the other. The strokes should be smooth and firm. If the rod goes in and out easily, or unevenly, the shock is defective and must be replaced.
11   Install the shocks in the reverse order of removal, but don't tighten the mounting bolts and nuts yet.
12   Bounce the rear of the vehicle a couple of times to settle the bushings, then tighten the nuts and bolts securely.

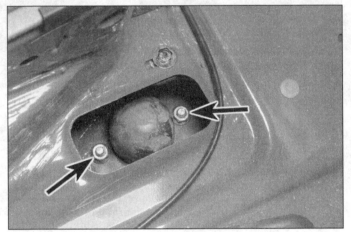

15.9 Shock absorber upper mounting nuts (arrows)

Rear coil springs

Removal

Note: *Although coil springs don't always wear out simultaneously, replace both left and right springs at the same time to prevent handling peculiarities and abnormal ride quality.*
14   Loosen the wheel lug bolts. Raise the rear of the vehicle and support it securely on jackstands. Make sure the stands don't interfere with the rear suspension when it's lowered and raised during this procedure. Remove the wheels.
15   On some models, disconnect the hangers and brackets which support the rear portion of the exhaust system and temporarily lower the exhaust system. Lower the exhaust system only enough to lower the suspension and remove the springs. Suspend it with a piece of wire.
16   Support the differential with a floor jack, remove the differential rear mounting bolt, push the differential down and wedge it into this lowered position with a block of wood. This reduces the drive angle, preventing damage to the CV joints when the trailing arms are lowered to remove the springs.
17   Place a floor jack under the trailing arm.

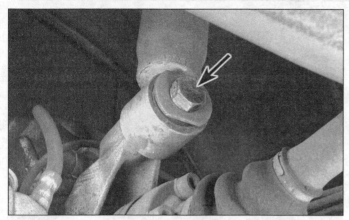

15.23 Remove the shock absorber lower mounting bolt (arrow)

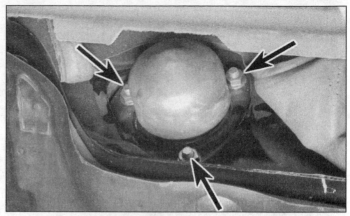

15.24 Shock absorber upper mounting nuts (arrows) (coil-over model shown)

18 If the vehicle is equipped with a rear stabilizer bar, disconnect the bar from its connecting links, or disconnect the links from the trailing arms.

19 Loop a chain through the coil spring and bolt the chain together to prevent the coil spring from popping out when the trailing arm is lowered. Be sure to leave enough slack in the chain to allow the spring to extend completely.

20 Disconnect the shock absorber lower mounting bolt, carefully lower the trailing arm and remove the coil spring.

21 Installation is the reverse of removal. As the trailing arm is raised back up, make sure the spring seats properly.

Rear shock absorber/coil spring assembly

Removal

Refer to illustrations 15.23 and 15.24

Note: *Although shock absorbers don't always wear out simultaneously, replace both left and right shocks at the same time to prevent handling peculiarities and abnormal ride quality.*

22 Loosen the wheel lug bolts, raise the vehicle and support it securely on jackstands. Remove the wheels.

23 Remove the shock lower mounting bolt **(see illustration)**.

24 On some models the trim inside the trunk must be peeled back far enough to access the upper mounting nuts. To get at the upper mounting nuts. Support the trailing arm with a jack and remove the upper mounting nuts **(see illustration)**. Lower the jack and remove the shock and the gasket.

25 Installation is the reverse of removal. Don't forget to install the gasket between the upper end of the shock and the body. Tighten the upper nuts securely. Don't tighten the lower bolt until the vehicle is lowered.

26 Lower the vehicle, allowing it to sit at normal ride height, and tighten the lower bolt securely.

Strut or shock absorber/coil spring

Replacement

Refer to illustrations 15.29, 15.30 and 15.35

27 If the struts, shock absorbers or coil springs exhibit the

telltale signs of wear (leaking fluid, loss of damping capability, chipped, sagging or cracked coil springs) explore all options before beginning any work. Strut or shock absorber assemblies complete with springs may be available on an exchange basis, which eliminates much time and work. Whichever route you choose to take, check on the cost and availability of parts before disassembling the vehicle. **Warning:** *Disassembling a strut or coil-over shock absorber assembly is a potentially dangerous undertaking and utmost attention must be directed to the job, or serious injury may result. Use only a high quality spring compressor and carefully follow the manufacturer's instructions furnished with the tool. After removing the coil spring from the strut assembly, set it aside in a safe, isolated area.*

28 Remove the strut or shock absorber assembly. Mount the assembly in a vise. Line the vise jaws with wood or rags to prevent damage to the unit and don't tighten the vise excessively.

29 Following the tool manufacturer's instructions, install the spring compressor (which can be obtained at most auto parts stores or equipment yards on a daily rental basis) on the spring and compress it sufficiently to relieve all pressure

15.29 Following the tool manufacturer's instructions, install the spring compressor on the spring and compress it sufficiently to relieve all pressure from the suspension support

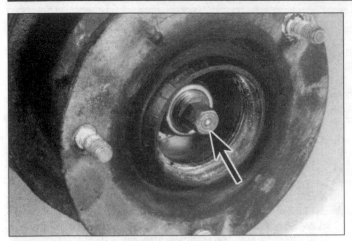

15.30 Pry the protective cap off the damper rod nut and remove the large nut (arrow) - to prevent the damper rod from turning, place a wrench on the hex-shaped end of the shaft

15.35 Make sure you align the end of the coil spring with the shoulder of the rubber ring and with the spring retainer

from the suspension support **(see illustration)**. This can be verified by wiggling the spring.

30 Pry the protective cap off the damper rod self-locking nut. Loosen the nut **(see illustration)** with an offset box end wrench while holding the damper rod stationary with another wrench.

31 Remove the nut, the strut bearing, the insulator and the large washer. Inspect the bearing for smooth operation. If it doesn't turn smoothly, replace it. Check the rubber insulator for cracking and general deterioration. If there is any separation of the rubber, replace the insulator.

32 Lift off the upper spring retainer and the rubber ring at the top of the spring. Check the rubber ring for cracking and hardness. Replace it if necessary.

33 Carefully lift the compressed spring from the assembly and set it in a safe place, such as a steel cabinet. **Warning:** *Never place your head near the end of the spring!*

34 Slide the protective tube and rubber bumper off the damper rod. If either of them is damaged or worn, replace it.

35 Installation is otherwise the reverse of removal. Tighten the threaded collar securely. Make sure you align the end of the coil spring with the shoulder of the rubber ring and with the spring retainer **(see illustration)**. Tighten the damper rod nut securely.

36 Install the strut or shock absorber assembly.

Rear trailing arms (with coil springs)

Refer to illustrations 15.39 and 15.42

37 Loosen the wheel lug bolts, raise the rear of the vehicle and support it securely on jackstands. Remove the wheel(s).

38 Remove the driveaxle (see Chapter 5).

39 Disconnect the rear brake hose from the metal brake line at the bracket on the trailing arm **(see illustration)**.

40 Disconnect the parking brake cable.

41 Disconnect the lower end of the shock absorber from the trailing arm and lower the trailing arm.

42 Remove the trailing arm pivot bolts **(see illustration)** and remove the trailing arm.

43 Inspect the pivot bolt bushings. If they're cracked,

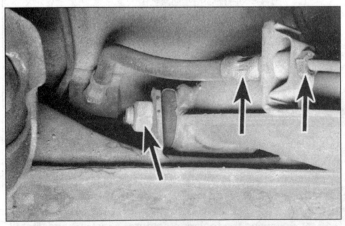

15.39 Disconnect the rear brake hose (arrow) from the metal brake line fitting (arrow) at this bracket on the trailing arm, then plug the line and hose immediately to prevent brake fluid leaks; the other arrow points to the nut for the inner pivot bolt

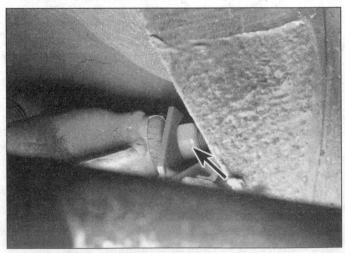

15.42 Nut (arrow) for the outer pivot bolt

dried out or torn, take the trailing arm to an automotive machine shop and have them replaced. Each bushing has a larger diameter shoulder on one end. Make sure this larger diameter shoulder on each bushing faces away from the trailing arm, i.e. the inner bushing shoulder faces the center of the vehicle and the outer bushing shoulder faces away from the vehicle.

44 Installation is the reverse of removal. Support the trailing arm with a floor jack and raise it to simulate normal ride height, then tighten the fasteners securely. Be sure to bleed the brakes.

Rear trailing arm (with coil-over shock absorbers)

Refer to illustration 15.47

45 Loosen the wheel lug bolts, raise the rear of the vehicle and support it securely on jackstands. Remove the wheel(s).

46 Remove the driveaxle (see Chapter 5).

47 Disconnect the rear brake hose from the metal brake line at the bracket on the trailing arm **(see illustration)**.

48 Disconnect the parking brake cable from the parking brake actuator and unclip the parking brake cable from the trailing arm.

49 Remove the ABS wheel speed sensor, if equipped, from the trailing arm, and unclip the sensor wire harness from the arm. Position the sensor aside so it won't be damaged during removal of the trailing arm.

50 If you're removing the right trailing arm, unplug the connector for the brake pad wear sensor.

51 Disconnect the rear stabilizer bar from the trailing arm.

52 Disconnect the shock absorber lower mounting bolt.

15.47 Disconnect the rubber brake hose (arrow) from the fitting on the metal brake line (arrow) at this bracket

53 Remove the two trailing arm pivot bolts and nuts and remove the trailing arm from the vehicle.

54 Inspect the pivot bolt bushings. If they're cracked, dried out or torn, take the trailing arm to an automotive machine shop and have them replaced. The bushing inner sleeve is longer on side. Make sure the bushings are installed with the longer side of the bushing sleeve facing toward the center of the vehicle.

55 Installation is the reverse of removal. Install the inner pivot bolt first. Don't tighten the nuts on the pivot bolts or the shock absorber yet.

56 Bleed the brakes.

57 Support the trailing arm with a floor jack and raise it to simulate normal ride height, then tighten the bolts and nuts securely.

16 Multi-link type

Typical multi-link type rear suspension

1	Stabilizer bar	4	Toe adjuster	7	Knuckle
2	Stabilizer bar bushing clamp	5	Front lateral arm	8	Strut/coil spring assembly
3	Rear lateral arm	6	Trailing arm (radius rod)		

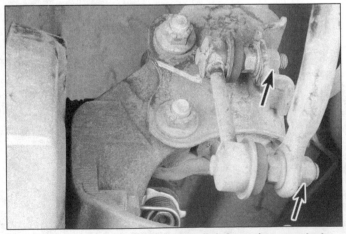

16.2 To disconnect the stabilizer link from the rear strut, remove the nut (indicated by the upper arrow). To disconnect the stabilizer link from the stabilizer bar, remove the nut (indicated by the lower arrow)

The multi-link type rear suspension system **(see illustration)** uses strut/coil springs, a pair of lateral suspension arms and a trailing arm, or radius rod, at each corner. The upper ends of the struts are attached to the vehicle body and their lower ends are attached to the upper ends of the rear knuckles. The lower ends of the knuckles are attached to the outer ends of the lateral arms; the strut/knuckle/hub assemblies are positioned laterally by the trailing arms. A stabilizer bar is attached to the vehicle by a pair of brackets and to the struts by link rods.

Rear stabilizer bar

Refer to illustrations 16.2 and 16.3

1   Loosen the rear wheel lug nuts, raise the rear of the vehicle, support it securely on jackstands and remove the wheels.

2   Remove the nuts from the link rods that attach the stabilizer to the struts **(see illustration)**.

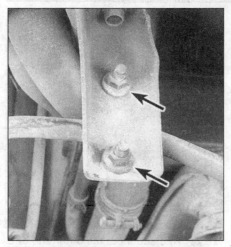

16.3 To disconnect the rear stabilizer bushing clamps, remove these nuts (arrows)

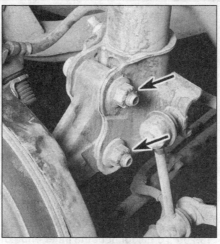

16.8 To disconnect the strut assembly from the rear knuckle, remove these two nuts and bolts (arrows)

16.11a Pry off this rubber cover . . .

3   Remove the bushing clamp nuts (see illustration) and remove the stabilizer.
4   Inspect all clamp bushings. If they're cracked or torn, replace them.
5   Installation is the reverse of removal. Be sure to tighten all fasteners securely.

Rear strut/coil spring assembly

Removal

Refer to illustrations 16.8, 16.11a and 16.11b

6   Loosen the rear wheel lug nuts, raise the rear of the vehicle and support it securely on jackstands. Remove the wheels.
7   Unbolt the brake hose bracket from the strut. If the vehicle is equipped with ABS, detach the speed sensor wiring harness from the strut by removing the clamp bracket bolt.
8   Remove the strut-to-knuckle nuts (see illustration) and knock the bolts out with a hammer and punch.
9   Separate the strut from the knuckle. Don't allow the knuckle to fall outward, as this may damage the brake hose.
10 Remove the back seat and the parcel shelf to get at the upper mounting nuts.
11 Remove the strut rubber cover (see illustration). Have an assistant support the strut and spring assembly while you remove the three strut-to-shock tower nuts (see illustration). Remove the assembly through the fenderwell.

Inspection

12 Inspect the strut. If any undesirable conditions exist, replace the strut (see Section 7, steps 28 thru 42).

Installation

13 Guide the strut assembly up into the fenderwell and insert the upper mounting studs through the holes in the shock tower. Make sure the strut upper spring seat is correctly aligned (the studs are not equidistant - they'll only fit through the holes in the shock tower one way). Once the

16.11b . . . and remove the strut upper mounting nuts (arrows)

studs protrude from the shock tower, install the nuts so the strut won't fall back through. This is most easily accomplished with the help of an assistant, as the strut is quite heavy and awkward.
14 Slide the steering knuckle into the strut flange and insert the two bolts. Install the nuts and tighten them securely.
15 Connect the brake hose bracket to the strut and tighten the bolt securely. If the vehicle is equipped with ABS, install the speed sensor wiring harness bracket.
16 Install the wheel and lug nuts, then lower the vehicle and tighten the lug nuts securely.
17 Tighten the upper mounting nuts securely.
18 Drive the vehicle to an alignment shop to have the wheel alignment checked, and if necessary, adjusted.

Rear suspension arms

19 Loosen the rear wheel lug nuts, raise the rear of the vehicle, support it securely on jackstands and remove the wheels.

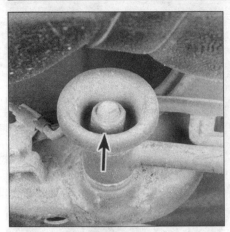

16.20a To disconnect the inner ends of a front lateral arm from the crossmember, remove this nut (arrow)

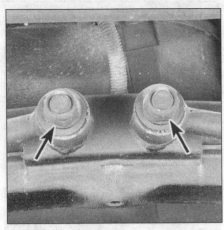

16.20b To disconnect the inner ends of the rear lateral arms, remove these nuts (arrows)

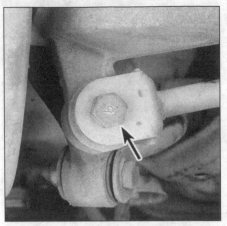

16.21 To disconnect the outer end of a lateral arm from the knuckle, remove the nut and bolt (arrow)

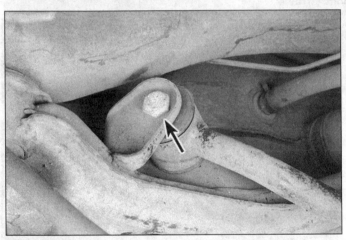

16.25 To disconnect the forward end of the trailing arm from the body, remove this nut and pivot bolt (arrow)

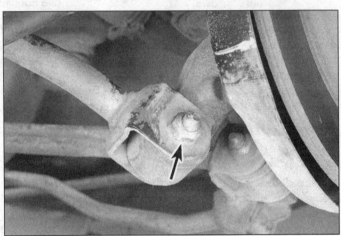

16.28 Remove the nut (arrow) and bolt that attach the rear end of the trailing arm to the knuckle

Lateral arms

Refer to illustrations 16.20a, 16.20b and 16.21

20   Remove the inner pivot nuts **(see illustrations)**.

21   Remove the bolts and nuts that attach the lateral arms to the knuckle **(see illustration)**.

22   Remove the arms.

23   Installation is the reverse of removal. Be sure to tighten all fasteners securely. **Note:** *Raise the rear suspension with a floor jack to simulate normal ride height before tightening the fasteners.*

24   When you're done, drive the vehicle to an alignment shop and have the rear-wheel toe adjusted.

Trailing arms

Refer to illustrations 16.25 and 16.26

25   Remove the pivot bolt and nut from the forward end of the trailing arm **(see illustration)**.

26   Remove the bolt and nut that attach the rear end of the arm to the knuckle **(see illustration)**.

27   Remove the trailing arm.

28   Installation is the reverse of removal. Place a floor jack under the rear knuckle and raise the suspension to simulate normal ride height, then tighten the fasteners securely.

All arms

29   Install the wheels and lug nuts, lower the vehicle and tighten the lug nuts securely.

30   When you're done, drive the vehicle to an alignment shop and have the rear-wheel toe adjusted.

17 Rear hub and bearing assembly - removal and installation

Non sealed bearings

This procedure is covered in Chapter 4 as part of *Wheel bearing check, repack and adjustment* procedure.

Sealed bearing

Refer to illustrations 17.3, 17.4 and 17.5

Note: *The rear hub and bearing are combined into a single assembly. The bearing is sealed for life and requires no lubrication or attention. If the bearing is worn or damaged, replace the entire hub and bearing assembly.*

1   Loosen the rear wheel lug nuts, raise the vehicle, place it securely on jackstands and remove the rear wheel.

2   Remove the brake drum or caliper and disc.

3   Remove the dust cover **(see illustration)**.

4   Unstake the hub retaining nut **(see illustration)**, unscrew the nut and remove the thrust washer then remove the hub assembly. **Caution:** *Some models use left hand threaded nuts.*

5   Install the new hub assembly and thrust washer, tighten the new nut securely, then stake its edge into the groove in the spindle **(see illustration)**.

17.3 Using a hammer and chisel, remove the dust cover

6   Install the dust cover by tapping lightly around the edge until it is seated.

7   The remainder of installation is the reverse of removal.

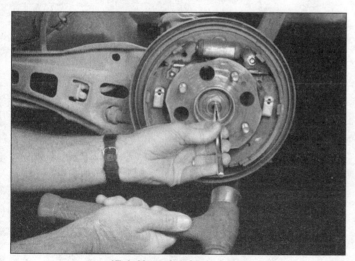

17.4 Unstake the hub nut

17.5 Stake the hub nut back into place

7 Steering systems

Contents

	Section
General information	1
Steering wheel - removal and installation	2
Power steering pump - removal and installation	3
Steering rack-and-pinion - removal and installation	4
Steering gearbox - removal and installation	5
Steering linkage - inspection, removal and installation	6
Steering knuckle - removal and installation	7

	Section
Hub and bearing assembly (front-wheel drive models) - removal and installation	8
Power steering system - bleeding	9
Wheels and tires - general information	10
Wheel alignment - general information	11
Airbag system - general information	12

1 General information

Refer to illustrations 1.1a, 1.1b, 1.1c and 1.1d

The two most common types of steering systems in use today are the gearbox system and the rack-and-pinion system **(see illustrations)**.

On gearbox steering systems, the gearbox transmits turning force through the steering linkage to the steering knuckles. The steering linkage consists of a Pitman arm, an idler arm, a center link (or drag link) and a pair of tie rods. Each tie-rod assembly consists of an inner tie rod, an adjuster sleeve and an outer tie-rod. Some models may be

1.1a Typical gearbox-type steering system components

1	*Steering gearbox*	*4*	*Idler arm*	*7*	*Outer tie-rods*
2	*Pitman arm*	*5*	*Inner tie-rods*	*8*	*Stabilizer bar*
3	*Center link*	*6*	*Tie-rod adjuster sleeves*		

1.1b Typical rack-and-pinion-type steering system components

1	Rack-and-pinion	4	Tie-rod jam nut	7	Power steering hydraulic lines
2	Rack-and-pinion boots	5	Tie-rod ends	8	Rack-and-pinion mounting
3	Inner tie-rods	6	Steering knuckle		bolts

equipped with a steering damper. The steering damper is located between the frame and the center link or drag link. It reduces unwanted "bump steer" (the slight turning or steering of a wheel away from its normal direction of travel as it moves through its suspension travel).

The rack-and-pinion unit is usually mounted to the fire-wall or to a lower crossmember **(see illustration)**. The rack-and-pinion actuates the tie-rods, which are attached to the steering knuckles. The inner ends of the tie-rods are protected by rubber boots which should be inspected periodically for secure attachment, tears and leaking lubricant.

The power assist system consists of a belt-driven hydraulic pump and associated lines and hoses **(see illustration)**. The fluid level in the power steering pump reservoir should be checked periodically (see Chapter 4).

The steering wheel operates the steering shaft, which actuates the gearbox or rack-and-pinion through universal joints. Looseness in the steering can be caused by wear in the steering shaft universal joints, the steering gear or rack-and-pinion, the tie-rod ends and loose retaining bolts.

The steering column, on most modern vehicles, is designed to collapse in the event of an accident. A small U-joint connects the steering shaft to the steering gearbox.

Frequently, when working on the suspension or steer-ing system components, you may come across fasteners which seem impossible to loosen. These fasteners on the underside of the vehicle are continually subjected to water, road grime, mud, etc., and can become rusted or "frozen," making them extremely difficult to remove. In order to unscrew these stubborn fasteners without damaging them (or other components), be sure to use lots of penetrating oil and allow it to soak in for a while. Using a wire brush to clean exposed threads will also ease removal of the nut or bolt and prevent damage to the threads. Sometimes a sharp blow with a hammer and punch will break the bond between a nut and bolt threads, but care must be taken to prevent the punch from slipping off the fastener and ruining the threads. Heating the stuck fastener and surrounding area with a torch sometimes helps too, but isn't recommended because of the obvious dangers associated with fire. Long breaker bars will increase leverage. Sometimes tightening the nut or bolt first will help to break it loose. Fasteners that require drastic measures to remove should always be replaced with new ones.

Since most of the procedures dealt with in this Chapter involve jacking up the vehicle and working underneath it, a good pair of jackstands will be needed. A hydraulic floor jack is the preferred type of jack to lift the vehicle, and it can

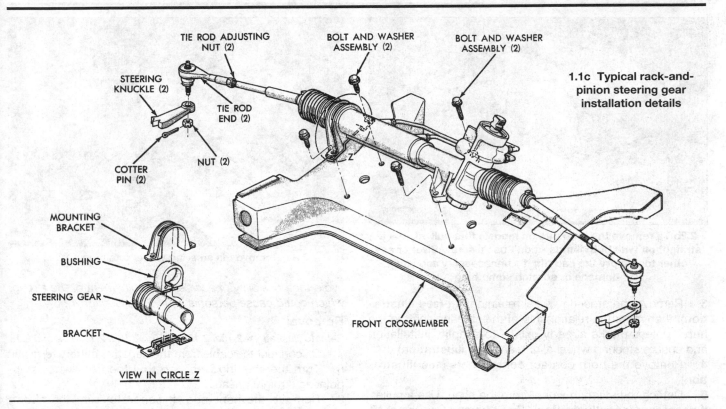

1.1c Typical rack-and-pinion steering gear installation details

VIEW IN CIRCLE Z

also be used to support certain components during various operations. **Warning:** *Never, under any circumstances, rely on a jack to support the vehicle while working on it. Whenever any of the suspension or steering fasteners are loosened or removed they must be inspected and, if necessary, replaced with new ones of the same part number or of original equipment quality and design. Never attempt to heat or straighten any suspension or steering components. Instead, replace any bent or damaged part with a new one.*

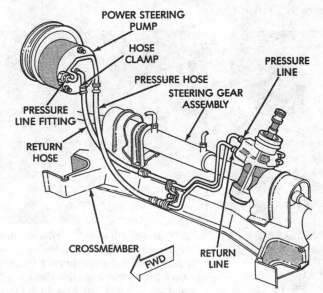

1.1d Typical power steering system components (rack-and-pinion type)

2 Steering wheel - removal and installation

Models without airbags

Removal

Refer to illustrations 2.2a, 2.2b, 2.3a, 2.3b, 2.4, 2.5 and 2.6

1   Disconnect the cable from the negative terminal of the battery. Make sure the front wheels are in the straight-ahead position.

2   Detach the horn pad from the steering wheel **(see illustrations)**.

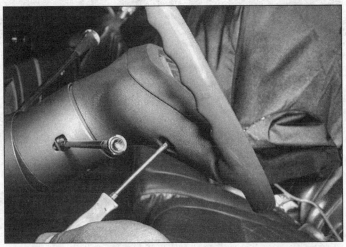

2.2a On models with the horn pad retained by screws, remove the screws from the back-side of the steering wheel

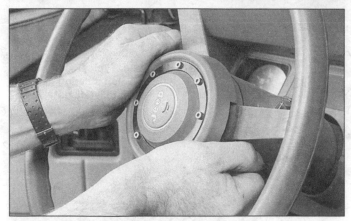

2.2b To remove the horn pad on models that pull off, pull it straight off with both hands - don't use a screwdriver or any other tool to pry the pad off; it's unnecessary and you'll damage or scratch something

2.3a Remove the steering wheel retaining nut . . .

3   Remove the steering wheel retaining nut (see illustration), then mark the relationship of the steering shaft to the hub - unless marks already exist - to simplify installation and ensure steering wheel alignment (see illustration).
4   Remove the horn contact components (see illustration).
5   Detach the steering wheel from the shaft. Use a puller, if necessary (see illustration). Don't hammer on the shaft to dislodge the steering wheel.

Installation
6   Installation is the reverse of removal (see illustration). When you slip the wheel onto the shaft, make sure you align the mark on the steering wheel hub with the mark on the shaft and tighten the nut securely.

Models with airbags
Warning: *On models equipped with airbags, always disable the airbag system before working in the vicinity of the impact sensors, steering column or instrument panel to*

avoid the possibility of accidental deployment of the airbag, which could cause personal injury (see Section 12).

Removal
Refer to illustrations 2.10a, 2.10b, 2.10c, 2.10d, 2.11, 2.12 and 2.13
7   Disconnect the cable from the negative battery terminal.
8   Turn the steering wheel so that the front wheels are pointing straight ahead.
9   Remove the horn pad, if separate from the airbag module.
10   Disarm the airbag system and locate the airbag module electrical connector. As a standard throughout the industry the airbag system wiring harness is covered in a yellow conduit and connectors are yellow. The airbag module connector may be located under a trim panel at the base of the steering wheel or steering column. Remove the airbag module from the steering wheel (see illustrations). **Warning:** *Handle the airbag module with care, carry the module with the trim cover side facing away from your body and store it in a safe location with the trim side facing up. See the precautions in Section 12.*

2.3b . . . then paint or scribe a mark (arrow) on the steering shaft and hub to ensure proper alignment during reassembly - remove the horn contact retaining screws (arrows)

2.4 Remove the horn contact components

2.5 You may be able to separate the steering wheel from the steering shaft by hand, but if the splines are frozen, remove the wheel from the shaft with a puller (available from your local auto parts store) - DO NOT HAMMER ON THE SHAFT!

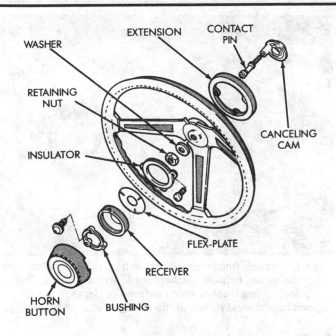

WASHER
EXTENSION
CONTACT PIN
RETAINING NUT
INSULATOR
CANCELING CAM
HORN BUTTON
BUSHING
RECEIVER
FLEX-PLATE

2.6 Exploded view of a typical steering wheel assembly

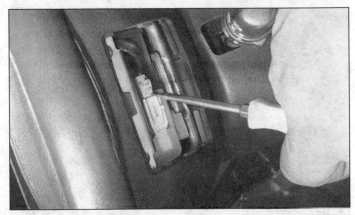

2.10a After disarming the airbag system, locate the airbag module electrical connector (usually a clearly marked yellow connector) and carefully separate the connector

11 Remove the steering wheel retaining nut, then mark the relationship of the steering wheel to the steering shaft **(see illustration)**.

12 Disconnect the electrical connector for the cruise control wiring harness, if equipped **(see illustration)**.

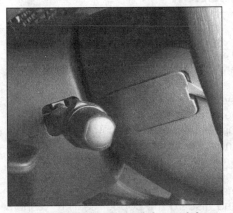

2.10b If the airbag module retaining screws are not clearly visible on the backside of the steering wheel, look for access covers and pry them off

2.10c Loosen and remove the airbag module retaining screws from each side of the steering wheel

2.10d Carefully lift the airbag module from the steering wheel

2.11 Remove the steering wheel retaining nut and apply match-marks on the steering wheel and shaft (arrows) so the steering wheel can be installed in it's original position

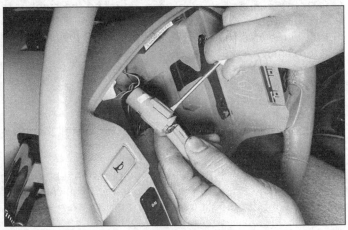

2.12 On models with the cruise control switch on the steering wheel, disconnect the electrical connector

2.13 Use a steering wheel puller to separate the steering wheel from the steering shaft

2.15 On models equipped with an airbag, note the alignment of the airbag module clockspring alignment marks. The clockspring must remain centered or damage to the clockspring and failure of the airbag system may occur

13 Use a steering wheel puller to separate the steering wheel from the steering shaft **(see illustration)**. When removing the wheel, make sure the electrical leads for the airbag module and the cruise control system don't snag on the wheel. **Warning:** *Do not turn the steering shaft while the steering wheel is removed.*

Installation

Refer to illustration 2.15

14 Verify that the front wheels are pointing straight ahead.

15 Pull the electrical leads for the airbag module and the cruise control system (if equipped) through the steering wheel and install the steering wheel. Make sure the clockspring is properly centered **(see illustration)**. As long as the clockspring is not separated from the steering shaft, it will remain centered. If for some reason you remove the clockspring, center the clockspring as follows:

a) *Turn the inner hub counterclockwise while holding the outer body until the inner hub reaches its stop. Do not force the hub against the stop.*

b) *Turn the inner hub clockwise, counting the number of turns required to reach the opposite stop.*

c) *Turn the inner hub clockwise exactly 1/2 the number of turns recorded above.*

d) *Align the alignment marks.*

e) *Install the clockspring.*

16 Install the steering wheel retaining nut.

17 Install the airbag module.

18 Plug in the airbag module connector. Install the lower cover.

19 Connect the negative battery cable.

20 Verify that the airbag circuit is operational by turning the ignition key to the On or Start position. The "AIR BAG" warning light should illuminate for about seven seconds, then turn off.

3 Power steering pump - removal and installation

Removal

Refer to illustrations 3.3, 3.5a, 3.5b and 3.6

1 Disconnect the cable from the negative battery terminal.

2 Using a large syringe or suction gun, remove as much fluid from the power steering fluid reservoir as possible. Place a drain pan under the vehicle to catch any fluid that spills out when the hoses are disconnected.

3 Loosen the clamp and disconnect the fluid return hose from the pump. Remove the pressure line-to-pump fitting or banjo bolt, then detach the line from the pump **(see illustration)**. Remove and discard the copper sealing washers (if equipped). They must be replaced when installing the pump.

3.3 Remove the fluid return hose (arrow) and the pressure line (arrow) from the power steering pump

3.5a If it's necessary to remove the power steering pump pulley to reach the mounting bolts, a special power steering pump puller is usually needed. Holding the puller with one wrench while tightening the center bolt pulls the pulley from the pump shaft

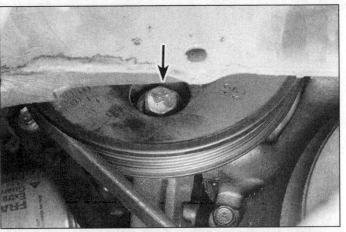

3.5b On some models holes will be provided in the pulley to access the power steering pump mounting bolts. Turn the pulley until the hole exposes one of the mounting bolts and use a socket to remove the bolt

4   Loosen the tension and remove the drivebelt.
5   On some models it may be necessary to remove the power steering pump pulley to access the mounting bolts **(see illustration)**. On other models, holes in the pulley may be provided to reach the mounting bolts **(see illustration)**.
6   Remove the pump mounting bolts **(see illustration)**, then remove the pump from the vehicle.

Installation

Refer to illustrations 3.7a, 3.7b and 3.8

7   If necessary, remove the pulley from the pump and install it on the replacement unit **(see illustrations)**.
8   Install the power steering pump. Install the pulley if it is

3.6 The power steering pump mounting bolts retain the power steering pump to the mounting bracket, on many models the bracket holes are elongated to allow for belt tension adjustment

3.7a If you're replacing the pump, remove the pulley using a special pulley removal tool . . .

3.7b . . . and install it on the replacement pump using a different special tool to press the pulley on the shaft

3.8 To improvise a pulley installation tool, use a long bolt with the same thread pitch as the internal threads of the power steering pump shaft, a nut, washer and a socket with the same diameter as the pulley hub

4.2 Disconnect the power steering pressure and return line fittings from the rack-and-pinion

of the type that must be installed after mounting the pump **(see illustration)**.

9   Adjust the drivebelt tension. Top up the fluid level in the reservoir and bleed the system.

4 Steering rack-and-pinion - removal and installation

Warning: *On models equipped with airbags, always disable the airbag system before working in the vicinity of the impact sensors, steering column or instrument panel to avoid the possibility of accidental deployment of the airbag, which could cause personal injury (see Section 12).*

Removal

Refer to illustrations 4.2, 4.3a, 4.3b, 4.4 and 4.5

1   Loosen the front wheel lug nuts, raise the front of the vehicle and support it securely on jackstands. Apply the parking brake and remove the wheels. Remove the engine splash shields.

2   Place a drain pan under the rack-and-pinion unit. Detach the power steering pressure and return lines **(see illustration)** and cap the ends to prevent excessive fluid loss and contamination.

3   Remove the universal joint cover **(see illustration)**. Mark the relationship of the lower universal joint to the steering gear input shaft **(see illustration)**. Remove the lower intermediate shaft pinch bolt.

4   Separate the tie-rod ends from the steering knuckle arms **(see illustration)**.

5   Support the rack-and-pinion and remove the mounting bolts **(see illustration)**. Separate the intermediate shaft from the rack-and-pinion input shaft and remove the rack-and-pinion assembly. **Warning:** *Do not allow the steering wheel to turn while the rack-and-pinion is removed on a model equipped with an airbag or damage to the airbag clockspring could occur resulting in airbag system failure and personal*

4.3a On most front-wheel drive models the steering column-to-rack-and-pinion universal joint is located inside the vehicle. Remove the nuts (arrows) or detach the clips retaining the two cover halves together and remove the cover

4.3b Mark the relationship of the universal joint to the rack-and-pinion input shaft and loosen the U-joint pinch bolt (arrow)

4.4 Separate the tie-rod end from the steering knuckle or strut using a small puller to press the balljoint stud out of the steering arm

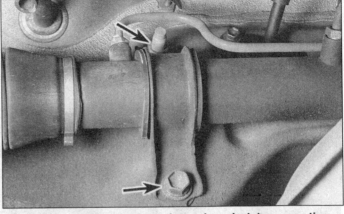

4.5 Remove the bolts from the rack-and-pinion mounting clamps (there are two clamps, one on each side)

injury. To prevent the steering wheel from turning, turn the ignition to the lock position and remove the key.

6   Check the rack-and-pinion rubber mounts for excessive wear or deterioration, replacing them if necessary.

Installation

7   Raise the rack-and-pinion into position and connect the U-joint, aligning the marks.

8   Install the mounting brackets and bolts.

9   Connect the tie-rod ends to the steering knuckle arms.

10   Install the U-joint pinch bolt.

11   If equipped with power steering, connect the power steering pressure and return hoses to the rack-and-pinion and fill the power steering pump reservoir with the recommended fluid.

12   Lower the vehicle and if equipped with power steering, bleed the steering system.

5 Steering gearbox - removal and installation

Warning: *On models equipped with airbags, always disable the airbag system before working in the vicinity of the impact sensors, steering column or instrument panel to avoid the possibility of accidental deployment of the airbag, which could cause personal injury (see Section 12).*

Removal

Refer to illustrations 5.2, 5.3, 5.4, 5.5 and 5.6

1   Place the front wheels in a straight-ahead position. Raise the front of the vehicle and support it securely on jackstands. Apply the parking brake.

2   Place a drain pan under the steering gearbox. Unscrew the fittings and detach the hoses **(see illustration)**, then cap the ends to prevent excessive fluid loss and contamination.

3   Remove the stone shield from the steering shaft-to-steering gearbox input shaft coupler **(see illustration)**.

4   Mark the relationship of the steering coupler to the

5.2 Disconnect the power steering fluid pressure and return lines from the gearbox. Use a flare-nut wrench on the fittings to prevent damage to the nuts

5.3 Most models use a stone shield to protect the steering coupler. Detach the clips or remove the nuts, separate the two halves and remove the stone shield

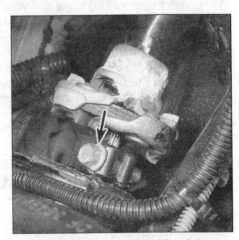

5.4 Mark the relationship of the steering coupler-to-steering gearbox input shaft, remove the pinch bolt and separate the coupler from the gearbox input shaft

5.5 Use a Pitman arm puller to separate the Pitman arm from the gearbox

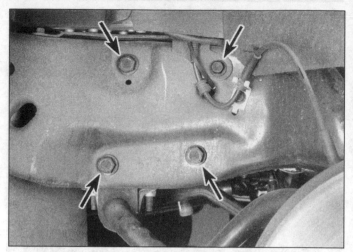

5.6 Remove the bolts retaining the gearbox to the frame

steering gearbox input shaft. Remove the steering coupler pinch bolt **(see illustration)**.

5   Detach the Pitman arm from the steering gearbox shaft **(see illustration)**.

6   Support the steering gearbox and remove the mounting bolts (see illustration). Lower the unit, separate the steering gearbox input shaft from the coupler and remove the steering gearbox from the vehicle. **Warning:** *Do not allow the steering wheel to turn while the steering gearbox is removed on a model equipped with an airbag or damage to the airbag clockspring could occur, resulting in airbag system failure and personal injury. To prevent the steering wheel from turning, turn the ignition to the lock position and remove the key.*

Installation

7   Raise the steering gearbox into position and connect the steering coupler to the steering gearbox input shaft. Be sure to match up the alignment marks.

8   Install the mounting bolts and washers.

9   Make sure the gearbox is centered and the front wheels are pointing straight ahead. Slide the Pitman arm onto the shaft. Install the washer and nut.

10   Install the steering coupler pinch bolt.

11   Connect the power steering hoses/lines to the steering gearbox and fill the power steering pump reservoir with the recommended fluid.

12   Lower the vehicle and bleed the steering system.

6   Steering linkage - inspection, removal and installation

Inspection

Refer to illustration 6.1

1   The steering linkage connects the steering gear to the front wheels and keeps the wheels in proper relation to each other **(see illustration)**. The linkage consists of a Pitman arm fastened to the steering gear shaft, which moves a

center link back and forth. The back-and-forth motion of the center link is transmitted to the steering knuckles through a pair of tie rods. Each tie-rod assembly consists of a pair of inner and outer tie rods connected by a threaded adjuster sleeve. An idler arm, connected between the center link and the frame reduces shimmy and unwanted forces to the steering gear. On some models, a steering damper may be mounted between the steering linkage and the frame to dampen unwanted oscillations in the steering linkage. Wheel toe adjustments are made at an adjuster sleeve on the tie-rod.

2   Set the wheels in the straight ahead position and lock the steering wheel.

3   Raise one side of the vehicle until the tire is approximately one inch off the ground.

4   Mount a dial indicator with the needle resting on the outside edge of the wheel. Grasp the front and rear of the tire and, using light pressure, wiggle the wheel back-and-forth and note the dial indicator reading. As a general rule, the gauge reading should be less than 0.125 (1/8-inch). If the play in the steering system is more than specified, inspect each steering linkage pivot point and ballstud for looseness and replace parts if necessary.

5   Raise the vehicle and support it on jackstands. Check for torn ballstud boots, frozen joints and bent or damaged linkage components.

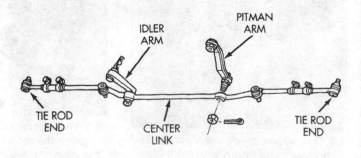

6.1 Typical steering linkage components

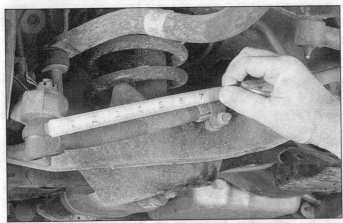

6.7a Before loosening the adjuster sleeve clamp and unscrewing the tie-rod end, measure the distance from the end of the adjuster sleeve to the center of the ballstud and record your measurement . . .

Removal and installation - gearbox type

6   Apply the parking brake. Loosen the front wheel lug nuts, raise the vehicle and support it securely on jackstands. Remove the front wheels.

Tie-rod

Refer to illustrations 6.7a, 6,7b, 6.8a, 6.8b, 6.9a, 6.9b, 6.9c, 6.12 and 6.14

Note: *This procedure covers replacing the tie-rod ends as well as the entire tie-rod. If you're only replacing a tie-rod end, ignore the Steps that don't apply. The procedure depicts a typical tie-rod end assembly, it applies to many similar models.*

7   If the inner or outer tie-rod end must be replaced, measure the distance between the adjuster sleeve and the center line of the ballstud or count the number of threads showing outside the adjuster sleeve (see illustrations). Record this measurement or number to maintain correct toe-in during reassembly.

8   Remove the cotter pin and loosen, but do not remove, the castle nut from the ballstud (see illustrations). If only

6.7b . . . or count the number of threads showing outside the adjuster sleeve

6.8a Remove the cotter pin from the tie-rod ballstud . . .

the outer tie-rod is to be replaced, loosen the outer nut; if only the inner tie-rod is to be replaced, loosen the inner nut; if the entire tie-rod is to be replaced, loosen both nuts.

9   If the outer tie-rod end or the entire tie-rod assembly is to be replaced, use a small puller to separate the outer tie-

6.8b . . . then loosen, but don't remove, the ballstud nut

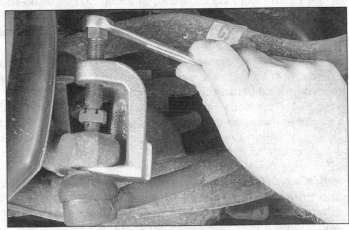

6.9a Install a small puller on the tie-rod end ballstud and separate the ballstud from the steering knuckle; leave the nut in place to prevent the parts from violently separating

6.9b To separate the inner tie-rod end from the center link, loosen (but do not remove) the nut . . .

6.9c . . . and press the ballstud out with a small puller

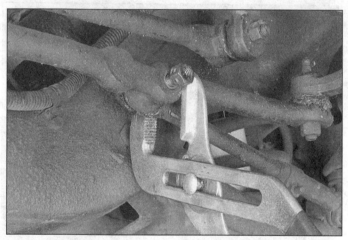

6.12 It may be necessary to force the ball stud into the tapered hole to keep it from turning while tightening the nut

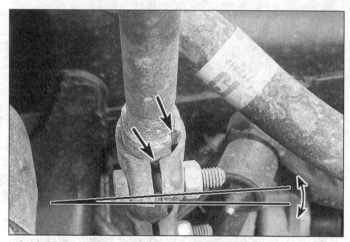

6.14 Make sure the adjuster sleeve clamp bolts are placed nearly horizontal and the slot in the adjuster sleeve is not aligned with the gaps in the clamps

rod end from the steering knuckle (see illustration), then remove the castellated nut and detach the outer tie-rod end ballstud from the knuckle. If the inner tie-rod end or the entire tie-rod is to be replaced, separate the inner tie-rod end from the center link (see illustrations).

10   Loosen the adjuster sleeve clamp bolts and unscrew the outer and or inner tie-rod ends.

11   Lubricate the threaded portion of the tie-rod end with chassis grease. Screw the new tie-rod end into the adjuster sleeve and adjust the distance from the sleeve to the ball-stud to the previously measured dimension. The number of threads showing on the inner and outer tie-rod ends should be equal within three threads. Don't tighten the adjuster sleeve clamp at this time.

12   Insert the tie-rod ballstud into the steering knuckle and/or center link. Make sure the ballstud is fully seated and tighten the nuts securely. If a ballstud spins when attempting to tighten the nut, force it into the tapered hole with a large pair of pliers (see illustration).

13   Continue tightening the nut slightly to align a slot in the nut with the hole in the ballstud - DO NOT loosen the nut to

align the slot with the hole. Install a new cotter pin.

14   Tighten the adjuster sleeve clamp nuts. The center of the adjuster sleeve clamp bolts should be nearly horizontal and the adjuster sleeve slot must not align with the gap in the clamps (see illustration).

15   Install the wheel and lug nuts, lower the vehicle and tighten the lug nuts. Drive the vehicle to an alignment shop to have the front end alignment checked and, if necessary, adjusted.

Center link/drag link

16   Loosen, but do not remove, the nuts securing the center link ballstuds to the tie-rod assemblies, the idler arm, the steering damper (if equipped) and the Pitman arm. Separate the ballstuds with a two-jaw puller, then remove the nuts.

17   If the tie-rod end is in need of replacement, refer to Steps 7 through 15.

18   Install the center link. Make sure the ballstuds are fully seated and tighten the nuts securely. If a ballstud spins when attempting to tighten the nut, force it into the tapered

6.19 Separate the center link from the Pitman arm with a small puller

6.20 Mark the relationship of the Pitman arm to the steering gearbox shaft

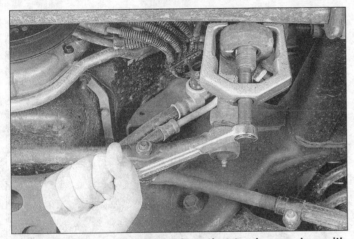

6.21 Separate the Pitman arm from the steering gearbox with a Pitman arm puller

6.24 To replace an idler arm, loosen the nut and use a small puller to separate the ballstud from the center link (arrow), then remove the mounting bolts from the frame (arrows) and remove the idler arm

hole with a large pair of pliers. If equipped with a castellated nut, continue tightening the nut slightly to align a slot in the nut with the hole in the ballstud - DO NOT loosen the nut to align the slot with the hole. Install a new cotter pin.

Pitman arm

Refer to illustrations 6.19, 6.20 and 6.21

Warning: *Do not allow the steering wheel to turn while the Pitman arm is disconnected from the steering gearbox on a model equipped with an airbag or damage to the airbag clockspring could occur resulting in airbag system failure and personal injury. To prevent the steering wheel from turning, turn the ignition to the lock position and remove the key.*

19 Loosen the Pitman arm-to-center link ballstud nut. Using a small puller, separate the center link from the Pitman arm **(see illustration)**.

20 Loosen the Pitman arm-to-steering gearbox nut and washer. Mark the relationship of the Pitman arm to the steering gearbox shaft to ensure proper alignment on reassembly **(see illustration)**.

21 Using a Pitman arm puller, separate the Pitman arm from the steering gearbox shaft **(see illustration)**.

22 Install the Pitman arm onto the steering gearbox. Most steering gearbox shafts and Pitman arms are keyed so the Pitman arm can only be installed one way. Regardless, make sure the front wheels are pointed straight ahead and the steering gearbox is centered when connecting the Pitman arm to the steering gearbox. Tighten the Pitman arm nut securely. Insert the ballstud into the center link. Make sure the ballstud is fully seated and tighten the nut securely. If a ballstud spins when attempting to tighten the nut, force it into the tapered hole with a large pair of pliers. If equipped with a castellated nut, continue tightening the nut slightly to align a slot in the nut with the hole in the ballstud - DO NOT loosen the nut to align the slot with the hole. Install a new cotter pin.

Idler arm

Refer to illustration 6.24

23 Remove the cotter pin and loosen the idler arm-to-center link ballstud nut. Separate the idler arm from the center link with a small puller.

24 Unbolt the idler arm from the frame **(see illustration)**.

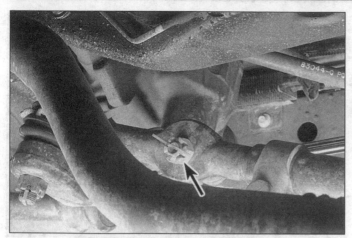

6.26 To replace the steering damper, remove the cotter pin (arrow), loosen the castle nut and use a small puller to separate the damper shaft from the center link . . .

6.27 . . . remove the nut and bolt from the frame bracket and remove the steering damper

6.30 Loosen the jam nut and mark the position of the tie-rod end in relation to the threads

6.31a Remove the cotter pin . . .

25 Install the idler arm to the frame and tighten the nuts/bolts securely. Insert the ballstud into the center link. Make sure the ballstud is fully seated and tighten the nut securely. If a ballstud spins when attempting to tighten the nut, force it into the tapered hole with a large pair of pliers. If equipped with a castellated nut, continue tightening the nut slightly to align a slot in the nut with the hole in the ball-stud - DO NOT loosen the nut to align the slot with the hole. Install a new cotter pin.

Steering damper

Refer to illustrations 6.26 and 6.27

26 To detach the steering damper from the center link, remove the cotter pin (if equipped) **(see illustration)**. Loosen the nut and use a small puller to separate the damper shaft from the center link.

27 To detach the damper from the frame bracket, remove the nut and bolt **(see illustration)**.

28 Install the damper onto the frame bracket and tighten nut/bolt securely. Insert the ball stud into the center link. Make sure the ballstud is fully seated and tighten the nut

securely. If a ballstud spins when attempting to tighten the nut, force it into the tapered hole with a large pair of pliers. If equipped with a castellated nut, continue tightening the nut slightly to align a slot in the nut with the hole in the ball-stud - DO NOT loosen the nut to align the slot with the hole. Install a new cotter pin.

Removal and installation - rack-and-pinion type

29 Loosen the wheel lug nuts. Raise the front of the vehicle, support it securely on jackstands, block the rear wheels and set the parking brake. Remove the front wheels.

Tie-rod ends

Refer to illustrations 6.30, 6.31a, 6.31b and 6.32

30 Loosen the jam nut enough to mark the position of the tie-rod end in relation to the threads **(see illustration)**.

31 Remove the cotter pin and loosen, but don't remove, the nut on the tie-rod end stud **(see illustrations)**.

32 Disconnect the tie-rod end from the steering knuckle arm with a puller **(see illustration)**. Remove the nut and separate the tie-rod.

6.31b . . . loosen, but don't remove, the tie-rod ballstud nut

6.32 Separate the tie-rod end from the steering knuckle arm using a small puller

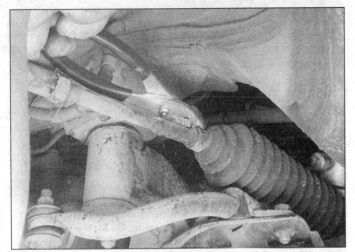

6.39a Remove the outer clamp from the rack-and-pinion boot

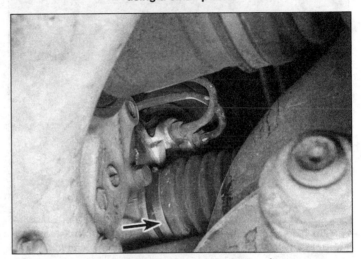

6.39b Cut off and discard the inner clamp

33 Unscrew the tie-rod end from the steering rod.

34 Thread the tie-rod end on to the marked position and insert the tie-rod stud into the steering knuckle arm. Tighten the jam nut securely.

35 Install the castle nut on the ballstud. Make sure the ballstud is fully seated and tighten the nut securely. If a ballstud spins when attempting to tighten the nut, force it into the tapered hole with a large pair of pliers. Continue tightening the nut slightly to align a slot in the nut with the hole in the ballstud - DO NOT loosen the nut to align the slot with the hole. Install a new cotter pin.

36 Install the wheel and lug nuts. Lower the vehicle and tighten the lug nuts.

37 Have the alignment checked by a dealer service department or an alignment shop.

Rack-and-pinion steering gear boots

Refer to illustrations 6.39a and 6.39b

38 Remove the tie-rod end and jam nut.

39 Remove the outer rack-and-pinion boot clamp (see illustration) with a pair of pliers. Cut off the inner boot clamp (see illustration) with a pair of diagonal cutters. Slide the boot off.

40 Before installing the new boot, wrap the threads on the end of the steering rod with a layer of tape so the small end of the new boot isn't damaged.

41 Slide the new boot into position on the rack-and-pinion until it seats in the groove in the steering rod and install new clamps.

42 Remove the tape and install the tie-rod end.

43 Install the wheel and lug nuts. Lower the vehicle and tighten the lug nuts.

7 Steering knuckle - removal and installation

Rear-wheel drive models

2WD models

Refer to illustrations 7.2, 7.4, 7.5a, 7.5b and 7.5c

1 Loosen the wheel lug nuts, raise the vehicle and sup-

7.2 Remove the mounting bolts (arrows) to detach the brake backing plate from the steering knuckle

7.4 Support the lower control arm with a jackstand or floor jack positioned under the lower control arm spring seat

7.5a To separate the upper control arm from the steering knuckle, remove the cotter pin and loosen the castle nut . . .

7.5b . . . use a puller or picklefork-type balljoint separator to separate the arm from the knuckle. Be aware that the picklefork may damage the balljoint boot - something you don't want to happen

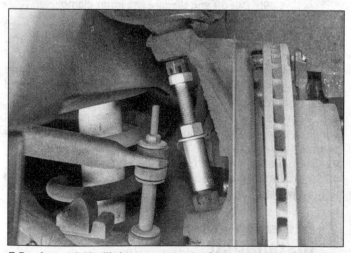

7.5c A special balljoint separator tool is available to press the balljoints out of the knuckle without damaging the balljoint boots. An alternative, such as the tool shown here, can be fabricated from a large bolt, nut, washer and deep socket - position the tool set-up as shown and turn the bolt to force the balljoint studs apart (be sure to leave the nuts loosely installed)

port it securely on jackstands. Remove the wheel.

2   Remove the disc brake caliper and brake disc. Use wire or a coat hanger to suspend the caliper aside. Remove the brake backing plate **(see illustration)**.

3   Disconnect the tie-rod from the steering knuckle (see Section 6).

4   Support the lower control arm with a jackstand or floor jack **(see illustration)**. The weight of the front of the vehicle should bear on the jackstand or floor jack to ensure the coil spring remains in position when the steering knuckle is removed. The jack must remain in this position throughout the entire procedure.

5   Separate the upper control arm from the steering knuckle **(see illustrations)**.

6   Separate the lower control arm from the steering knuckle and lift the knuckle off the balljoint studs. **Warning:** *DO NOT remove the jackstand or floorjack support from under the lower control arm with the steering knuckle removed or the lower control arm will be forced down by the coil spring with considerable force, possibly causing personal injury.*

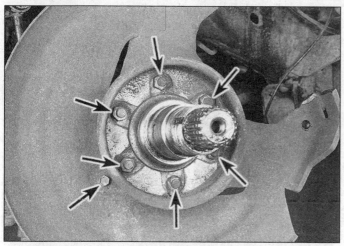

7.11a If equipped with a typical solid axle-type 4WD system, remove the nuts securing the spindle to the steering knuckle, remove the brake backing plate . . .

7.11b . . . and remove the spindle

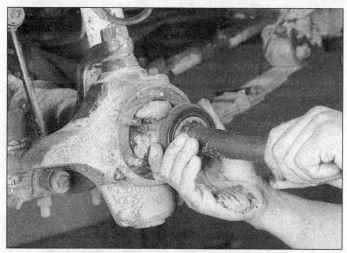

7.11c Pull the axleshaft straight out of the housing

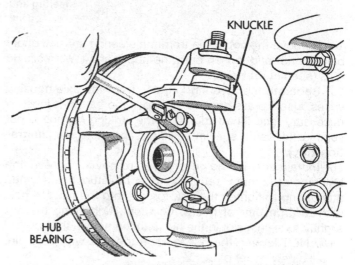

7.11d On some solid axle-type and most independent front suspension models, remove the bolts securing the hub/bearing assembly to the steering knuckle and remove the hub/bearing assembly

7   Carefully inspect the steering knuckle for cracks, especially around the steering arm and spindle mounting area. Also inspect the balljoint stud holes. If they're elongated, or if you find any cracks in the knuckle, replace the steering knuckle.

8   To install the steering knuckle, place it onto the lower balljoint stud and install the nut. Insert the upper balljoint stud and install the nut. Tighten the nuts securely. If a ball-stud spins when attempting to tighten the nut, force it into the tapered hole with a large pair of pliers. Continue tightening the nut slightly to align a slot in the nut with the hole in the ballstud - DO NOT loosen the nut to align the slot with the hole. Install a new cotter pin. The remainder of installation is the reverse of removal.

4WD models

Refer to illustrations 7.11a, 7.11b, 7.11c, 7.11d, 7.14, 7.15a and 7.15b

9   Loosen the wheel lug nuts, raise the vehicle and support it securely on jackstands. Remove the front wheels.

10   Remove the disc brake caliper and disc. Use wire or a coat hanger to suspend the caliper aside. Remove the brake backing plate.

11   Depending on the model and 4WD type, remove the spindle and axleshaft or hub/bearing assembly **(see illustrations)**. If equipped with an independent front suspension, it may not be necessary to remove the driveaxle.

12   Disconnect the tie rod or drag link from the steering knuckle (see Section 6).

13   If equipped with independent front suspension, support the lower control arm with a jackstand or floor jack **(see illustration 7.4)**. **Warning:** *DO NOT remove the jackstand or floorjack support from under the lower control arm during the entire procedure or the lower control arm will be forced down by the spring/torsion bar with considerable force, possibly causing personal injury.*

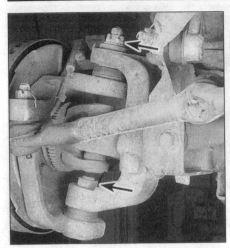

7.14 Remove the cotter pins from the ballstuds and loosen the nuts

7.15a After loosening the nuts, the balljoints may be separated from the knuckle by aggressively striking the steering knuckle in the area of the ballstud

7.15b Using a large hammer, strike the knuckle in this area to separate the lower ballstud

14   Remove the cotter pin from the steering knuckle upper ballstud nut and loosen both the upper and lower ballstud nuts **(see illustration)**.

15   Separate the steering knuckle from the axle housing yokes. Use a brass hammer to knock the ballstuds loose, if necessary **(see illustrations)**. If the striking method is not successful, use a special balljoint separator **(see illustration 7.5c)**.

16   Installation is the reverse of removal. Tighten the balljoint stud nuts securely. If a ballstud spins when attempting to tighten the nut, force it into the tapered hole with a large pair of pliers. Continue tightening the nut slightly to align a slot in the nut with the hole in the ballstud - DO NOT loosen the nut to align the slot with the hole. Install a new cotter pin.

Front-wheel drive models

Refer to illustrations 7.20, 7.22a, 7.22b, 7.22c, 7.22d and 7.23

Removal

17   Loosen the wheel lug nuts, raise the vehicle and sup-port it securely on jackstands. Remove the wheel.

18   Remove the brake caliper and the brake disc, and disconnect the brake hose from the strut. Using wire or a coat hanger, support the caliper aside.

19   Break the driveaxle loose from the hub and bearing assembly (see Section 8). Remove the hub and bearing assembly, if necessary.

20   Remove the strut-to-steering knuckle nuts, but don't remove the bolts yet **(see illustration)**.

21   Separate the tie-rod end from the steering knuckle arm (see Section 6).

22   Separate the balljoint from the steering knuckle **(see illustrations)**.

23   Push the driveaxle from the hub **(see illustration)**. Support the end of the driveaxle with a section of wire.

24   Remove the strut-to-knuckle bolts and separate the knuckle from the strut.

Installation

25   Guide the knuckle and hub assembly into position, inserting the driveaxle into the hub.

7.20 To remove the strut assembly from the steering knuckle, remove these two nuts and bolts (arrows)

7.22a Remove the cotter pin . . .

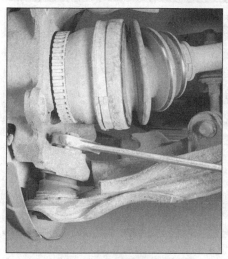

7.22b . . . loosen the balljoint stud nut and back it off several threads

7.22c Separate the balljoint from the steering knuckle with a picklefork-type separator - apply grease to the forks to prevent splitting the balljoint boot

7.22d Pry the lower control arm down to remove the ballstud from the knuckle

26 Push the knuckle into the strut flange and install the bolts and nuts, but don't tighten them yet.

27 Attach the control arm to the steering knuckle. Tighten the ballstud nut securely. If the ballstud spins when attempting to tighten the nut, force it into the tapered hole with a large pair of pliers. Continue tightening the nut slightly to align a slot in the nut with the hole in the ballstud - DO NOT loosen the nut to align the slot with the hole. Install a new cotter pin.

28 Attach the tie-rod end to the steering knuckle arm (see Section 6). Tighten the strut-to-knuckle nuts.

29 Place the brake disc on the hub and install the caliper.

30 Install the driveaxle/hub nut (see Section 8).

31 Install the wheel and tighten the lug nuts but don't tighten them yet.

32 Lower the vehicle and tighten the wheel lug nuts.

8 Hub and bearing assembly (front-wheel drive models) - removal and installation

Refer to illustrations 8.3a, 8.3b, 8.3c, 8.4a, 8.4b, 8.5 and 8.6

Note: *On some models the hub and bearings are pressed into the steering knuckle. The steering knuckle must be removed before the hub and bearings can be pressed out. Due to the special tools and expertise required to press the hub and bearing from the steering knuckle, this job should be left to a professional. However, the steering knuckle and hub may be removed and the assembly taken to an automotive repair facility or machine shop. See Section 7 for the steering knuckle and hub removal procedure.*

1 Loosen the front wheel lug nuts, raise the vehicle, support it securely on jackstands and remove the front wheels.

2 Remove the brake caliper and the brake disc. Using wire or a coat hanger, support the caliper aside.

3 Remove the cotter pin and the nut lock, if equipped **(see illustrations)**. Some models may be equipped with a

7.23 Pull out on the steering knuckle and detach the driveaxle from the hub

8.3a Using a pair of pliers or side cutters, remove the cotter pin . . .

8.3b . . . and the nut lock

8.3c If equipped with a staked nut, use a hammer and chisel to unstake the nut

8.4a Use a large prybar to hold the hub while loosening the driveaxle hub nut

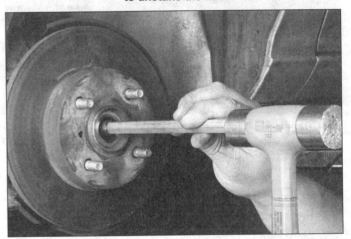

8.4b Using a brass drift or soft faced hammer, strike the end of the driveaxle sharply to break it free of the hub splines

self-locking nut or a nut that is staked to the axleshaft (see illustration).

4   Remove the driveaxle hub nut (see illustration). Using a brass drift or soft-faced hammer, lightly strike the end of

the driveaxle to loosen it from the hub splines (see illustration). If striking the driveaxle fails to break it loose a special tool designed especially for this task may be necessary.

5   Remove the bolts retaining the hub and bearing assem-

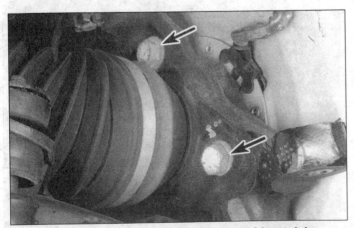

8.5 Remove the hub and bearing assembly retaining nuts/bolts (arrows) - there are typically three or four nuts or bolts and they're located on the backside of the knuckle or accessible from the front side through holes in the hub

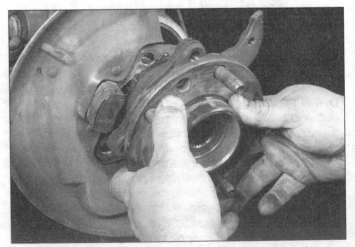

8.6 Remove the hub and bearing assembly from the steering knuckle

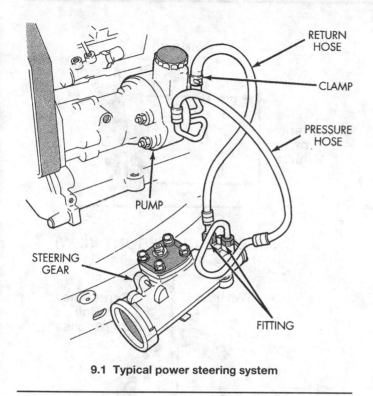

9.1 Typical power steering system

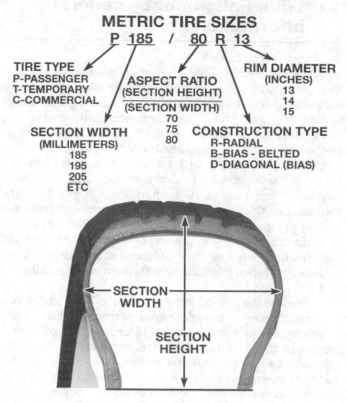

10.1 Metric tire size code

bly to the steering knuckle **(see illustration)**.

6   Remove the hub and bearing assembly **(see illustration)**.

7   Installation is the reverse of removal.

9 Power steering system - bleeding

Refer to illustration 9.1

1   Following any operation in which the power steering fluid lines have been disconnected, the power steering system must be bled to remove all air and obtain proper steering performance **(see illustration)**.

2   With the front wheels in the straight ahead position, check the power steering fluid level and, if low, add fluid until it reaches the Cold mark on the dipstick.

3   Start the engine and allow it to run at fast idle. Recheck the fluid level and add more if necessary to reach the Cold mark on the dipstick.

4   Bleed the system by turning the wheels from side to side, without hitting the stops. This will work the air out of the system. Keep the reservoir full of fluid as this is done.

5   When the air is worked out of the system, return the wheels to the straight ahead position and leave the vehicle running for several more minutes before shutting it off.

6   Road test the vehicle to be sure the steering system is functioning normally and noise free.

7   Recheck the fluid level to be sure it is up to the Hot mark on the dipstick while the engine is at normal operating temperature. Add fluid if necessary.

10 Wheels and tires - general information

Refer to illustration 10.1

1   Modern vehicles are equipped with metric-sized steel belted radial tires **(see illustration)**. Use of other size or type of tires on a vehicle designed for radial tires may affect the ride and handling of the vehicle. Don't mix different types of tires, such as radials and bias belted, on the same vehicle as handling may be seriously affected. It's recommended that tires be replaced in pairs on the same axle, but if only one tire is being replaced, be sure it's the same size, structure and tread design as the other.

2   Because tire pressure has a substantial effect on handling and wear, the pressure on all tires should be checked at least once a month or before any extended trips.

3   Wheels must be replaced if they are bent, dented, leak air, have elongated bolt holes, are heavily rusted, out of vertical symmetry or if the lug nuts won't stay tight. Wheel repairs that use welding or peening are not recommended.

4   Tire and wheel balance is important in the overall handling, braking and performance of the vehicle. Unbalanced wheels can adversely affect handling and ride characteristics as well as tire life. Whenever a tire is installed on a wheel, the tire and wheel should be balanced by a shop with the proper equipment.

11 Wheel alignment - general information

Refer to illustration 11.1

A wheel alignment refers to the adjustments made to the wheels so they are in proper angular relationship to the suspension and the ground. Wheels that are out of proper alignment not only affect vehicle control, but also increase tire wear. The front end angles normally measured are camber, caster and toe-in **(see illustration)**.

Getting the proper wheel alignment is a very exacting process, one in which complicated and expensive machines are necessary to perform the job properly. Because of this, you should have a technician with the proper equipment perform these tasks. We will, however, use this space to give you a basic idea of what is involved with a wheel alignment so you can better understand the process and deal intelligently with the shop that does the work.

Toe-in is the turning in of the wheels. The purpose of a toe specification is to ensure parallel rolling of the wheels. In a vehicle with zero toe-in, the distance between the front edges of the wheels will be the same as the distance between the rear edges of the wheels. The actual amount of toe-in is normally only a fraction of an inch. On the front end, toe-in is controlled by the tie-rod end position on the tie-rod. On the rear end, it's controlled by a cam on the inner end of the rear (number two) suspension arm. Incorrect toe-in will cause the tires to wear improperly by making them scrub against the road surface.

Camber is the tilting of the wheels from vertical when viewed from one end of the vehicle. When the wheels tilt out at the top, the camber is said to be positive (+). When the wheels tilt in at the top the camber is negative (-). The amount of tilt is measured in degrees from vertical and this measurement is called the camber angle. This angle affects the amount of tire tread which contacts the road and compensates for changes in the suspension geometry when the vehicle is cornering or traveling over an undulating surface.

Caster is the tilting of the front steering axis from the vertical. A tilt toward the rear is positive caster and a tilt toward the front is negative caster.

12 Airbag system - general information

Note: *Refer to your owners manual for specific information on the airbag system installed in your vehicle.*

Some models are equipped with a Supplemental Restraint System (SRS), more commonly known as an airbag. This system is designed to protect the driver, and in some models, the front seat passenger, from serious injury in the event of a head-on or frontal collision. It consists of an airbag module in the center of the steering wheel and the right side of the instrument panel and a sensing/diagnostic module mounted in the center of the vehicle.

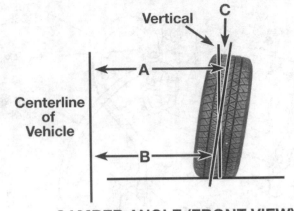

CAMBER ANGLE (FRONT VIEW)

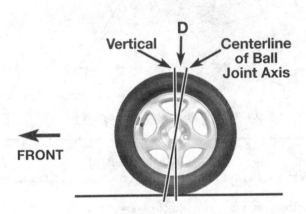

CASTER ANGLE (SIDE VIEW)

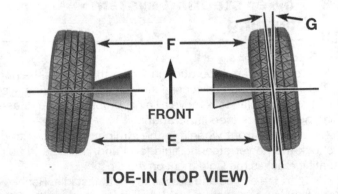

TOE-IN (TOP VIEW)

11.1 Camber, caster and toe-in angles

Airbag module

Steering wheel-mounted

The airbag inflator module contains a housing incorporating the cushion (airbag) and inflator unit, mounted in the center of the steering wheel The inflator assembly is mounted on the back of the housing over a hole through which gas is expelled, inflating the bag almost instantaneously when an electrical signal is sent from the system. A coil assembly on the steering column under the module carries this signal to the module.

This coil assembly can transmit an electrical signal regardless of steering wheel position. The igniter in the air bag converts the electrical signal to heat and ignites the sodium azide/copper oxide powder, producing nitrogen gas, which inflates the bag.

Instrument panel-mounted

The airbag is mounted above the glove compartment and is usually designated by the letters SRS (Supplemental Restraint System). It consists of an inflator containing an igniter, a bag assembly, a reaction housing and a trim cover.

The air bag is considerably larger that the steering wheel-mounted unit and is supported by the steel reaction housing. The trim cover is textured and painted to match the instrument panel and has a molded seam which splits when the bag inflates. As with the steering-wheel-mounted air bag, the igniter electrical signal converts to heat, converting sodium azide/iron oxide powder to nitrogen gas, inflating the bag.

Sensing and diagnostic module

The sensing and diagnostic module supplies the current to the airbag system in the event of the collision, even if battery power is cut off. It checks this system every time the vehicle is started, causing the "AIR BAG" light to go on then off, if the system is operating properly. If there is a fault in the system, the light will go on and stay on, flash, or the dash will make a beeping sound. If this happens, the vehicle should be taken to your dealer immediately for service.

Precautions

Warning: *Failure to follow these precautions could result in accidental deployment of the airbag and personal injury.*

Whenever working in the vicinity of the steering wheel, steering column or any of the other SRS system components, the system must be disarmed. To disarm the system:

a) *Point the wheels straight ahead and turn the key to the Lock position. Remove the key.*
b) *Disconnect the cable from the negative battery terminal.*
c) *Wait at least 5 minutes for the back-up power supply to be depleted.*

Whenever handling an airbag module, always keep the airbag opening pointed away from your body. Never place the airbag module on a bench of other surface with the airbag opening facing the surface. Always place the airbag module in a safe location with the airbag opening facing up.

Never measure the resistance of any SRS component. An ohmmeter has a built-in battery supply that could accidentally deploy the airbag.

Never use electrical welding equipment on a vehicle equipped with an airbag without first disconnecting the yellow airbag connector, usually located under the steering column near the combination switch connector (driver's airbag) and behind the glove box (passenger's airbag).

Never dispose of a live airbag module. Return it to your dealer for safe deployment, using special equipment, and disposal.

Notes

8 Modifications

How to improve your vehicle's handling

Introduction

While it is safe to say that all passenger vehicles on the road are equipped from the manufacturer with adequate steering and suspension systems that perform their designed functions well enough, there is always the driver who enjoys attacking remote canyon roads, putting himself and his vehicle to the ultimate test. Speed is craved by a certain few, and there will always be those who just can not resist modifying their vehicle. They must extract every ounce of performance from the original design as possible.

This vehicle has been modified with components specially tuned to provide positive suspension handling and a new low profile appearance (photo courtesy of Ground Force Inc.)

But it does not do any good to be able to attack that canyon road if your vehicle can not handle the curves the road throws at it. Improving your vehicle's handling to some extent can be accomplished by the relative simple, inexpensive replacement of parts. All-out modifications can be very costly, require some fabrication, render the vehicle unsuitable for everyday driving and should not be performed on a vehicle intended for street use.

Of course some vehicles will out perform others is stock form, but impressive gains can usually be expected from any vehicle by adding a sticky set of tires and performance shock absorbers, lowering the center of gravity with the proper components and either adding stabilizer bars, if not equipped, or increasing the size of the existing stabilizer bars. A modern performance vehicle, properly modified can easily obtain a cornering force of 0.9g, or better! Although our goal here is not to design an all-out race car, but only to describe how to modify your favorite vehicle to increase its cornering speed and overall braking and handling performance.

The key point to remember is this; use the proper components that are specifically tuned to work together in your vehicle. The proper set-up will produce an incredibly enjoyable driving experience without compromising the inherent ride qualities of your vehicle. On the other hand, the use of improper components, or shoddy workmanship, could render your vehicle unsafe to drive. Be very careful when replacing suspension components, removing control arms and replacing coil springs can be a dangerous job requiring specific tools and equipment to perform safely.

Lastly, if your vehicle is covered by the manufacturers new vehicle warranty or an aftermarket service contract, modifications to the vehicle may void the warranty. This is another good reason to use quality aftermarket products, a reputable company will warranty their components - some as long as the lifetime of the original owner.

Tires

The tire is the connection between your vehicle and the road. It is the most significant item effecting overall traction. Improvement of the tires ability to grip the road is the key to improving your vehicle's handling and the following modifications listed in this section will strive to increase the tire's grip on the road surface. If the vehicle is not equipped with a suitable set of tires, many of the improvement gains of the suspension system will not be fully realized.

Tire technology is a continually improving process, what may be the greatest tire on the road today, will surely be replaced by a better tire or new technology. The invention of the steel belted radial-ply tire started a revolution in the tire industry. The performance gains of the radial tire as compared to the bias-ply tire, are so significant that bias-

The wide stance and low profile of a high-performance tire increases handling performance by allowing more rubber to contact the road

ply tires are now obsolete on modern street driven vehicles. High-performance tires are available from a wide variety of manufacturers and in a wide price range. Not unlike most components in the high-performance industry, the higher the level of performance in a given tire - the higher the price.

What constitutes a modern high-performance tire?

The tread of a high-performance tire is manufactured with a soft rubber compound and is the single most contributing factor in improving the tire's traction capabilities. The soft rubber compound allows the tread to mold around the minute irregularities of the road surface, creating gripping forces not obtainable in a standard tire with harder compounds. The trade off is tread wear; the softer compounds will wear faster as more friction is generated at the road surface.

The tire size and series, or aspect ratio, also denotes a tire as being a high-performance tire. If you compare the tire size of a high-performance tire with an economy or

Most high-performance tires are designed with a unique tread pattern and asymmetric shoulders to combine a quiet running tire with excellent handling characteristics in wet or dry weather conditions

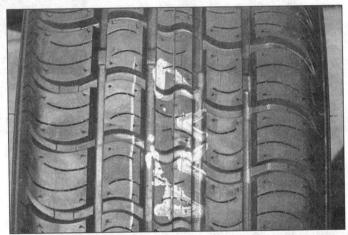

The wide grooves and cross-slots of this tread pattern will improve wet weather traction and resist hydroplaning by allowing the water to flow through the tread to the sides of the tire

The latest ultra high-performance tread design incorporates large tread blocks in a V shaped pattern and are designed to rotate in a specific direction

A uni-directional tire will have an arrow on the sidewall indicating the recommended tire rotational direction

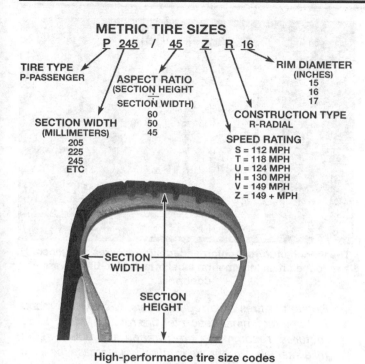

METRIC TIRE SIZES

P 245 / 45 Z R 16

TIRE TYPE
P-PASSENGER

ASPECT RATIO
(SECTION HEIGHT ÷ SECTION WIDTH)
60
50
45

SECTION WIDTH
(MILLIMETERS)
205
225
245
ETC

RIM DIAMETER
(INCHES)
15
16
17

CONSTRUCTION TYPE
R-RADIAL

SPEED RATING
S = 112 MPH
T = 118 MPH
U = 124 MPH
H = 130 MPH
V = 149 MPH
Z = 149 + MPH

High-performance tire size codes

TREADWEAR 100 200 300	THE TREADWEAR GRADE IS A COMPARATIVE RATING BASED ON THE WEAR RATE OF THE TIRE WHEN TESTED UNDER CONTROLLED CONDITIONS-BASED ON 100 (A TIRE WITH A 200 RATING WOULD WEAR TWICE AS WELL)
TRACTION A B C	THE TRACTION GRADE REPRESENTS THE TIRES ABILITY TO STOP ON WET PAVEMENT UNDER CONTROLLED CONDITIONS WITH **A** BEING THE BEST
TEMPERATURE A B C	THE TEMPERATURE GRADE REPRESENTS THE TIRES RESISTANCE TO THE GENERATION OF HEAT AND ITS ABILITY TO DISSIPATE HEAT WHEN TESTED UNDER CONTROLLED CONDITIONS WITH **A** BEING THE HIGHEST

Explanation of the Federal Uniform Tire Quality Grading codes found on the tire sidewall

Before mounting a set of wide tires, determine the available clearance between the wheelwells and suspension components

high-mileage tire, you will notice a big difference in size between the two. The high performance tire is wider and has a lower profile than a standard tire. The wider tread increases the tire's "contact patch" (actual rubber-to-road surface contact area). Increasing the contact patch has a direct relationship with traction. A larger contact patch will increase traction proportionally, simply because there is more rubber available to grip the road. A high-performance tire has a low profile or short section height (distance between the rim bead and the tread). The sidewalls flex less in a low profile tire, making the tire more responsive to steering input.

Choosing the correct tire for your vehicle

Choosing the correct tire for your vehicle will depend largely on three things; the intended use of the vehicle, the level of performance desired and the price of the tire. If the vehicle is driven daily, the wear rating of the tire will be an important consideration; if price is not an object, you may wish to reach for the highest level of performance obtainable. Generally, a balance between performance, wear and value is best in most cases. Tread design must also be considered. Ultra high-performance tires, such as those used in racing, are designed to grip dry pavement and are not suitable for wet conditions. Unless you plan on never taking your vehicle out in the rain or snow you will probably settle on an all-weather tread design. Some of the latest unidirectional tread designs (tread that is designed to rotate in a specific direction) offer an excellent compromise between high-performance handling and wet traction. Talk to a local tire dealer about your needs, they are experienced in these matters and may offer several options for your consideration.

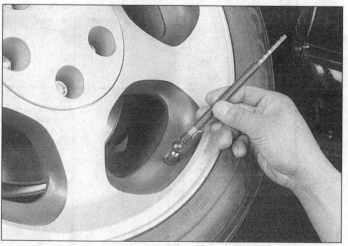

To extend the life of the tires, check the air pressure at least once a week with an accurate gauge

The statement made earlier, "increasing the contact patch increases traction," may lead one to believe in the "bigger-is-better" theory. But the truth is, the tire must be balanced with the total vehicle package. As tire size increases, so does rolling resistance, vehicle performance will suffer if the tire is too large. Also, the advantages of the wider tire might not be realized unless the suspension is able to keep the contact patch flat on the road. If the engine has enough horsepower to handle a larger tire and the suspension has been modified to improve handling, then an increase in tire size may be justified. Tire-to-wheel well clearance may also be a factor in tire choice. The tire must fit in the wheel well without rubbing during suspension travel or steering input or damage to the tire or vehicle bodywork could occur.

Tire maintenance

A quality set of high-performance tires is a sizable investment. Properly maintaining the set will insure a longer lifespan and maximum performance. Following these simple guidelines will allow your tire to reach its highest potential:

1) *Check the tire pressures regularly - Maintain the tire at the manufactures recommended air pressure. An under inflated tire will wear excessively on the shoulders, while an over inflated tire will wear in the center of the tread. Operating the vehicle with a severely under inflated tires can cause sidewall damage, not to mention overheating which could lead to a blow-out. It is imperative that the tire be at the correct pressure for maximum performance.*

The manufacturers recommended air pressures are generally found on an information label attached to the drivers door post

2) *Rotate the tires on a regular basis - Follow the manufactures recommendation for tire rotation. Most manufacturers recommend tire rotation at 5,000 to 6,000 mile intervals. Proper tire rotation will increase the life of the set of tires, especially if the front or rear tires tend to wear at a faster rate.*

3) *Keep the wheels aligned and balanced - Wheels that are out-of-alignment will wear the tires quickly. Excessive positive camber wears the tire on the outside edge. Excessive negative camber wears the tire on the inside edge. An incorrect toe setting causes the edges of the tire tread to scrub against the road surface and acceler-*

Condition	Probable cause	Corrective action	Condition	Probable cause	Corrective action
Shoulder wear	• Underinflation (both sides wear) • Incorrect wheel camber (one side wear) • Hard cornering • Lack of rotation	• Measure and adjust pressure. • Repair or replace axle and suspension parts. • Reduce speed. • Rotate tires.	Feathered edge Toe wear	• Incorrect toe	• Adjust toe-in.
Center wear	• Overinflation • Lack of rotation	• Measure and adjust pressure. • Rotate tires.	Uneven wear	• Incorrect camber or caster • Malfunctioning suspension • Unbalanced wheel • Out-of-round brake drum • Lack of rotation	• Repair or replace axle and suspension parts. • Repair or replace suspension parts. • Balance or replace. • Turn or replace. • Rotate tires.

This chart will help determine the condition of a vehicle's tires, as well as the possible cause of improper wear patterns

A set of lightweight alloy wheels, properly designed fit your vehicle, will enhance the performance of the tires and suspension system

When selecting custom wheels, measure the wheel offset to determine if they are compatible with your vehicle

ated wear. An out-of-balance tire causes the tread to wear unevenly and may be evident by a vibration at highway speeds. If any abnormal wear is noticed, have the alignment and balance checked immediately by a properly equipped facility.

4) Inspect the tires for damage and punctures - Periodically inspect the tire sidewalls for cuts or damage. Check the tire around its total circumference for imbedded objects, such as nails or screws, in the tire tread. Repair or replace the tire immediately if damage is found.

Tire repair

Punctured tires may be repaired using a variety of methods, if the puncture is not too big. Most punctures are caused by nails, screws and similar sized metal objects and create only a small hole in the tire tread. If the tire has been cut, replacement will most likely be required. For a puncture to meet the repair criteria it must be in the tire tread area. No attempt to repair punctures in the sidewall area should be made, the sidewall is far to flexible to repair and any attempt will surely leak.

Plugging small punctures is the most common repair practice in use by the average tire outlet. In this method a small rubber "plug" is inserted at the point of puncture using a special tool and rubber adhesive as a lubricate. A far superior method of repair involves removing and demounting the tire from the rim and repairing the puncture from the inside with a radial tire patch and the proper adhesive. Tires used in a high performance application should be repaired by patching or replaced.

Wheels

How about mounting those sticky high-performance tires on a new set of lightweight alloy wheels? If your are lucky enough to afford this luxury, the choices in wheels are astounding. But don't just chose a set of wheels that look good, chose a set that will enhance the performance

aspects of your suspension and tires. Wide tires need wide wheels and lightweight alloy wheels reduce unsprung weight, which in turn increases suspension responsiveness.

Most modern performance vehicles are equipped with a decent set of alloy wheels from the factory - but they may be improved upon. If your vehicle is riding on standard steel wheels, then upgrading to a set of wide alloy wheels is a must if installing a set of wide high-performance tires. Every tire manufacturer lists an approved wheel width for each tire size, for maximum performance, use the maximum wheel width allowable. The wider wheel will allow more tire tread to contact the road surface and increase traction.

Some research should be done before the final purchase to ensure an intelligent decision is made. The wider tire/wheel combination must fit inside the wheelwell without interference. Make some careful measurements to determine how much clearance is available with your existing tires, then calculate the added width of the new set-up into the formula to determine if the new wider tire will rub on the steering linkage, suspension components or inner fender panel. Even more importantly the correct wheel offset must be maintained to ensure the tires don't rub, the steering geometry remains the same and no extra load is placed on the wheel bearings. Consult the tire and wheel manufacturers or the experts at your local tire store for advice, if necessary, and purchase a set of tires that fit your wheelwells and have the correct wheel offset.

Shock absorbers

Installing high-performance shock absorbers is probably the most common suspension performance upgrade. Replacing standard shock absorbers with a high-performance set increases handling responsiveness and weight transfer. The most popular performance shock is the gas-pressurized shock absorber. Gas-pressurized shock absorbers dissipate heat faster than a standard shock and they resist oil foaming.

For best overall performance the shock absorber must be compatible with the spring. If a shorter and stiffer spring

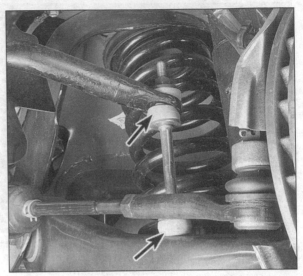

High-performance shock absorbers, such as these Ground Force nitrogen gas charged shocks, are specially valved to maintain tire-to-road contact by reducing aeration. A precision matched shock absorber is necessary in a lowered vehicle to maintain positive suspension control

The front stabilizer bar is connected to the suspension arm by bushings (arrows)

has been installed, then stiffer shock absorbers with less travel should be installed as well.

Stabilizer bars

A stabilizer bar is a type of torsion bar that resists the side-to-side movement, or body roll, of the suspended body. Excessive body roll has a large effect on vehicle handling. As a vehicle enters a corner centrifugal force causes the weight of the body to roll to the outside, this in turn causes a portion of the tires to lose contact with the road.

Many vehicles are equipped with stabilizer bars in production. The stock bars do an adequate job in resisting

Notice how the stabilizer bar on this high-performance vehicle, connects both suspension arms with the vehicle frame/body - the stabilizer bar resists body roll by twisting, much like a spring

| 1 | Front stabilizer bar | 2 | Stabilizer bar-to-frame bushing clamp | 3 | Stabilizer bar-to-lower control arm link |

A vehicle designed for performance and handling is equipped with a stout stabilizer bar (arrow) on the rear suspension as well as the front

body roll and are designed to work in conjunction with the stock suspension. If an increase in performance is desired, a stabilizer bar with a higher roll resistance may be installed. Usually this is accomplished by installing a replacement stabilizer bar with a larger diameter than original. A high-performance vehicle or sports car may be equipped with substantially-sized stabilizer bars and would probably not benefit from a change unless the total suspension system is upgraded.

Stabilizer bars may be installed at the front and rear of the vehicle, but care must be taken to maintain proper balance as excessive roll resistance at the front or rear will cause understeer or oversteer accordingly. Excessively stiff bars could have a negative effect on handling as well, causing the inside rear wheel to unload during cornering. Purchase a set of stabilizer bars from a reputable performance suspension company designed specifically for your vehicle. Consult their representative to discuss your options, if necessary.

Polyurethane bushings

The vehicle manufacturer's standard suspension bushing material of choice is rubber. Their goals are to create a soft ride and isolate noise. Rubber bushings conform to suspension component movement, absorbing vibration and shock. While all is well and good when they are new, as rubber bushings wear they allow the alignment settings to change as the bushings distort under torque and load.

Cracked suspension bushings and leaking shock absorbers (arrows) indicate this vehicle would be a good candidate for a suspension upgrade

The material of choice in a high-performance application is polyurethane. Polyurethane bushings lack of compliance transmits movement of the suspension components without distortion, which in turn, increases responsiveness. On the negative side, the polyurethane bushing material is harder than rubber, so they may exhibit an increase in ride harshness and possibly a small increase in vibration and noise.

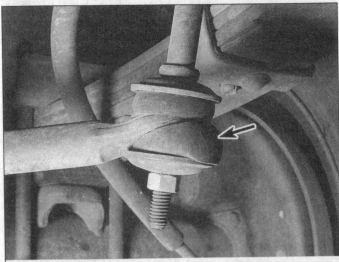

Polyurethane stabilizer bar bushings, such as these Prothane bushings, allow the stabilizer bar to react quicker to suspension movement. Replacing your original rubber stabilizer bar bushings with polyurethane bushings will decrease unwanted body roll

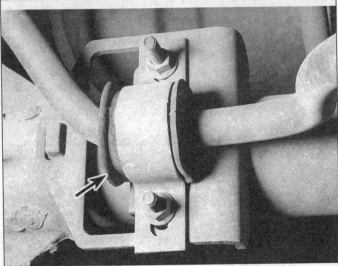

Notice how these rubber stabilizer bar bushings are deformed

On a street driven vehicle it may be impractical to replace all the suspension bushings with polyurethane. The stabilizer bar, however is one area where significant gains can be made by changing bushings. Installing polyurethane stabilizer bar bushings will effectively increase the stabilizer bar rate and they are simple to install.

Replacing the original rubber control arm bushings with polyurethane control arm bushings reduce bushing compliance and increase suspension responsiveness

An often overlooked, but necessary, component of the suspension system is the bump stop. Replacing your worn-out rubber bump stops with polyurethane bump stops increases suspension travel and is a necessary modification when installing lowering springs

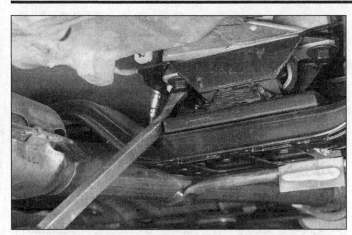

A separated transmission mount can cause excessive driveline movement, to check the transmission mount, pry between the crossmember and the mount - there should be very little movement

A polyurethane transmission mount reduces excessive movement in the driveline

Springs

The most significant gain in vehicle handling can be accomplished by lowering the vehicle's center of gravity with a balanced set of performance springs. Lowering the center of gravity with a shorter, stiffer spring reduces weight transfer and body roll. The reduced body roll allows more of the tire tread to remain in contact with the road surface during cornering, therefore the vehicle can obtain a higher cornering speed.

Installing a shorter, stiffer set of springs, in most cases, will have some negative effects as well. The stiffer spring will increase ride harshness and when a vehicle is lowered, ground clearance is reduced. Some part of the vehicle undercarriage or bodywork may contact the pavement, especially entering driveways. The vehicle suspension may bottom against the bump stops over some relatively minor bumps. The steering linkage may also be effected and require modification to restore the original steering performance.

Springs, and suspension, are a very important vehicle component. Many hours are spent by the manufacturer designing the suspension system of a vehicle to create a balance of good handling and comfortable ride characteristics. Much thought and research should be done before deciding to modify your vehicle's suspension, making a poor decision here could mean disaster. Do not cut coils or

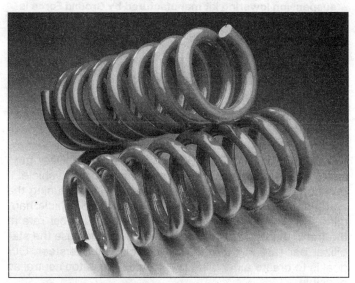

Precision engineered coil springs, such as these Ground Force coils, are designed to lower the vehicle a specific amount while still maintaining good ride characteristics. As a further benefit, they are either powder coated or PCV coated to prevent corrosion

remove leafs from your existing springs. Instead purchase a balanced set of springs for a reputable company - they too have invested many hours designing their product to give the best ride quality while offering an increased level of performance.

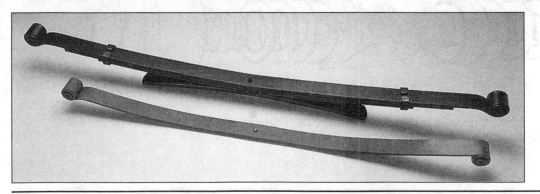

Precision engineered leaf springs are available to properly lower the rear of a vehicle with a leaf spring type suspension. They are available in both multi-leaf or mono-leaf designs

Assembling a critically balanced sport-tuned suspension system isn't something you can accomplish by chance. This suspension lowering kit manufactured by Ground Force is precision matched to give your vehicle the effect you're looking for

Complete suspension kits

For a suspension to work properly the components must be designed to work together. A properly balanced front and rear system must work together and not oppose one another or the handling characteristics of the vehicle will be thrown off. Changing springs and stabilizer bars have a huge effect on the overall balance of the vehicle. If one component in a system is changed without tuning the complete system, it has a negative effect on vehicle handling. For example; if we increase the stabilizer bar rate at the front, the vehicle will understeer, if we increase the stabilizer bar rate at the rear, the vehicle will oversteer. Our goal is to create a handling package as close too neutral as possible.

One way to ensure a balanced system is to install one of the complete and "tuned" suspension systems offered by a performance suspension company. Typically a complete system will consist of springs, shocks, stabilizer bars and polyurethane bushings. The components of the kit are designed to work together to achieve the correct balance. Accompanied by a good set of tires, the handling improvement of a complete suspension upgrade is incredible.

Limited-slip differential

The inherent design of a standard differential causes it to deliver most of the power to the tire with the least amount of traction. A limited-slip differential attempts to deliver power equally to both driving tires in a low traction situation. Having a limited-slip differential installed is most beneficial in an off-road driven vehicle, where maximum traction on slippery or loose surfaces is a huge advantage. Straight-line acceleration, such as that found in drag racing, also improves greatly with the addition of a limited-slip differential.

Installing a limited-slip differential in your street driven vehicle will improve its overall traction characteristics. Traction under less than desirable conditions will improve, making your vehicle safer to drive in rain, snow or ice. In a cornering situation, traction will improve coming out of the corner under hard acceleration when a limited-slip differential has been installed.

If you have recently purchase a vehicle and you are unsure if it's equipped with a limited-slip differential there are a couple of easy ways to determine the type presently installed. Block the front wheels, raise the rear of the vehicle and support it securely on jackstands. With the transmission in neutral and the parking brake released, rotate one wheel while watching the other, if the other wheel spins in the opposite direction the differential is "open" and not a limited-slip type. If both wheels spin the same direction, have an assistant attempt to hold the opposite wheel while you rotate one wheel, if both wheels are locked together the differential is a limited-slip type. A further inspection can be made by removing the rear cover (if equipped) and inspecting the differential unit. If clutch discs or cones are apparent in the area of the side gears, the differential is a limited-slip type.

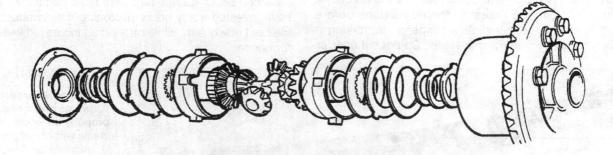

Exploded view of a typical limited slip differential

To determine if your vehicle is equipped with a limited slip differential, remove the rear cover and inspect the area between the side gears (arrow), if clutches, springs and/or cones are visible it's a limited slip unit

A aerodynamic spoiler adds mainly to the appearance of a street driven vehicle, but may create some downforce at high speeds

Aerodynamics

No discussion of vehicle handling would be complete without a discussion of aerodynamics. The aerodynamic down force created by wings, spoilers and ground effect bodyworks contributes to the incredible handling characteristics of an all-out racing vehicle. On a street driven vehicle, spoilers and ground effects are mainly cosmetic improvements, although they may provide some downforce at high speed (over 60 mph).

The addition of front air dams and rear spoilers are a very popular option on today's production vehicles, especially those equipped with some sort of "sport package." The front airdam reduces airflow under the vehicle. Substantial airflow under a vehicle at high speed creates lift, reducing the load on the tires. If the lift is great enough, the vehicle may feel unstable and certainly would be unsafe to operate at high speed. A well-designed rear spoiler may create some downforce on the rear tires. Any increase in downforce increases traction. If you choose to add aerodynamic components to your vehicle, buy only high quality components that fit well and are designed by the manufacturer to improve handling.

Off-road suspension modifications

Introduction

When driving off-road, your main objective will be attempting to maintain traction under all conditions. The original factory tires and suspension system will perform adequately under most conditions, but will not perform as well as a properly modified vehicle when the "going gets rough." Even though some vehicles may be equipped with a factory off-road package, the design engineers probably feel these vehicles will be operated on the pavement far more than off-road. Therefore the suspension system is designed primarily for good highway handling with improved off-road performance, as limitations allow.

An improvement in Off-road performance can easily be obtained by installing a set of off-road performance gas-charged shock absorbers and upgrading the tires to a slightly larger diameter with a more aggressive tread design. For those seeking a higher level of off-road performance more extensive modifications may be desired.

A properly modified 4X4 vehicle sports a slightly taller stance, aggressive tires and matched wheels

Tires

Preparing your vehicle to meet the challenges of off-road driving will surely include a new set of tires. There are three basic areas to consider when selecting tires for on/off-road use; tire construction, tire size and tread design.

Tire construction

Radial tire construction is by far the most popular type of construction with bias-type available for serious off-road duty. Radial tires are superior tires for highway duty and will probably be the most popular choice. Radial tires offer excellent traction both on and off road, the only drawback for off-road use is the soft, flexible sidewall - it's more prone to damage from rocks. Bias-ply and bias-belted tires are becoming hard to locate in today's market because of the popularity of radial tires. Bias tires are generally reserved for serious off-road duty as they offer the maximum level of durability.

Look for the designation "LT" on the side of the tire, it's incorporated with the tire size. The LT designations means

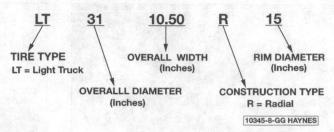

LT	31	10.50	R	15
TIRE TYPE LT = Light Truck	OVERALLL DIAMETER (Inches)	OVERALL WIDTH (Inches)	CONSTRUCTION TYPE R = Radial	RIM DIAMETER (Inches)

10345-8-GG HAYNES

Light truck tire size codes

it's a tire designed for Light Truck use. A tire designed for passenger car use may not stand up to the rigors of off-road driving.

Tire size

Tire size is a very controversial issue. Some succumb to the "bigger is better" theory, but in reality tire size must be balanced with the suspension, steering and driveline systems of the individual vehicle. The two most important specification to look at when selecting tires, is the overall diameter and width of the tire. Increasing tire diameter over the stock diameter, increases vehicle ground clearance, which is a plus in an off-road situation. If you're satisfied with an increase of one or two tire sizes and the chosen tire fits in your wheel well without rubbing, all is well, but if you're intent on installing a much larger tire then many factors must be considered. Most full-size pick-ups and utility vehicles will accept a 31 to 33 inch diameter tire with no, or minor, modifications. A tire larger than 35 inches will, in most cases, require extensive modifications to the steering, suspension, braking and driveline systems. In reality a 31 to 33 inch diameter tire will perform admirably in most off-road situations. Huge tires with a diameter of 36 to 44 inches should be left to specialty, off-road only or competition vehicles.

Tread design

The number one factor to consider when choosing a tread design is the intended use of the vehicle. The different tread designs available for light truck and utility vehicles

An all-terrain type tread pattern is designed to offer a compromise in good off-road traction while maintaining highway manners and a quiet ride

The large blocks of this tread pattern are designed for exceptional traction when off-road . . .

. . . and the tread pattern is directional, as indicated by the arrows on the sidewall

Large off-road tires require a wheel specifically designed for light trucks and utility vehicles

Adjustable shock absorbers (as indicated by the knob near the lower mounting ring) allow multiple settings for compression/rebound damping control. They can be adjusted on the vehicle from Firm, for off-road, to Soft, for highway use

range from a passenger car type tread to a serious off-road mud and snow type tread incorporating large tread blocks and open space. Obviously, if you plan to operate the vehicle primarily on the highway then a standard passenger car type tread design will do. If maximum traction for serious off-road driving is desired, then opt for a very aggressive tread design. The drawback to an aggressive tread design is its inherent highway manners, the large tread blocks of an aggressive off-road tire are very noisy and bumpy when driven on the highway. A compromise, and probably the most popular light truck tire sold, is the all-terrain type tread design. All terrain tires are available from all major tire manufacturers. The tread is designed to give decent traction off-road while maintaining good highway performance and low noise levels.

Wheels

There are a number of factors to consider when selecting new wheels for off-road use. Primarily, the wheel must be designed to fit the vehicle properly. They must have the proper bolt pattern and offset. Refer to the wheel manufacturers catalog to ensure the wheel in question is properly designed for your vehicle. Secondly, the wheel must be the proper width to accept the tire mounted to it. Again, every tire manufacturer publishes the approved wheel width in their tire brochure, so make sure the wheel falls into the recommended width range.

Lightweight alloy wheels look great and reduce unsprung weight (which increases suspension responsiveness) but are not mandatory for off-road use on all models. Rugged steel wheels have long been available for light trucks and utility vehicles and are the choice of many serious off-roaders who prefer the durability of steel. Generally speaking, if your vehicle is equipped with independent front suspension and/or state-of-the art four-link coil sprung suspension it will benefit from lightweight alloy wheels. While a heavy-duty vehicle equipped with solid axles may not reap the benefits of lightweight wheels.

Shock absorbers

Good quality shock absorbers, designed specifically for off-road use are available from a variety of manufacturers. Gas-charged shocks have become the standard for replacement shocks in most all applications and the same holds true in the off-road arena. Compare shock specifications between several manufacturers. The twin-tube shock with a piston bore diameter of 1-3/8 inch and a 5/8 inch chrome hardened piston rod is the most common performance shock. Other areas to look at are; reserve tube size, piston seal type and valving. Most performance shocks are equipped with urethane shock bushings, but you may have to purchase the shock boot extra.

Suspension lift-kits

Warning: *Excessively lifting a vehicle affects vehicle stability as well as safe handling and braking characteristics. Before*

The extended length of this coil spring raises the vehicle frame off the suspension providing the lift - note the extra long shock absorber has been relocated to provide added suspension travel

If possible, use a lift kit including a complete spring pack instead of riser blocks or add-a-leafs

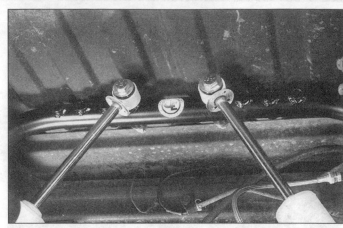

The rear shock mounts on this Bronco have been relocated to improve ride quality and handling

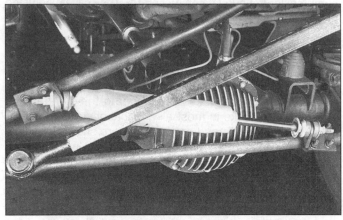

When lifting a vehicle it may be necessary to replace some steering components with longer components to compensate for the added lift - the steering stabilizer shock is designed to restore control and resist front end shimmy due to the added stress of larger tires on the steering linkage

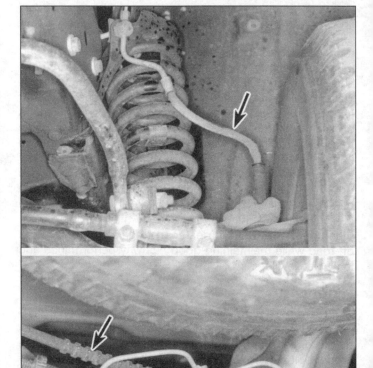

It may be necessary to replace the brake hoses at each front wheel and the rear axle with longer hoses - check this important item before lifting a vehicle

considering a suspension lift, check your state regulations regarding lifted vehicles. Most states have current laws in effect regulating vehicle height and oversized tires.

Modifying a stock suspension system has an affect on steering geometry, center of gravity, roll center and braking performance. With so many variables and safety issues, it makes sense to use only a well engineered, quality suspension lift kit available from one of the major aftermarket manufacturers. A complete kit will include all the necessary components to properly lift the vehicle, as well as other necessary components such as longer brake hoses, a dropped Pitman arm, extended stabilizer bar brackets, bump stops and longer shock absorbers.

Make sure the kit addresses driveline length, pinion angle and driveaxle CV joint angles on independent front axle models. Consult the kit manufacturer or an independent four-wheel drive shop before selecting a kit to ensure the vehicle can be properly modified in all areas to restore the original suspension travel, steering geometry and braking performance.

Whichever lift kit you chose, it is recommended you lift your vehicle only the amount necessary for adequate tire clearance and retain full suspension travel. Use common sense, safe methods and proper tools when installing the kit or have the kit installed by a professional. Perform any other necessary modifications (steering, braking, etc.) to ensure your vehicle will be road worthy and safe to drive.

Load capacity and trailer towing modifications

Introduction

Most vehicles, pick-up trucks especially, are designed to carry or tow a specific amount of weight. When that weight is exceeded, modifications to "beef-up" the rear suspension may be necessary. The component or system you select to increase the load carrying or trailer towing capacity of your vehicle will ultimately depend on the amount of weight to be carried or towed. Since most suspension modifications tend to stiffen the spring rate don't go overboard in this area. Stiffer spring rates mean the vehicle ride will be much harsher when unloaded. Choose a system that will meet your demands but not cause the ride quality to become unbearable.

Tires/wheels

When choosing a set of tires and wheels for a vehicle destined to carry a heavy load or pull a trailer, the rated load capacity of the tire must be taken into consideration. All light truck tires have the maximum load rating per tire, at the recommended air pressure, molded into the sidewall of the tire. You'll also find the load range letter designation (A, B, C, D, etc.) and the ply rating on the sidewall. All of this information is available to help you determine how much weight can safely be loaded on a given tire.

The first thing you must do is determine the actual weight resting on each axle of your vehicle. To do this, you'll have to locate a public scales, look in the telephone book yellow pages or contact a freight office or moving/storage facility to locate one in your area. Drive the vehicle, fully loaded, onto the scales stopping just before the rear tires reach the scales and record the weight on the front axle, then drive the front wheels just off the scales and

Large load-carrying vehicles require special wheels and tires

record the weight on the rear axle (its important you perform this operation with the vehicle fully loaded, just as you would normally use it). Divide the axle weight by two and you have the actual weight on each tire. You can now chose, from the tire manufacturers chart, a tire with a maximum load capacity greater than the actual weight that will be placed on the tire at any given time.

As with any other application, when choosing new wheels, chose the correct wheels for the job at hand. If the wheels are to be installed on a heavy laden vehicle, make sure the manufacturer guarantees they are rated for the increased weight. Check the wheel offset (see previous Section) to ensure there will be no interference and make sure the wheel is the correct width for the tire.

Shock absorbers

Coil-over-shocks

A coil-over-shock, or "load-leveler shock" as they're

Coil-over-shock absorbers can be installed to increase the load-carrying capabilities of a vehicle

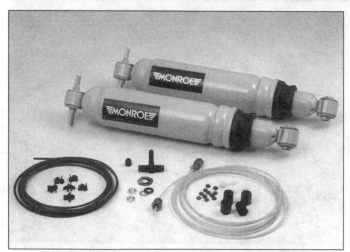

Air shock kits include the air shocks and all the necessary tubing and fittings to install them

Air bags are installed between the axle and the frame rail - when inflated with air they support a portion of the vehicle's weight, supplementing the springs

sometimes called, is a conventional shock absorber with a coil spring attached the shock body. The coil spring increases the shock rate, resulting in an increased load carrying capability. Coil-over-shocks can be installed on most vehicles in place of the original shock absorber. They are a great addition to passenger cars and station wagons carrying extra weight or towing a small trailer. They will raise the ride height and cause some degradation of ride quality in an unloaded vehicle.

Air shocks

An air shock is a conventional oil filled shock absorber with an internal air chamber. A fitting on the side of the air shock is connected to a Schrader valve fitting with small diameter tubing (usually plastic). The Schrader valve is then mounted in a convenient location which will allow it to be reached with an air hose. Pressurizing the air shock with compressed air increases the shock rate allowing an increased load to be placed on the vehicle suspension.

Air shocks are installed in place of the original shock absorbers, making their installation rather straight forward. More attention should be placed on installing and routing the air lines. Routing the air lines away from moving suspension components and hot exhaust pipes will ensure many years of leak-free service.

When compared to other suspension modifications, air shocks should only be considered for a light-to-medium duty application. But if that's all you need, they are great for adding a little extra support to the rear suspension of your passenger car or mini-van when pulling a small to medium trailer such as a boat or personal watercraft. The beauty of air shocks is they can be fine-tuned by varying the air pressure in the shock. The vehicle can be operated with the air pressure set at the minimum level until the need arises to increase the pressure.

Airbags

An air bag is a large, heavy duty rubber bladder with a Schrader valve fitting which allows it to be pressurized with

compressed air in the same manner as a tire. They are designed to be installed in the rear suspension in a location which will allow the weight of the vehicle to rest on the airbag. When inflated with compressed air, the air bag provides additional support to the rear springs, thereby increasing the load-carrying capacity of the vehicle. By varying the air pressure inside the bag, an airbag can be fine-tuned to your specific needs.

There are typically two different applications for an air bag, one used with a leaf spring type suspension and one used with a coil spring type suspension. On a leaf spring type, the air bag is mounted on top of the axle housing or spring mount, the frame rail rests on top of the air bag. On a coil spring type suspension, the air bag is placed inside the coil spring. The air pressure in the air bags may be adjusted to meet your needs and when inflated to their maximum pressure are able to support a tremendous amount of weight.

An overload spring leaf (arrow) is incorporated into the existing spring pack to increase the spring rate

Overload springs

Overload springs, "add-a-leaf" or "helper springs", as they're sometimes called, are an additional spring leaf designed to be incorporated into the existing rear leaf spring pack of a specific vehicle. Adding the additional leaf to the rear spring pack increases the spring rate and the load carrying capacity of the vehicle. Generally they are easy to install and are available for pick-up trucks and passenger cars equipped with rear leaf springs. Once installed the overload spring becomes a permanent part of the suspension system, unless removed. Therefore the overload spring is generally installed to compensate for an increase in load such as a large tool box or the addition of a camper shell in a light duty pick-up.

Sources

California Tire
1180 Newbury Rd.
Newbury Park, CA 91320
Phone: 805-499-4811
Retail tire outlet

Cross Enterprises
3833 Old Conejo Rd.
Newbury Park, CA 91320
Phone: 805-499-8169
Suspension components and accessories for early model Ford Bronco

Garry Moore, Distributor
2610 Ave. de Las Plantas
Thousand Oaks, CA 91360
Phone: 805-492-2146
MAC Tool Distributor

Ground Force Inc.
P.O. Box 149
Mount Braddock, PA 15465
Phone: 724-430-2068
www.groundforce.com
Suspension lowering kits, springs, spindles, shock absorbers and related components

Lisle Corp.
807 E. Main St.
Clarinda, IA 51632
Phone: 712-542-5101
Specialty automotive tools and equipment

Prothane Inc.
3560 Cadillac Ave.
Costa Mesa, CA 92626
Phone: 714-979-4990
Polyurethane bushings, bump stops, spring pads, motor and transmission mounts

Specialty Products Co.
4045 Specialty Place
Longmont, CO 80502
Phone: 800-525-6505
Wheel alignment tools, adjusters and accessories

Wheel alignment specifications

ACURA

Integra

1986 thru 1989
  Front
    Caster... 2.20-degree(s)
    Camber .... -0.50-degree(s)
    Toe-in .... -3/64-inch
  Rear
    Camber .... -0.75-degree(s)
    Toe-in 3/32-inch

1990 thru 1993
  Front
    Caster... 1.50-degree(s)
    Camber .... 0-degree(s)
    Toe-in 0-inch
  Rear
    Camber .... -0.66-degree(s)
    Toe-in 3/32-inch

1994 thru 1996
  Front
    Caster... 1.16-degree(s)
    Camber .... -0.16-degree(s)
    Toe-in 0-inch
  Rear
    Caster... -0.75-degree(s)
    Camber .... 0.08-degree(s)
    Toe-in 5/32-inch

Legend

1986 thru 1990
  Front
    Caster... 1.66-degree(s)
    Camber .... 0-degree(s)
    Toe-in 0-inch
  Rear
    Camber .... 0-degree(s)
    Toe-in 3/32-inch

1991 thru 1995
  Front
    Caster... 3.75-degree(s)
    Camber .... 0-degree(s)
    Toe-in 0-inch
  Rear
    Caster... -0.33-degree(s)
    Camber .... -0.08-degree(s)
    Toe-in 5/32-inch

Vigor

1992 thru 1994
  Front
    Caster... 1.66-degree(s)
    Camber .... 0-degree(s)
    Toe-in 0-inch
  Rear
    Caster... -0.50-degree(s)
    Camber .... 0.12-degree(s)
    Toe-in 1/4-inch

2.5 TL

1995 and 1996
  Front
    Caster... 2.10-degree(s)
    Camber .... 0-degree(s)
    Toe-in 0-inch
  Rear
    Caster... -0.50-degree(s)
    Camber .... 0.08-degree(s)
    Toe-in 5/32-inch

3.2 TL

1996
  Front
    Caster... 3.66-degree(s)
    Camber .... 0-degree(s)
    Toe-in .... -3/64-inch
  Rear
    Caster... -0.50-degree(s)
    Camber .... 0.12-degree(s)
    Toe-in 1/4-inch

AMERICAN MOTORS

Alliance and Encore

1983 thru 1987
  Front
    Caster... 1.25-degree(s)
    Camber .... 0.06-degree(s)
    Toe-in .... -1/8-inch
  Rear
    Camber .... -0.75-degree(s)
    Toe-in 0-inch

Concord

1983
  Front
    Caster... 4.50-degree(s)
    Camber
      Left wheel .... 0.38-degree(s)
      Right wheel .... 0.12-degree(s)
    Toe-in 1/8-inch

Eagle

1983
  Front
    Caster... 4.00-degree(s)
    Camber .... 0.38-degree(s)
    Toe-in 1/8-inch

1984 thru 1988
  Front
    Caster... 2.50-degree(s)
    Camber .... 0.38-degree(s)
    Toe-in 1/8-inch

GTA

1987
  Front
    Caster... 1.25-degree(s)
    Camber .... -1.33-degree(s)
    Toe-in .... -1/8-inch
  Rear
    Camber .... -1.38-degree(s)
    Toe-in 3/16-inch

Medallion

1988 and 1988
  Front
    Caster... 1.00-degree(s)
    Camber .... 0.50-degree(s)
    Toe-in 3/32-inch

Rear
    Camber .... -0.70-degree(s)
    Toe-in .... -19/64-inch

Spirit
1983
Front
    Caster .... 4.50-degree(s)
    Camber
        Left wheel .... 0.38-degree(s)
        Right Wheel .... 0.12-degree(s)
    Toe-in 1/8-inch

AUDI

4000
1983 thru 1987 (except Quatro)
Front
    Caster .... 0.50-degree(s)
    Camber .... -0.70-degree(s)
    Toe-in 5/64-inch
Rear
    Camber .... -1.00-degree(s)
    Toe-in 5/32-inch
1984 thru 1987 (4000S Quattro)
Front
    Caster .... 1.25-degree(s)
    Camber .... -0.75-degree(s)
    Toe-in 3/64-inch
Rear (up to Chassis No. FA095223)
    Camber .... -0.50-degree(s)
    Toe-in 5/32-inch
Rear (trom Chassis No. FA095224)
    Camber .... -1.00-degree(s)
    Toe-in 5/32-inch

5000
1983
Front
    Caster .... 0.70-degree(s)
    Camber .... -0.50-degree(s)
    Toe-in 0-inch
Rear
    Camber .... -0.50-degree(s)
    Toe-in 1/8-inch
1984 thru 1988 (except Quattro)
Front
    Caster .... 1.00-degree(s)
    Camber .... -0.50-degree(s)
    Toe-in 0-inch
Rear (up to VIN EA085288, EA082448)
    Camber .... -0.80-degree(s)
    Toe-in 1/8-inch
Rear (from VIN EA085289, EA082449)
    Camber .... -0.80-degree(s)
    Toe-in 5/64-inch
1983 thru 1986 Quattro
Front
    Caster .... 1.50-degree(s)
    Camber .... -0.80-degree(s)
    Toe-in 0-inch
Rear
    Camber .... -0.50-degree(s)
    Toe-in 0-inch

1986 thru 1988 (5000CS Quattro)
Front
    Caster .... 1.00-degree(s)
    Camber .... -0.50-degree(s)
    Toe-in 3/64-inch
Rear
    Camber .... -0.25-degree(s)
    Toe-in 5/64-inch

80 and 90
1988 thru 1992 (except Quattro)
Front
    Caster .... 1.25-degree(s)
    Camber .... -0.75-degree(s)
    Toe-in 5/64-inch
Rear
    Camber .... -1.00-degree(s)
    Toe-in 5/32-inch
1988 thru 1992 Quattro sedan (except DOHC)
Front
    Caster .... 1.25-degree(s)
    Camber .... -0.80-degree(s)
    Toe-in 5/64-inch
Rear
    Camber .... -0.75-degree(s)
    Toe-in 5/32-inch
1990 and 1991 Quattro DOHC
Front
    Caster .... 1.44-degree(s)
    Camber .... -0.80-degree(s)
    Toe-in 5/64-inch
Rear
    Camber .... -0.75-degree(s)
    Toe-in 5/32-inch
1990 and 1991 Quattro coupe
Front
    Caster .... 2.25-degree(s)
    Camber .... -0.80-degree(s)
    Toe-in 5/64-inch
Rear
    Camber .... -0.75-degree(s)
    Toe-in .... 0.16-degree(s)
1993 thru 1995 (except Quattro)
Front
    Caster .... 2.16-degree(s)
    Camber .... -0.75-degree(s)
    Toe-in 5/64-inch
Rear
    Camber .... -1.50-degree(s)
    Toe-in 1/4-inch
1992 thru 1995 Quattro
Front
    Caster .... 2.16-degree(s)
    Camber .... -0.75-degree(s)
    Toe-in 5/64-inch
Rear
    Camber .... -0.66-degree(s)
    Toe-in 5/64-inch

100, 200 and A6
1989 thru 1991 (2WD)
Front
    Caster .... 1.00-degree(s)
    Camber .... -0.50-degree(s)
    Toe-in 0-inch
Rear (up to VIN 44JN201030)
    Camber .... -0.70-degree(s)
    Toe-in 5/32-inch

Rear (from VIN 44JN201031)
- Camber .... -0.80-degree(s)
- Toe-in 5/32-inch

1989 thru 1991 Quattro (except DOHC)
Front
- Caster...................................... 1.00-degree(s)
- Camber .... -0.50-degree(s)
- Toe-in 3/64-inch

Rear
- Camber .... -0.25-degree(s)
- Toe-in 5/64-inch

1990 and 1991 Quattro DOHC
Front
- Caster...................................... 1.25-degree(s)
- Camber .... -0.50-degree(s)
- Toe-in .... 5/64 inch

Rear
- Camber .... -0.25-degree(s)
- Toe-in 3/64-inch

1990 thru 1994 Quattro V8
Front
- Caster...................................... 1.25-degree(s)
- Camber .... -0.50-degree(s)
- Toe-in 5/64-inch

Rear
- Camber .... -0.25-degree(s)
- Toe-in 5/64-inch

1992 thru 1996 (except Quattro, S4 and S6)
Manual transmission
Front
- Caster...................................... 1.16-degree(s)
- Camber .... -0.80-degree(s)
- Toe-in 1/0-inch

Automatic transmission
Front
- Caster...................................... 1.16-degree(s)
- Camber .... -0.66-degree(s)
- Toe-in 1/8-inch

Rear
- Camber .... -0.80-degree(s)
- Toe-in 5/32-inch

1992 thru 1996 Quattro
Front
- Caster...................................... -0.90-degree(s)
- Camber .... -0.66-degree(s)
- Toe-in 1/8-inch

Rear
- Camber .... -0.25-degree(s)
- Toe-in 5/64-inch

1992 thru 1995 S4 and S6
Front
- Caster...................................... 0.90-degree(s)
- Camber .... -0.80-degree(s)
- Toe-in 1/8-inch

Rear
- Camber .... -0.25-degree(s)
- Toe-in 5/64-inch

Cabriolet
1994 thru 1996
Front
- Caster...................................... 2.16-degree(s)
- Camber .... -0.75-degree(s)
- Toe-in 5/64-inch

Rear
- Camber .... -1.00-degree(s)
- Toe-in 5/32-inch

BMW

3-series
1983 320i
Front
- Caster...................................... 8.90-degree(s)
- Camber .... 0-degree(s)
- Toe-in 1/16-inch

Rear
- Camber .... -2.33-degree(s)
- Toe-in 5/32-inch

1983 thru 1991 (all others, except 325iX and M3)
Front
- Caster...................................... 8.50-degree(s)
- Camber .... -0.65-degree(s)
- Toe-in 5/32-inch

Rear
- Camber .... -2.00-degree(s)
- Toe-in 5/32-inch

1988 thru 1991
325iX
Front
- Caster...................................... 1.55-degree(s)
- Camber .... -1.00-degree(s)
- Toe-in 0-inch

Rear
- Camber .... -2.00-degree(s)
- Toe-in 13/64-inch

M3
Front
- Caster...................................... 9.10-degree(s)
- Camber .... -0.67-degree(s)
- Toe-in 9/64-inch

Rear
- Camber .... -2.33-degree(s)
- Toe-in 15/64-inch

1992 thru 1994
Front
- Caster...................................... 3.75-degree(s)
- Camber .... -0.67-degree(s)
- Toe-in 5/32-inch

Rear
- Camber .... -1.50-degree(s)
- Toe-in 13/64-inch

1995 and 1996
Except M3
Front
- Caster...................................... 3.75-degree(s)
- Camber .... -0.50-degree(s)
- Toe-in 5/32-inch

Rear
- Camber .... -2.00-degree(s)
- Toe-in 13/64-inch

M3
Front
- Caster...................................... 6.65-degree(s)
- Camber .... -0.92-degree(s)
- Toe-in 5/64-inch

Rear
- Camber .... -1.75-degree(s)
- Toe-in 1/4-inch

5-series
1983 thru 1988
Front
- Caster...................................... 8.00-degree(s)
- Camber .... -0.33-degree(s)
- Toe-in 5/32-inch

Rear
    Camber .... -2.33-degree(s)
    Toe-in... 5/32-inch

1989 thru 1996
  Except M5
    Front
      Caster .... 7.88-degree(s)
      Camber .... -0.22-degree(s)
      Toe-in... 5/32-inch
    Rear
      Camber .... -2.33-degree(s)
      Toe-in... 13/64-inch
  M5
    Front
      Caster .... 8.15-degree(s)
      Camber .... -0.22-degree(s)
      Toe-in... 5/32-inch
    Rear
      Camber .... -2.25-degree(s)
      Toe-in... 1/8-inch

6-series
1983 thru 1989
  Front
    Caster .... 8.00-degree(s)
    Camber .... -0.33-degree(s)
    Toe-in... 5/32-inch
  Rear
    Camber .... -2.33-degree(s)
    Toe-in... 5/32-inch

7-series
1983 thru 1987
  Front
    Caster .... 8.95-degree(s)
    Camber .... 0-degree(s)
    Toe-in... 1/32-inch
  Rear
    Camber .... -2.33-degree(s)
    Toe-in... 5/32-inch

1988 thru 1994
  Front
    Caster .... 8.00-degree(s)
    Camber .... -0.18-degree(s)
    Toe-in... 5/32-inch
  Rear
    Camber .... -2.33-degree(s)
    Toe-in... 5/32-inch

1995 and 1996
  Front
    Caster .... 6.00-degree(s)
    Camber .... -0.22-degree(s)
    Toe-in... 7/64-inch
  Rear
    Camber .... -1.50-degree(s)
    Toe-in... 5/32-inch

8-series
1991 thru 1996
  840Ci and 850Ci
    Front
      Caster .... 8.00-degree(s)
      Camber .... -0.18-degree(s)
      Toe-in... 5/32-inch
    Rear
      Camber .... -1.16-degree(s)
      Toe-in... 5/32-inch

  850CSi
    Front
      Caster .... 8.12-degree(s)
      Camber .... -0.40-degree(s)
      Toe-in... 7/32-inch
    Rear
      Camber .... -1.38-degree(s)
      Toe-in... 5/32-inch

BUICK

Century
1983
  Front
    Caster... 2.00-degree(s)
    Camber .... 0-degree(s)
    Toe-in... 0-inch
  Rear
    Camber .... 0-degree(s)
    Toe-in... 1/32-inch

1984 thru 1992
  Front
    Caster... 1.75-degree(s)
    Camber .... 0-degree(s)
    Toe-in... 0-inch
  Rear
    Camber... 0-degree(s)
    Toe-in... 0-inch

1993 thru 1996
  Front
    Caster... 3.00-degree(s)
    Camber .... 0-degree(s)
    Toe-in... 0-inch
  Rear
    Camber .... 0-degree(s)
    Toe-in... 0-inch

Skyhawk
1983 thru 1989
  Front
    Caster... 1.70-degree(s)
    Camber .... 0.80-degree(s)
    Toe-in... -1/16-inch
  Rear
    Camber .... -0.25-degree(s)
    Toe-in... 1/16-inch

Skylark and Sommerset
1983 Skylark
  Front
    Caster... 2.00-degree(s)
    Camber .... 0-degree(s)
    Toe-in... 0-inch
  Rear
    Camber .... 0-degree(s)
    Toe-in... 1/32-inch

1984 and 1985 Skylark
  Front
    Caster... 1.88-degree(s)
    Camber .... 0-degree(s)
    Toe-in... 0-inch
  Rear
    Camber .... 0-degree(s)
    Toe-in... 0-inch

1985 Somerset and 1986 thru 1989 Somerset and Skylark
  Front
    Caster... 1.70-degree(s)
    Camber .... 0.80-degree(s)
    Toe-in... -1/16-inch

Rear
    Camber .... -0.25-degree(s)
    Toe-in 1/8-inch

1990 thru 1996 Skylark
  Front
    Caster .... 1.50-degree(s)
    Camber .... 0-degree(s)
    Toe-in 0-inch
  Rear
    Camber .... -0.25-degree(s)
    Toe-in 7/64-inch

Lesabre, Electra and Park Avenue

1983 thru 1990 (RWD)
  Front
    Caster .... 3.00-degree(s)
    Camber .... 0.80-degree(s)
    Toe-in 5/32-inch

1985 and 1996 (FWD)
  Front
    Caster .... 3.00-degree(s)
    Camber .... 0.20-degree(s)
    Toe-in 0-inch
  Rear
    Camber .... -0.30-degree(s)
    Toe-in 3/64-inch

Regal

1983 thru 1988 Regal (RWD)
  Caster .... 3.00-degree(s)
  Camber .... 0.50-degree(s)
  Toe-in 5/32-inch

1988 and 1995 Regal (FWD)
  Front
    Caster .... 2.00-degree(s)
    Camber .... 0.75-degree(s)
    Toe-in 0-inch
  Rear
    Camber .... 0.06-degree(s)
    Toe-in .... -3/64-inch

1996 Regal
  Front
    Caster .... 1.80-degree(s)
    Camber .... 0.70-degree(s)
    Toe-in 0-inch
  Rear
    Camber .... -0.35-degree(s)
    Toe-in 0-inch

Reatta

1987 and 1990
  Front
    Caster .... 2.80-degree(s)
    Camber .... 0-degree(s)
    Toe-in 0-inch
  Rear
    Camber .... 0.30-degree(s)
    Toe-in 0-inch

1991
  Front
    Caster .... 2.80-degree(s)
    Camber .... 0-degree(s)
    Toe-in 3/32-inch
  Rear
    Camber .... -0.10-degree(s)
    Toe-in 3/32-inch

Riviera

1983 thru 1986
  Caster .... 2.50-degree(s)
  Camber .... 0-degree(s)
  Toe-in 0-inch
  Rear
    Camber .... 0-degree(s)
    Toe-in 3/32-inch

1987 and 1988
  Front
    Caster .... 2.30-degree(s)
    Camber .... 0-degree(s)
    Toe-in 0-inch
  Rear
    Camber .... -0.70-degree(s)
    Toe-in 3/32-inch

1989 thru 1991
  Front
    Caster .... 2.30-degree(s)
    Camber .... 0-degree(s)
    Toe-in 7/64-inch
  Rear
    Camber .... 0-degree(s)
    Toe-in 7/64-inch

1992
  Front
    Caster .... 2.30-degree(s)
    Camber .... -0.50-degree(s)
    Toe-in 3/32-inch
  Rear
    Camber .... 0-degree(s)
    Toe-in 3/32-inch

1993 and 1994
  Front
    Caster .... 2.30-degree(s)
    Camber .... 0-degree(s)
    Toe-in 7/64-inch
  Rear
    Camber .... 0-degree(s)
    Toe-in 7/64-inch

1995 and 1996
  Front
    Caster .... 5.50-degree(s)
    Camber .... 0.20-degree(s)
    Toe-in 0-inch
  Rear
    Camber .... -0.30-degree(s)
    Toe-in 7/64-inch

Roadmaster

1991 thru 1996
  Front
    Caster .... 3.50-degree(s)
    Camber .... 0-degree(s)
    Toe-in 5/64-inch

CADILLAC

Allante

1987 and 1988
  Front
    Caster .... 2.80-degree(s)
    Camber .... 0-degree(s)
    Toe-in 0-inch
  Rear
    Camber .... 0.20-degree(s)
    Toe-in 3/32-inch

1989 and 1990
- Front
  - Caster .... 2.80-degree(s)
  - Camber .... 0.20-degree(s)
  - Toe-in 3/32-inch
- Rear
  - Camber .... 0.20-degree(s)
  - Toe-in 3/32-inch

1991
- Front
  - Caster .... 2.30-degree(s)
  - Camber .... 0-degree(s)
  - Toe-in 3/32-inch
- Rear
  - Camber .... 0.70-degree(s)
  - Toe-in 3/32-inch

1992
- Front
  - Caster .... 2.30-degree(s)
  - Toe-in 3/32-inch
  - Left wheel
    - Camber .... -0.50-degree(s)
  - Right wheel
    - Camber .... 0.50-degree(s)
  - Rear
    - Camber .... 0.70-degree(s)
    - Toe-in 3/32-inch

1993 thru 1996
- Front
  - Caster .... 2.50-degree(s)
  - Camber .... 0-degree(s)
  - Toe-in 7/64-inch
- Rear
  - Camber .... 0-degree(s)
  - Toe-in 7/64-inch

Cimarron
1983 thru 1988
- Front
  - Caster .... 1.70-degree(s)
  - Camber .... 0.80-degree(s)
  - Toe-in .... -1/16-inch
- Rear
  - Camber .... -0.25-degree(s)
  - Toe-in 1/16-inch

Deville and Fleetwood
1985 thru 1996 (FWD)
- Front
  - Caster .... 3.00-degree(s)
  - Camber
    - Left wheel .... -0.50-degree(s)
    - Right wheel .... 0.50-degree(s)
  - Toe-in 0-inch
- Rear
  - Camber .... -0.30-degree(s)
  - Toe-in 3/64-inch

1983 and 1984 (RWD)
- Front
  - Caster .... 3.00-degree(s)
  - Camber .... 0.50-degree(s)
  - Toe-in 1/8-inch

1985 thru 1992 (RWD)
- Front
  - Caster .... 3.00-degree(s)
  - Camber .... 0-degree(s)
  - Toe-in 0-inch

1993 thru 1996 (RWD)
- Front
  - Caster .... 3.50-degree(s)
  - Camber .... 0-degree(s)
  - Toe-in 5/64-inch

Eldorado and Seville
1983 thru 1986
- Front
  - Caster .... 2.50-degree(s)
  - Camber .... 0-degree(s)
  - Toe-in 0-inch
- Rear
  - Camber .... 0-degree(s)
  - Toe-in 3/32-inch

1987 thru 1991
- Front
  - Caster .... 2.30-degree(s)
  - Camber .... 0-degree(s)
  - Toe-in 0-inch
- Rear
  - Camber .... -0.10-degree(s)
  - Toe-in 3/32-inch

1992 thru 1996
- Front
  - Caster .... 2.30-degree(s)
  - Camber
    - Left wheel .... -0.50-degree(s)
    - Right wheel .... 0.50-degree(s)
  - Toe-in 3/32-inch
- Rear
  - Camber .... 0-degree(s)
  - Toe-in 3/32-inch

CHEVROLET

Camaro
1983 and 1986
- Front
  - Caster .... 3.00-degree(s)
  - Camber .... 1.00-degree(s)
  - Toe-in 3/64-inch

1987 thru 1990
- Front
  - Caster .... 4.70-degree(s)
  - Camber .... 0.30-degree(s)
  - Toe-in 0-inch

1991 thru 1995
- Front
  - Caster .... 4.80-degree(s)
  - Camber .... 0.30-degree(s)
  - Toe-in 0-inch

1996
- Front
  - Caster .... 5.00-degree(s)
  - Camber .... 0-degree(s)
  - Toe-in 0-inch

Caprice and Impala
1983 thru 1986
- Front
  - Caster .... 3.00-degree(s)
  - Camber .... 0.80-degree(s)
  - Toe-in 5/32-inch

1987 thru 1990
- Front
  - Caster .... 2.80-degree(s)
  - Camber .... 0.80-degree(s)
  - Toe-in 3/64-inch

1991 thru 1996
Front
    Caster..3.50-degree(s)
    Camber ....0-degree(s)
    Toe-in..5/64-inch

Cavalier
1983 and 1984
Front
    Caster..1.70-degree(s)
    Camber ....0.80-degree(s)
    Toe-in..-1/16-inch
Rear
    Camber ....-0.25-degree(s)
    Toe-in..1/16-inch
1985 and 1986
Front
    Caster..1.70-degree(s)
    Camber ....0.80-degree(s)
    Toe-in..-1/16-inch
Rear
    Camber ....-0.25-degree(s)
    Toe-in..5/32-inch
1987 thru 1989
Front
    Caster..1.70-degree(s)
    Camber ....0.80-degree(s)
    Toe-in..0-inch
Rear
    Camber ....-0.25-degree(s)
    Toe-in..1/8-inch
1990 thru 1992
Front
    Caster..1.70-degree(s)
    Camber ....0-degree(s)
    Toe-in..0-inch
Rear
    Camber ....-0.25-degree(s)
    Toe-in..1/8-inch
1993 and 1994
Front
    Caster..1.30-degree(s)
    Camber ....-0.15-degree(s)
    Toe-in..0-inch
Rear
    Camber ....-0.25-degree(s)
    Toe-in..1/8-inch
1995 and 1996
Front
    Caster..4.30-degree(s)
    Camber ....-0.20-degree(s)
    Toe-in..3/64-inch
Rear
    Camber ....-0.25-degree(s)
    Toe-in..7/64-inch

Celebrity
1983
Front
    Caster..2.00-degree(s)
    Camber ....0-degree(s)
    Toe-in..0-inch
Rear
    Camber ....0-degree(s)
    Toe-in..1/32-inch
1984 thru 1986
Front
    Caster..1.88-degree(s)
    Camber ....0-degree(s)
    Toe-in..0-inch

Rear
    Camber ....0-degree(s)
    Toe-in..0-inch
1987 thru 1990
Front
    Caster..1.70-degree(s)
    Camber ....0-degree(s)
    Toe-in..0-inch
Rear
    Camber ....0-degree(s)
    Toe-in..0-inch

Chevette
1983 thru 1986
Front
    Caster..5.00-degree(s)
    Camber ....0.20-degree(s)
    Toe-in
        2-door..1/8-inch
        4-door..1/16-inch
1987
Front
    Caster..4.70-degree(s)
    Camber ....0.20-degree(s)
    Toe-in..1/16-inch

Citation
1983 thru 1985
Front
    Caster..1.88-degree(s)
    Camber ....0-degree(s)
    Toe-in..0-inch
Rear
    Camber ....0-degree(s)
    Toe-in..0-inch

Corsica and Beretta
1987 thru 1990
Front
    Caster..1.16-degree(s)
    Camber ....0.50-degree(s)
    Toe-in..0-inch
Rear
    Camber ....-0.25-degree(s)
    Toe-in..5/32-inch
1991 and 1992
Front
    Caster..1.16-degree(s)
    Camber ....0.10-degree(s)
    Toe-in..0-inch
Rear
    Camber ....-0.25-degree(s)
    Toe-in..5/32-inch
1993 and 1994
Front
    Caster..1.20-degree(s)
    Camber ....0-degree(s)
    Toe-in..0-inch
Rear
    Camber ....-0.25-degree(s)
    Toe-in..0-inch
1995 and 1996
Front
    Caster..1.20-degree(s)
    Camber ....0-degree(s)
    Toe-in..0-inch
Rear
    Camber ....-0.40-degree(s)
    Toe-in..7/64-inch

Haynes suspension, steering and driveline manual

Corvette

1984 and 1985
  Front
    Caster.. 3.00-degree(s)
    Camber .... 0.80-degree(s)
    Toe-in.. 5/32-inch
  Rear (1984)
    Camber .... 0-degree(s)
    Toe-in.. 5/32-inch
  Rear (1985)
    Camber .... 0.38-degree(s)
    Toe-in.. 5/32-inch

1986
  Front
    Caster.. 6.00-degree(s)
    Camber .... 0.80-degree(s)
    Toe-in.. 0-inch
  Rear
    Camber .... 0.38-degree(s)
    Toe-in.. 0-inch

1987 and 1988
  Front
    Caster.. 5.50-degree(s)
    Camber .... 0.80-degree(s)
    Toe-in.. 0-inch
  Rear
    Camber .... 0.38-degree(s)
    Toe-in.. 0-inch

1989
  Front
    Caster.. 5.80-degree(s)
    Camber .... 0.50-degree(s)
    Toe-in.. 0-inch
  Rear
    Camber .... 0.20-degree(s)
    Toe-in.. 0-inch

1990 thru 1996
  Front
    Caster.. 6.00-degree(s)
    Camber .... 0.50-degree(s)
    Toe-in.. 0-inch
  Rear
    Camber .... 0-degree(s)
    Toe-in.. 0-inch

Lumina and Monte Carlo (FWD)

1990
  Front
    Caster.. 2.00-degree(s)
    Camber .... 0.70-degree(s)
    Toe-in.. 0-inch
  Rear
    Camber .... 0.10-degree(s)
    Toe-in.. 0-inch

1991
  Front
    Caster.. 1.80-degree(s)
    Camber .... 0.70-degree(s)
    Toe-in.. 0-inch
  Rear
    Camber .... -0.16-degree(s)
    Toe-in.. -3/64-inch

1992 thru 1994
  Front
    Caster.. 2.00-degree(s)
    Camber .... 0.70-degree(s)
    Toe-in.. 0-inch
  Rear
    Camber .... 0.10-degree(s)
    Toe-in.. -3/64-inch

1995 and 1996 Lumina
  Front
    Caster.. 1.80-degree(s)
    Camber .... 0.70-degree(s)
    Toe-in.. 0-inch
  Rear
    Camber .... -0.35-degree(s)
    Toe-in.. 0-inch

Malibu, Monte Carlo (RWD) and El Camino

1983 thru 1986
  Front
    Caster.. 3.00-degree(s)
    Camber .... 0.50-degree(s)
    Toe-in.. 5/32-inch

1987 and 1988
  Front
    Caster.. 2.80-degree(s)
    Camber .... 0.50-degree(s)
    Toe-in.. 3/64-inch

Nova

1985 thru 1988
  Front
    Caster.. 0.80-degree(s)
    Camber .... -0.25-degree(s)
    Toe-in.. 1/32-inch
  Rear
    Camber .... -0.50-degree(s)
    Toe-in.. 5/32-inch

Spectrum

1985 thru 1988
  Front
    Caster.. 2.25-degree(s)
    Camber .... 0.30-degree(s)
    Toe-in.. 0-inch
  Rear (1985 and 1986)
    Camber .... 0-degree(s)
    Toe-in.. 1/16-inch
  Rear (1987 and 1988)
    Camber .... -0.16-degree(s)
    Toe-in.. 1/16-inch

Sprint

1985 and 1986
  Front
    Caster.. 3.16-degree(s)
    Camber .... 1.00-degree(s)
    Toe-in.. 1/16-inch
  Rear
    Camber .... 0-degree(s)
    Toe-in.. 0-inch

1987 and 1988
  Front
    Caster.. 3.16-degree(s)
    Camber .... 0.25-degree(s)
    Toe-in.. 0-inch
  Rear
    Camber .... 0-degree(s)
    Toe-in.. 0-inch

1989 thru 1991 (Canada)
  Front
    Caster.. 3.00-degree(s)
    Camber .... 0-degree(s)
    Toe-in.. 0-inch
  Rear
    Camber .... 0-degree(s)
    Toe-in.. 3/32-inch

Chevrolet and GMC Trucks

Includes Chevrolet Lumina APV, Oldsmobile Bravada and Silhouette, Pontiac Trans Sport
Note: *Caster adjustments vary with the ride height, which is measured between the frame crossmember and the bracket mount for the lower control arm bumper, and is known as Dimension "A". Vehicle must be parked on a smooth, level surface when the measurement is taken.*

Astro and Safari

1985 thru 1987
    Caster .... 2.70-degree(s)
    Camber .... 0.90-degree(s)
    Toe-in 5/32-inch
1988 thru 1996
    2WD
        Caster .... 2.70-degree(s)
        Camber .... 0.80-degree(s)
        Toe-in 7/64-inch
    4WD (1988 thru 1990)
        Caster .... 3.00-degree(s)
        Camber .... 0.88-degree(s)
        Toe-in .... -3/64-inch
    4WD (1991 thru 1996)
        Caster .... 2.00-degree(s)
        Camber .... 0.88-degree(s)
        Toe-in .... -1/32-inch

Lumina APV, Silhouette and Trans Sport

1990 thru 1992
    Front
        Caster .... 1.70-degree(s)
        Camber .... 0-degree(s)
        Toe-in 0-inch
    Rear
        Camber .... -0.10-degree(s)
        Toe-in 0-inch
1993 thru 1996
    Front
        Caster .... 3.00-degree(s)
        Camber .... 0-degree(s)
        Toe-in 0-inch
    Rear
        Camber .... -0.10-degree(s)
        Toe-in 0-inch

S10, S15, Sonoma, S-Blazer, S-Jimmy and Bravada

1983 thru 1994
    2WD
        Caster .... 2.00-degree(s)
        Camber .... 0.80-degree(s)
        Toe-in 5/32-inch
    4WD
        Caster .... 2.00-degree(s)
        Camber .... 0.80-degree(s)
        Toe-in 5/32-inch
1995 and 1996
    2WD
        Caster .... 3.00-degree(s)
        Camber .... 0-degree(s)
        Toe-in 7/64-inch
    4WD
        Caster .... 3.00-degree(s)
        Camber .... 0-degree(s)
        Toe-in 7/64-inch
    Typhoon
        Caster .... 3.50-degree(s)
        Camber .... 0-degree(s)
        Toe-in 5/32-inch

C-R10, 20 and 30 Pickups; Suburban (2WD)

1983 thru 1988
    C-R10
        Caster
            Dimension A
                2-1/2 inches .... 3.70-degree(s)
                3 inches .... 3.20-degree(s)
                3-1/2 inches .... 2.63-degree(s)
                4 inches .... 2.12-degree(s)
                4-1/2 inches .... 1.50-degree(s)
                5 inches .... 1.00-degree(s)
        Camber .... 0.70-degree(s)
        Toe-in 3/16-inch
    C-R 20 and 30
        Caster
            Dimension A
                2-1/2 inches .... 1.50-degree(s)
                3 inches .... 0.88-degree(s)
                3-1/2 inches .... 0.30-degree(s)
                4 inches .... 0-degree(s)
                4-1/2 inches .... -0.70-degree(s)
                5 inches .... -1.20-degree(s)
        Camber .... 0.25-degree(s)
        Toe-in 3/16-inch
1989 thru 1991
    R10
        Caster
            Dimension A
                2-1/2 inches .... 3.70-degree(s)
                3 inches .... 3.20-degree(s)
                3-1/2 inches .... 2.63 degree(s)
                4 inches .... 2.12-degree(s)
                4-1/2 inches .... 1.50 degree(s)
                5 inches .... 1.00-degree(s)
        Camber .... 0.70-degree(s)
        Toe-in 5/64-inch
    R20 and 30
        Caster
            Dimension A
                2-1/2 inches .... 1.50-degree(s)
                3 inches .... 0.88-degree(s)
                3-1/2 inches .... 0.30-degree(s)
                4 inches .... 0-degree(s)
                4-1/2 inches .... -0.70-degree(s)
                5 inches .... -1.20-degree(s)
        Camber .... 0.25-degree(s)
        Toe-in 5/64-inch

K-V10, 20 and 30 series Pickups; Blazer, Jimmy and Suburban (4WD)

1983 and 1984
    Caster
        Dimension A
            4 inches .... 8.00-degree(s)
    Camber .... 1.00-degree(s)
    Toe-in 3/16-inch
1985 thru 1988
    Caster
        Dimension A
            4 inches .... 8.00-degree(s)
        Camber .... 1.50-degree(s)
        Toe-in 0-inch
1989
    Caster
        Dimension A
            4 inches .... 8.00-degree(s)
    Camber .... 1.50-degree(s)
    Toe-in 5/64-inch

1990 thru 1992

V10 and 20
  Caster
    Dimension A
      4 inches .... 8.00-degree(s)
    Camber .... 1.00-degree(s)
    Toe-in 5/64-inch

V3500
  Caster
    Dimension A
      4 inches .... 8.00-degree(s)
    Camber .... 0-degree(s)
    Toe-in 5/64-inch

C1500, 2500 and 3500 Pickup

1988 thru 1996

Except C3500HD
  Caster.. 3.75-degree(s)
  Camber .... 0.50-degree(s)
  Toe-in 1/8-inch
C3500HD (with I-beam front suspension)
  Caster.. 5.00-degree(s)
  Camber .... 1.25-degree(s)
  Toe-in 1/16-inch

K1500, K2500 LD, K2500 HD and K3500 Pickup

1988

K1500 and 2500
  Caster.. 4.00-degree(s)
  Camber .... 0.88-degree(s)
  Toe-in 1/16-inch
K3500
  Caster.. 4.00-degree(s)
  Camber .... 0.75-degree(s)
  Toe-in 1/16-inch

1989 thru 1996

Light duty
  Caster.. 3.00-degree(s)
  Camber .... 0.66-degree(s)
  Toe-in 1/8-inch
Heavy duty
  Caster.. 3.00-degree(s)
  Camber .... 0.50-degree(s)
  Toe-in 1/8-inch

Blazer, Jimmy, Tahoe, Yukon, Suburban (C-K Series)

1992 thru 1996

2WD
  Caster.. 3.75-degree(s)
  Camber .... 0.50-degree(s)
  Toe-in 1/8-inch
4WD (Light duty)
  Caster.. 3.00-degree(s)
  Camber .... 0.66-degree(s)
  Toe-in 1/8-inch
4WD (Heavy duty)
  Caster.. 3.00-degree(s)
  Camber .... 0.50-degree(s)
  Toe-in 1/8-inch

G10, 20 and 30 (Full-size Vans)

1983 and 1984

G10 and 20
  Caster
    Dimension A
      1-1/2 inches .... 3.50-degree(s)
      2 inches .... 3.12-degree(s)
      2-1/2 inches .... 2.70-degree(s)
      3 inches .... 2.38-degree(s)
      3-1/2 inches .... 2.12-degree(s)
      4 inches .... 1.80-degree(s)
      4-1/2 inches .... 1.38-degree(s)
    Camber .... 0.50-degree(s)
    Toe-in 3/16-inch
G30
  Caster
    Dimension A
      1-1/2 inches .... 2.80-degree(s)
      2 inches .... 2.20-degree(s)
      2-1/2 inches .... 1.63-degree(s)
      3 inches .... 1.00-degree(s)
      3-1/2 inches .... 0.50-degree(s)
      4 inches .... 0-degree(s)
    Camber .... 0.25-degree(s)
    Toe-in 3/16-inch

1985 thru 1988

G10 and 20
  Caster
    Dimension A
      1-1/2 inches .... 3.44-degree(s)
      2 inches .... 3.00-degree(s)
      2-1/2 inches .... 2.70-degree(s)
      3 inches .... 2.30-degree(s)
      3-1/2 inches .... 2.00-degree(s)
      4 inches .... 1.70-degree(s)
      4-1/2 inches .... 1.38-degree(s)
    Camber .... 0.50-degree(s)
    Toe-in 3/16-inch
G30
  Caster
    Dimension A
      1-1/2 inches .... 3.10-degree(s)
      2 inches .... 2.70-degree(s)
      2-1/2 inches .... 2.12-degree(s)
      3 inches .... 1.50-degree(s)
      3-1/2 inches .... 1.00-degree(s)
      4 inches .... 0.50-degree(s)
      4-1/2 inches .... -0.06-degree(s)
    Camber .... 0.25-degree(s)
    Toe-in 3/16-inch

1989 thru 1993

G10 and 20
  Caster
    Dimension A
      1-1/2 inches .... 3.40-degree(s)
      2 inches .... 3.00-degree(s)
      2-1/2 inches .... 2.70-degree(s)
      3 inches .... 2.30-degree(s)
      3-1/2 inches .... 2.00-degree(s)
      4 inches .... 1.70-degree(s)
      4-1/2 inches .... 1.40-degree(s)
      5 inches .... 1.20-degree(s)
    Camber .... 0.50-degree(s)
    Toe in 0-inch
G30
  Caster
    Dimension A
      1-1/2 inches .... 3.10-degree(s)
      2 inches .... 2.70-degree(s)
      2-1/2 inches .... 2.10-degree(s)
      3 inches .... 1.50-degree(s)
      3-1/2 inches .... 1.00-degree(s)
      4 inches .... 0.50-degree(s)
      4-1/2 inches .... -0.10-degree(s)
    Camber .... 0.25-degree(s)
    Toe in 0-inch

1994 thru 1996
  G1500,2500
    Caster
      Dimension A

1-1/2 inches	5.40-degree(s)
2 inches	5.00-degree(s)
2-1/2 inches	4.70-degree(s)
3 inches	4.30-degree(s)
3-1/2 inches	4.00-degree(s)
4 inches	3.70-degree(s)
4-1/2 inches	3.40-degree(s)
5 inches	3.20-degree(s)
Camber	0.50-degree(s)
Toe in	0-inch

  G30,G3500
    Caster
      Dimension A

1-1/2 inches	5.10-degree(s)
2 inches	4.70-degree(s)
2-1/2 inches	4.10-degree(s)
3 inches	3.50-degree(s)
3-1/2 inches	3.00-degree(s)
4 inches	2.50-degree(s)
4-1/2 inches	2.00-degree(s)
5 inches	1.50-degree(s)
Camber	0.25-degree(s)
Toe-in	0-inch

P20 and 30 (Cube Van)

1985 thru 1989
  Except with 4-wheel disc brakes, hydroboost, dual rear wheels,
  or 5,000 lb I-beam front axle
    Caster
      Dimension A

2 inches	3.00-degree(s)
2-1/2 inches	2.25-degree(s)
3 inches	1.75-degree(s)
3-1/2 inches	1.16-degree(s)
4 inches	0.63-degree(s)
4-1/2 inches	0.16-degree(s)
Camber	0.25-degree(s)
Toe-in	3/16-inch

  With 4-wheel disc brakes or hydroboost
    Caster
      Dimension A

2 inches	3.25-degree(s)
2-1/2 inches	2.63-degree(s)
3 inches	2.00-degree(s)
3-1/2 inches	1.50-degree(s)
4 inches	0.90-degree(s)
4-1/2 inches	0.44-degree(s)
Camber	0.25-degree(s)
Toe-in	3/16-inch

  With dual rear wheels
    Caster
      Dimension A

2 inches	2.50-degree(s)
2-1/2 inches	1.88-degree(s)
3 inches	1.30-degree(s)
3-1/2 inches	0.80-degree(s)
4 inches	0.25-degree(s)
4-1/2 inches	-0.20-degree(s)
Camber	0.25-degree(s)
Toe-in	3/16-inch

  With 5,000 lb I-beam front axle

Caster	5.00-degree(s)
Camber	1.50-degree(s)
Toe-in	1/8-inch

1990 thru 1992
  Except with 4-wheel disc brakes, hydroboost, dual rear wheels, or
  5,000 lb I-beam front axle
    Caster
      Dimension A

2 inches	3.00-degree(s)
2-1/2 inches	2.25-degree(s)
3 inches	1.75-degree(s)
3-1/2 inches	1.16-degree(s)
4 inches	0.63-degree(s)
4-1/2 inches	0.16-degree(s)
Camber	0.10-degree(s)
Toe in	3/16-inch

  With 4-wheel disc brakes or hydroboost
    Caster
      Dimension A

2 inches	3.25-degree(s)
2-1/2 inches	2.63-degree(s)
3 inches	2.00-degree(s)
3-1/2 inches	1.50-degree(s)
4 inches	0.90-degree(s)
4-1/2 inches	0.44-degree(s)
Camber	0.10-degree(s)
Toe in	3/16-inch

  With dual rear wheels
    Caster
      Dimension A

2 inches	2.50-degree(s)
2 1/2 inches	1.00 degree(s)
3 inches	1.30-degree(s)
3-1/2 inches	0.80-degree(s)
4 inches	0.30-degree(s)
4-1/2inches	-0.20-degree(s)
Camber	0.10-degree(s)
Toe-in	1/32-inch

  With 5,000 lb I-beam front axle

Caster	5.00-degree(s)
Camber	1.50-degree(s)
Toe-in	1/8-inch

1993 thru 1996
  Except with 4-wheel disc brakes, hydroboost, dual rear wheels, or
  5,000 lb I-beam front axle
    Caster
      Dimension A

2 inches	3.00-degree(s)
2-1/2 inches	2.30-degree(s)
3 inches	1.70-degree(s)
3-1/2 inches	0.20-degree(s)
4 inches	0.60-degree(s)
4-1/2 inches	0.20-degree(s)
Camber	0.10-degree(s)
Toe-in	1/32-inch

  With 4-wheel disc brakes or hydroboost
    Caster
      Dimension A

2 inches	3.20-degree(s)
2-1/2 inches	2.60-degree(s)
3 inches	2.10-degree(s)
3-1/2 inches	1.50-degree(s)
4 inches	1.00-degree(s)
4-1/2 inches	0.50-degree(s)
Camber	0.10-degree(s)
Toe-in	1/32-inch

With dual rear wheels
  Caster
    Dimension A
      2 inches .... 2.50-degree(s)
      2-1\2 inches .... 1.90-degree(s)
      3 inches .... 1.30-degree(s)
      3-1/2 inches .... 0.80-degree(s)
      4 inches .... 0.30-degree(s)
      4-1/2 inches .... -0.20-degree(s)
  Camber .... 0.10-degree(s)
  Toe-in 1/32-inch
With 5,000 lb I-beam front axle
  Caster .... 5.00-degree(s)
  Camber .... 1.50-degree(s)
  Toe-in 1/32-inch

CHRYSLER

Cirrus
1995 and 1996
  Front
    Caster .... 3.30-degree(s)
    Camber .... 0.30-degree(s)
    Toe-in 3/64-inch
  Rear
    Camber .... -0.20-degree(s)
    Toe-in 3/64-inch

Concorde
1993 thru 1996
  Front
    Caster .... 3.00-degree(s)
    Camber .... 0-degree(s)
    Toe-in 7/64-inch
  Rear
    Camber .... -0.10-degree(s)
    Toe-in 3/64-inch

Conquest
1987 thru 1989
  Front
    Caster .... 5.88-degree(s)
    Camber .... -0.50-degree(s)
    Toe-in 0-inch
  Rear (1987)
    Camber .... 0-degree(s)
    Toe-in 0-inch
  Rear (1988 and 1989)
    Camber .... -0.25-degree(s)
    Toe-in 0-inch

Cordoba and Imperial (RWD)
1983
  Front
    Caster .... 2.50-degree(s)
    Camber .... 0.50-degree(s)
    Toe-in 1/8-inch

Laser
1984 thru 1986
  Front
    Caster .... 1.20-degree(s)
    Camber .... 0.30-degree(s)
    Toe-in 1/16-inch
  Rear
    Camber .... -0.50-degree(s)
    Toe-in 0-inch

LeBaron
1983 thru 1990
  Front
    Caster .... 1.20-degree(s)
    Camber .... 0.30-degree(s)
    Toe-in 1/16-inch
  Rear
    Camber .... -0.50-degree(s)
    Toe-in 0-inch
1991 thru 1995
  Front
    Caster .... 2.80-degree(s)
    Camber .... 0.30-degree(s)
    Toe-in 1/16-inch
  Rear
    Camber .... -0.50-degree(s)
    Toe-in 0-inch

New Yorker, Dynasty, E-Class, Fifth Avenue, Imperial and LHS (FWD)
1983 thru 1990
  Front
    Caster .... 1.20-degree(s)
    Camber .... 0.30-degree(s)
    Toe-in 1/16-inch
  Rear
    Camber .... -0.50-degree(s)
    Toe-in 0-inch
1991 thru 1993
  Front
    Caster .... 2.80-degree(s)
    Camber .... 0.30-degree(s)
    Toe-in 1/16-inch
  Rear
    Camber .... -0.50-degree(s)
    Toe-in 0-inch
1994 thru 1996
  Front
    Caster .... 3.00-degree(s)
    Camber .... 0-degree(s)
    Toe-in 7/64-inch
  Rear
    Camber .... -0.10-degree(s)
    Toe-in 3/64-inch

New Yorker, Fifth Avenue and Newport (RWD)
1983 thru 1989
  Front
    Caster .... 2.50-degree(s)
    Camber .... 0.50-degree(s)
    Toe-in 1/8-inch

Sebring
1995 and 1996 (except convertibile)
  Front
    Caster .... 4.33-degree(s)
    Camber .... 0-degree(s)
    Toe-in 0-inch
  Rear
    Camber .... -1.33-degree(s)
    Toe-in 1/8-inch
1996 Convertible
  Front
    Caster .... 3.30-degree(s)
    Camber .... 0.30-degree(s)
    Toe-in 3/64-inch
  Rear
    Camber .... -0.20-degree(s)
    Toe-in 3/64-inch

Town and Country

1983 thru 1989
- Front
  - Caster .... 0.88-degree(s)
  - Camber .... 0.30-degree(s)
  - Toe-in 1/16-inch
- Rear
  - Camber .... -0.50-degree(s)
  - Toe-in 0-inch

1990 and 1991
- Front
  - Caster .... 1.70-degree(s)
  - Camber .... 0.30-degree(s)
  - Toe-in 1/16-inch
- Rear
  - Camber .... -0.20-degree(s)
  - Toe-in 0-inch

1992 thru 1995
- Front
  - Caster .... 1.30-degree(s)
  - Camber .... 0.30-degree(s)
  - Toe-in 1/16-inch
- Rear
  - Camber .... -0.20-degree(s)
  - Toe-in 0-inch

1996
- Front
  - Caster .... 1.40-degree(s)
  - Camber .... 0.10-degree(s)
  - Toe-in 3/64-inch
- Rear
  - Camber .... 0-degree(s)
  - Toe-in 0-inch

DODGE

Aries, 400 and 600

1983 thru 1989
- Front
  - Caster (except wagon) .... 1.20-degree(s)
  - Caster (wagon) .... 0.88-degree(s)
  - Camber .... 0.30-degree(s)
  - Toe-in 1/16-inch
- Rear
  - Camber .... -0.50-degree(s)
  - Toe-in 0-inch

Avenger

1995 and 1996
- Front
  - Caster .... 4.33-degree(s)
  - Camber .... 0-degree(s)
  - Toe-in 0-inch
- Rear
  - Camber .... -1.33-degree(s)
  - Toe-in 1/8-inch

Caravan and Mini Ram Van

1984 thru 1990
- Front
  - Caravan
    - Caster .... 0.70-degree(s)
    - Camber .... 0.30-degree(s)
    - Toe-in 1/16-inch
  - Mini Ram Van
    - Caster .... 0.44-degree(s)
    - Camber .... 0.30-degree(s)
    - Toe-in 1/16-inch

Rear (1984 thru 1989)
- Camber .... -0.63-degree(s)
- Toe-in .... 0- inch

Rear (1990)
- Camber .... -0.20-degree(s)
- Toe-in 0-inch

1991
- Front
  - Caster .... 1.70-degree(s)
  - Camber .... 0.30-degree(s)
  - Toe-in 1/16-inch
- Rear
  - Camber .... -0.20-degree(s)
  - Toe-in 0-inch

1992 thru 1995
- Front
  - Caster .... 1.30-degree(s)
  - Camber .... 0.30-degree(s)
  - Toe-in 1/16-inch
- Rear
  - Camber .... -0.20-degree(s)
  - Toe-in 0-inch

1996
- Front
  - Caster .... 1.40-degree(s)
  - Camber .... 0.10-degree(s)
  - Toe-in 3/64-inch
- Rear
  - Camber .... 0-degree(s)
  - Toe-in 0-inch

Challenger

1983
- Front
  - Caster .... 2.70-degree(s)
  - Camber .... 1.20-degree(s)
  - Toe-in 5/32-inch

Charger and Rampage

1983 thru 1987
- Front
  - Caster .... 1.88-degree(s)
  - Camber .... 0.30-degree(s)
  - Toe-in 1/16-inch
- Rear
  - Camber .... -0.75-degree(s)
  - Toe-in 3/32-inch

Colt and Champ (except Vista)

1983
- Front
  - Caster .... 0.80-degree(s)
  - Camber .... 0.50-degree(s)
  - Toe-in 1/16-inch

1984 Colt
- Front
  - Caster .... 0.80-degree(s)
  - Camber .... 0.50-degree(s)
  - Toe-in 1/16-inch

1985 and 1986 Colt
- Front
  - Caster .... 0.75-degree(s)
  - Camber .... 0-degree(s)
  - Toe-in 0-inch
- Rear
  - Camber .... -0.70-degree(s)
  - Toe-in 0-inch

1987 and 1988 Colt (except wagon)
Front
    Caster.. 1.00-degree(s)
    Camber .... 0-degree(s)
    Toe-in 0-inch
Rear
    Camber .... -0.70-degree(s)
    Toe-in 1/8-inch

1988 and 1990 Colt wagon
2WD
  Front
    Caster (w/o power steering)............... 0.70-degree(s)
    Caster (with power steering) .... 1.30-degree(s)
    Camber .... 0-degree(s)
    Toe-in 0-inch
  Rear
    Camber .... 0-degree(s)
    Toe-in 0-inch
4WD
  Front
    Caster .... 1.00-degree(s)
    Camber .... 0-degree(s)
    Toe-in 0-inch
  Rear
    Camber .... 0-degree(s)
    Toe-in 0-inch

1992 thru 1996 Colt wagon (Canada)
2WD
  Front
    Caster .... 2.16-degree(s)
    Camber .... 0.33-degree(s)
    Toe-in 0-inch
4WD
  Front
    Caster .... 2.08-degree(s)
    Camber .... 0.66-degree(s)
    Toe-in 0-inch
Rear
    Camber .... -0.50-degree(s)
    Toe-in 3/32-inch

1989 thru 1996 Colt (except wagon)
Front
    Caster .... 2.25-degree(s)
    Camber .... 0-degree(s)
    Toe-in 0-inch
Rear
    Camber .... -0.66-degree(s)
    Toe-in 1/8-inch

Colt Vista
1984 thru 1991
2WD
  Front
    Caster .... 0.80-degree(s)
    Camber .... 0.44-degree(s)
    Toe-in 0-inch
  Rear
    Camber .... -0.56-degree(s)
    Toe-in 0-inch
4WD
  Front
    Caster .... 0.80-degree(s)
    Camber .... 0.88-degree(s)
    Toe-in 0-inch
  Rear
    Camber .... 0-degree(s)
    Toe-in 0-inch

Conquest
1984 and 1985
Front
    Caster .... 5.30-degree(s)
    Camber .... 0-degree(s)
    Toe-in 0-inch
Rear
    Camber .... 0-degree(s)
    Toe-in 0-inch
1986
Front
    Caster .... 5.88-degree(s)
    Camber .... 0.50-degree(s)
    Toe-in 0-inch
Rear
    Camber .... 0-degree(s)
    Toe-in 0-inch

Daytona, Dynasty, Lancer, Shadow and Spirit
1984 thru 1990
  Front
    Caster .... 1.20-degree(s)
    Camber .... 0.30-degree(s)
    Toe-in 1/16-inch
  Rear
    Camber .... -0.50-degree(s)
    Toe-in 0-inch
1991 thru 1996
  Front
    Caster .... 2.80-degree(s)
    Camber .... 0.30-degree(s)
    Toe-in 1/16-inch
  Rear
    Camber .... -0.50-degree(s)
    Toe-in 0-inch

Diplomat
1984 thru 1989
  Front
    Caster .... 2.50-degree(s)
    Camber .... 0.50-degree(s)
    Toe-in 1/8-inch

Intrepid
1993 thru 1996
  Front
    Caster .... 3.00-degree(s)
    Camber .... 0-degree(s)
    Toe-in 7/64-inch
  Rear
    Camber .... -0.10-degree(s)
    Toe-in 3/64-inch

Mirada
1983
  Front
    Caster .... 2.50-degree(s)
    Camber .... 0.50-degree(s)
    Toe-in 1/8-inch

Monaco
1990 thru 1992
  Front
    Caster .... 2.00-degree(s)
    Camber .... -0.30-degree(s)
    Toe-in 0-inch
  Rear
    Camber .... -0.50-degree(s)
    Toe-in 0-inch

Neon

1995 and 1996
Front
 Caster.. 2.80-degree(s)
 Camber .... 0-degree(s)
 Toe-in 3/64-inch
Rear
 Camber .... -0.25-degree(s)
 Toe-in 3/64-inch

Omni

1983 thru 1990
Front
 Caster.. 1.44-degree(s)
 Camber .... 0.30-degree(s)
 Toc in 1/16-inch
Rear
 Camber .... -0.75-degree(s)
 Toe-in .... 0.09- inch

Stealth

1991 thru 1996
Front
 Caster.. 3.90-degree(s)
 Camber .... 0-degree(s)
 Toe-in 0-inch
Rear (FWD)
 Camber .... 0-degree(s)
 Toe-in 1/64-inch
Rear (4WD)
 Camber .... -0.15-degree(s)
 Toe-in 1/64-inch

Stratus

1995 and 1996
Front
 Caster.. 3.30-degree(s)
 Camber .... 0.30-degree(s)
 Toe-in 3/64-inch
Rear
 Camber .... -0.20-degree(s)
 Toe-in 3/64-inch

2000 GTX

1989 thru 1992
2WD
 Front
  Caster.. 2.00-degree(s)
  Camber .... 0.38-degree(s)
  Toe-in 0-inch
 Rear
  Camber .... -0.75-degree(s)
  Toe-in 0-inch
4WD
 Front
  Caster.. 1.90-degree(s)
  Camber .... 0.50-degree(s)
  Toe-in 0-inch
 Rear
  Camber .... -1.00-degree(s)
  Toe-in 1/8-inch

Dodge and Plymouth RWD Trucks

D100, 150, 250, 350 and Ramcharger (2WD)

1983 and 1984
Caster.. 0.50-degree(s)
Camber.. 0.50-degree(s)
Toe-in 1/8-inch

1985 thru 1987
Caster.. 0.50-degree(s)
Camber .... 0.50-degree(s)
Toe-in 7/32-inch

1988 thru 1993
Caster.. 0.50-degree(s)
Camber .... 0.50-degree(s)
Toe-in 1/8-inch

W100, 150, 250, 350 and Ramcharger (4WD)

1983 thru 1987
Caster.. 2.00-degree(s)
Camber .... 1.00-degree(s)
Toe-in 7/32-inch

1988 thru 1991
Caster.. 2.00-degree(s)
Camber .... 1.00-degree(s)
Toe-in 3/32-inch

1992 and 1993
Caster.. 2.00-degree(s)
Camber .... 0-degree(s)
Toe-in 3/32-inch

Pickups (Full-size)

1994 thru 1996
2WD
 1500,119-inch wheelbase
  Caster .... 3.65-degree(s)
  Camber .... 0.50-degree(s)
  Toe-in 1/8-inch
 1500,135-inch wheelbase
  Caster .... 3.80-degree(s)
  Camber .... 0.50-degree(s)
  Toe-in.. 1/8-inch
 1500,139-inch wheelbase
  Caster .... 3.70-degree(s)
  Camber .... 0.50-degree(s)
  Toe-in.. 1/8-inch
 1500, 155-inch wheelbase
  Caster .... 3.85-degree(s)
  Camber .... 0.50-degree(s)
  Toe-in.. 1/8-inch
 2500 LD
  Caster .... 3.55-degree(s)
  Camber .... 0.50-degree(s)
  Toe-in.. 1/8-inch
 2500 HD, 135-inch wheelbase
  Caster .... 3.45-degree(s)
  Camber .... 0.50-degree(s)
  Toe-in.. 1/8-inch
 2500 HD, 155-inch wheelbase
  Caster .... 3.65-degree(s)
  Camber .... 0.50-degree(s)
  Toe-in.. 1/8-inch
 3500,135-inch wheelbase
  Caster .... 3.25-degree(s)
  Camber .... 0.50-degree(s)
  Toe-in.. 1/8-inch
 3500,155-inch wheelbase
  Caster .... 3.45-degree(s)
  Camber .... 0.50-degree(s)
  Toe-in.. 1/8-inch
4WD
 1500,119-inch wheelbase
  Caster .... 3.35-degree(s)
  Camber .... 0.25-degree(s)
  Toe-in.. 1/8-inch

Haynes suspension, steering and driveline manual

1500, 135-inch wheelbase
Caster .... 3.45-degree(s)
Camber .... 0.25-degree(s)
Toe-in 1/8-inch

1500,139-inch wheelbase
Caster .... 3.40-degree(s)
Camber .... 0.25-degree(s)
Toe-in 1/8-inch

1500,155-inch wheelbase
Caster .... 3.55-degree(s)
Camber .... 0.25-degree(s)
Toe-in 1/8-inch

2500 LD
Caster .... 3.00-degree(s)
Camber .... 0.25-degree(s)
Toe-in 1/8-inch

2500 HD, 135-inch wheelbase
Caster .... 3.10-degree(s)
Camber .... 0.25-degree(s)
Toe-in 1/8-inch

2500, 155-inch wheelbase
Caster .... 3.25-degree(s)
Camber .... 0.25-degree(s)
Toe-in 1/8-inch

3500, 135-inch wheelbase
Caster .... 2.90-degree(s)
Camber .... 0.25-degree(s)
Toe-in 1/8-inch

3500,155-inch wheelbase
Caster .... 3.00-degree(s)
Camber .... 0.25-degree(s)
Toe-in 1/8-inch

B150-350

1984 thru 1987
Caster .... 2.50-degree(s)
Camber .... 0.38-degree(s)
Toe-in 1/8-inch

1988 thru 1996
Caster .... 2.50-degree(s)
Camber .... 0-degree(s)
Toe-in 0-inch

Dakota

1987 thru 1996
Caster .... 1.50-degree(s)
Camber .... 0.50-degree(s)
Toe-in 1/8-inch

Ram 50 and Arrow

1983
2WD
Caster .... 2.50-degree(s)
Camber .... 1.00-degree(s)
Toe-in 7/32-inch
4WD
Caster .... 2.00-degree(s)
Camber .... 1.00-degree(s)
Toe-in 7/32-inch

1984 thru 1986
2WD
Caster .... 2.50-degree(s)
Camber .... 1.00-degree(s)
Toe-in 7/32-inch
4WD
Caster .... 2.00-degree(s)
Camber .... 1.00-degree(s)
Toe-in 7/32-inch

1987 thru 1993
2WD
Caster .... 2.50-degree(s)
Camber .... 0.66-degree(s)
Toe-in 7/32-inch
4WD
Caster .... 2.00-degree(s)
Camber .... 1.00-degree(s)
Toe-in 7/32-inch

Raider

1987 and 1988
Caster .... 3.00-degree(s)
Camber .... 1.00-degree(s)
Toe-in 7/32-inch

1989
Caster .... 2.90-degree(s)
Camber .... 1.00-degree(s)
Toe-in 7/32-inch

EAGLE

Vista - see *Dodge*

Medallion

1988 and 1989
Front
Caster .... 4.00-degree(s)
Camber .... 0.50-degree(s)
Toe-in 3/32-inch
Rear
Camber .... -0.70-degree(s)
Toe-in .... -5/16-inch

Premier

1988 and 1989
Front
Caster .... 2.00-degree(s)
Camber .... -0.30-degree(s)
Toe-in 1/8-inch
Rear
Camber .... -0.70-degree(s)
Toe-in 9/32-inch

1990 thru 1992
Front
Caster .... 2.00-degree(s)
Camber .... -0.30-degree(s)
Toe-in 0-inch
Rear
Camber .... -0.50-degree(s)
Toe-in 0-inch

Summit

1989 thru 1996 (except wagon)
Front
Caster .... 2.25-degree(s)
Camber .... 0-degree(s)
Toe-in 0-inch
Rear
Camber .... -0.66-degree(s)
Toe-in 1/8-inch

1992 thru 1996 (wagon)
2WD
Front
Caster .... 2.16-degree(s)
Camber .... 0.33-degree(s)
Toe-in 0-inch

4WD
- Front
  - Caster .... 2.08-degree(s)
  - Camber .... 0.66-degree(s)
  - Toe-in... 0-inch
- Rear
  - Camber .... -0.50-degree(s)
  - Toe-in... 3/32-inch

Talon
1990 thru 1994
- 1.8L
  - Front
    - Caster .... 2.30-degree(s)
    - Camber .... 0.25-degree(s)
    - Toe-in... 0-inch
  - Rear
    - Camber .... -0.75-degree(s)
    - Toe-in... 0-inch
- 2.0L (2WD)
  - Front
    - Caster .... 2.40-degree(s)
    - Camber .... 0.10-degree(s)
    - Toe-in... 0-inch
  - Rear
    - Camber .... -0.75-degree(s)
    - Toe-in... 0-inch
- 2.0L (4WD)
  - Front
    - Caster .... 2.30-degree(s)
    - Camber .... 0.15-degree(s)
    - Toe-in... 0-inch
  - Rear
    - Camber .... -1.56-degree(s)
    - Toe-in... 5/32-inch

1995
- ESi, TSi (4WD)
  - Front
    - Caster .... 4.66-degree(s)
    - Camber .... -0.08-degree(s)
    - Toe-in... 0-inch
- TSi (2WD)
  - Front
    - Caster .... 4.66-degree(s)
    - Camber .... -0.33-degree(s)
    - Toe-in... 0-inch
  - Rear (2WD with 14-inch wheels or 4WD with A/T)
    - Camber .... -1.33-degree(s)
    - Toe-in... 1/8-inch
  - Rear (2WD with 16-inch wheels or 4WD with M/T)
    - Camber .... -1.66-degree(s)
    - Toe-in... 1/8-inch

1996
- Front (with 14-inch wheels)
  - Caster... 4.33-degree(s)
  - Camber .... 0-degree(s)
  - Toe-in... 0-inch
- Front (with 16-inch wheels)
  - Caster .... 4.66-degree(s)
  - Camber .... -0.08-degree(s)
  - Toe-in... 0-inch
- Rear
  - Camber .... -1.33-degree(s)
  - Toe-in... 1/8-inch

Vision
1993 thru 1996
- Front
  - Caster .... 3.00-degree(s)

- Camber .... 0-degree(s)
- Toe-in... 0.10 inch
- Rear
  - Camber .... -0.10-degree(s)
  - Toe-in... 0.05 inch

FORD

Aspire
1994 thru 1996
- Front
  - Caster .... 1.66-degree(s)
  - Camber .... 0.80-degree(s)
  - Toe-in... 9/64-inch
- Rear
  - Camber .... -0.25-degree(s)
  - Toe-in... 1/8-inch

Contour
1995 and 1996
- Except SE
  - Front
    - Caster .... 2.30-degree(s)
    - Camber .... -0.40-degree(s)
    - Toe-in... -11/64-inch
  - Rear
    - Camber .... -0.40-degree(s)
    - Toe-in... 5/32-inch
- Contour SE
  - Front
    - Caster .... 2.40-degree(s)
    - Camber .... -0.53-degree(s)
    - Toe-in... -11/64-inch
  - Rear
    - Camber .... -0.53-degree(s)
    - Toe-in... 5/32-inch

Crown Victoria and Country Squire
1983 and 1984
- Front
  - Caster... 3.00-degree(s)
  - Camber .... 0.50-degree(s)
  - Toe-in... 1/8-inch

1985 and 1986
- Front
  - Caster... 3.00-degree(s)
  - Camber .... 0.50-degree(s)
  - Toe-in... 1/16-inch

1987 thru 1991
- Front
  - Caster... 3.50-degree(s)
  - Camber .... -0.50-degree(s)
  - Toe-in... 1/16-inch

1992 thru 1996
- Front
  - Caster... 5.50-degree(s)
  - Camber .... -0.50-degree(s)
  - Toe-in... 1/16-inch

Escort and EXP
1983 and 1984
- Front
  - Caster... 1.38-degree(s)
  - Camber
    - Left wheel .... 2.20-degree(s)
    - Right wheel .... 1.70-degree(s)
  - Toe-in... -7/64-inch

Rear
    Camber .... -1.25-degree(s)
    Toe-in 3/16-inch

1985
  Except turbo
    Front
      Caster .... 1.30-degree(s)
      Camber
        Left wheel .... 1.88-degree(s)
        Right wheel .... 1.44-degree(s)
      Toe-in .... -7/64-inch
    Rear
      Camber .... -1.20-degree(s)
      Toe-in 3/16-inch
  Turbo
    Front
      Caster .... 1.00-degree(s)
      Camber
        Left wheel .... 1.20-degree(s)
        Right wheel .... 0.75-degree(s)
      Toe-in .... -7/64-inch
    Rear
      Camber .... -1.30-degree(s)
      Toe-in 3/16-inch

1986 thru 1989
  Front
    Caster .... 2.34-degree(s)
    Camber
      Left wheel .... 1.16-degree(s)
      Right wheel .... 0.75-degree(s)
    Toe-in .... -3/32-inch
  Rear
    Camber .... -0.44-degree(s)
    Toe-in 3/16-inch

1990
  Except GT
    Front
      Caster .... 2.30-degree(s)
      Camber
        Left wheel .... 1.40-degree(s)
        Right wheel .... 1.00-degree(s)
      Toe-in .... -3/32-inch
    Rear
      Camber .... -0.34-degree(s)
      Toe-in 3/32-inch
  GT
    Front
      Caster .... 2.30-degree(s)
      Camber
        Left wheel .... 1.00-degree(s)
        Right wheel .... 0.56-degree(s)
      Toe-in .... -3/32-inch
    Rear
      Camber .... -0.34-degree(s)
      Toe-in 3/32-inch

1991 thru 1996
  Front
    Caster .... 1.90-degree(s)
    Camber .... -0.06-degree(s)
    Toe-in 3/32-inch
  Rear
    Camber .... -0.33-degree(s)
    Toe-in 3/32-inch

Fairmont and Futura

1983
  Front
    Caster .... 1.12-degree(s)
    Camber .... 0.44-degree(s)
    Toe-in 3/16-inch

Festiva

1988 thru 1993
  Front
    Caster .... 1.56-degree(s)
    Camber .... 0.66-degree(s)
    Toe-in 9/64-inch
  Rear
    Camber .... -0.25-degree(s)
    Toe-in 1/8-inch

LTD

1983
  Except wagon
    Front
      Caster .... 1.12-degree(s)
      Camber .... 0.44-degree(s)
      Toe-in 3/16-inch
  Wagon
    Front
      Caster .... 0.88-degree(s)
      Camber .... 0.50-degree(s)
      Toe-in 3/16-inch

1984
  Except wagon
    Front
      Caster .... 1.00-degree(s)
      Camber .... 0.44-degree(s)
      Toe-in 3/16-inch
  Wagon
    Front
      Caster .... 1.00-degree(s)
      Camber .... 0.50-degree(s)
      Toe-in 3/16-inch

1985 and 1986
  Except wagon
    Front
      Caster .... 0.88-degree(s)
      Camber .... 0.38-degree(s)
      Toe-in 3/16-inch
  Wagon
    Front
      Caster .... 0.75-degree(s)
      Camber .... 0.44-degree(s)
      Toe-in 3/16-inch

Mustang

1983 and 1984
  Front
    Caster .... 1.25-degree(s)
    Camber .... 0-degree(s)
    Toe-in 3/16-inch

1985 and 1986
  Front
    Caster .... 1.00-degree(s)
    Camber .... 0-degree(s)
    Toe-in 3/16-inch

1987
  Except 5.0L and GT
    Front
      Caster .... 0.80-degree(s)
      Camber .... -0.03-degree(s)
      Toe-in 3/16-inch
  5.0L and GT
    Front
      Caster .... 1.28-degree(s)
      Camber .... -0.12-degree(s)
      Toe-in 3/16-inch

1988 and 1989
 Except 5.0L and GT
  Front
   Caster .... 1.16-degree(s)
   Camber .... -0.10-degree(s)
   Toe-in 3/16-inch
 5.0L and GT
  Front
   Caster .... 1.28-degree(s)
   Camber .... 0.14-degree(s)
   Toe-in 3/16-inch
1990 thru 1993
 4-cylinder
  Front
   Caster .... 1.88-degree(s)
   Camber .... -0.50-degree(s)
   Toe-in .... -1/8-inch
 V8 except Cobra
  Front
   Caster .... 1.88-degree(s)
   Camber .... -0.63-degree(s)
   Toe-in .... -1/8-inch
 Cobra
  Front
   Caster .... 1.88-degree(s)
   Camber .... -0.63-degree(s)
   Toe-in 0-inch
1994 thru 1996
 Except GT and Cobra
  Front
   Caster .... 3.40-degree(s)
   Camber .... -0.60-degree(s)
   Toe-in 1/8-inch
 GT and Cobra
  Front
   Caster .... 3.40-degree(s)
   Camber .... -0.50-degree(s)
   Toe-in 1/8-inch

Probe
1989
 Front
  Caster... 1.22-degree(s)
  Camber .... 0.28-degree(s)
  Toe-in 1/8-inch
 Rear
  Camber .... 0.50-degree(s)
  Toe-in 0-inch
1990
 Front
  Caster... 1.25-degree(s)
  Camber .... -0.25-degree(s)
  Toe-in 1/8-inch
 Rear
  Camber .... -0.44-degree(s)
  Toe-in 1/8-inch
1991 and 1992
 Front
  Caster... 1.70-degree(s)
  Camber .... -0.25-degree(s)
  Toe-in 0-inch
 Rear
  Camber .... -0.44-degree(s)
  Toe-in 1/8-inch
1993 and 1994
 4-cylinder
  Front
   Caster .... 3.00-degree(s)
   Camber .... -0.70-degree(s)
   Toe-in 1/8-inch

  Rear
   Camber .... -0.90-degree(s)
   Toe-in .... -1/8-inch
 V6
  Front
   Caster .... 3.00-degree(s)
   Camber .... -0.90-degree(s)
   Toe-in 1/8-inch
  Rear
   Camber .... -0.36-degree(s)
   Toe-in 1/8-inch
1995
 4-cylinder
  Front
   Caster .... 3.00-degree(s)
   Camber .... -0.70-degree(s)
   Toe-in 1/8-inch
  Rear
   Camber .... -0.33-degree(s)
   Toe-in 1/8-inch
 V6
  Front
   Caster .... 3.00-degree(s)
   Camber .... -0.90-degree(s)
   Toe-in 1/8-inch
  Rear
   Camber .... -0.45-degree(s)
   Toe-in 1/8-inch
1996
 4-cylinder
  Front
   Caster .... 2.88-degree(s)
   Camber .... -0.70-degree(s)
   Toe-in 1/8-inch
  Rear
   Camber .... -0.70-degree(s)
   Toe-in 1/8-inch
 V6
  Front
   Caster .... 2.90-degree(s)
   Camber .... -0.90-degree(s)
   Toe-in 1/8-inch
  Rear
   Camber .... -0.90-degree(s)
   Toe-in .... -1/8-inch

Taurus
1986 and 1987
 Except wagon
  Front
   Caster .... 4.00-degree(s)
   Camber .... -0.50-degree(s)
   Toe-in .... -3/32-inch
  Rear
   Camber .... -0.90-degree(s)
   Toe-in 1/16-inch
 Wagon
  Front
   Caster .... 3.80-degree(s)
   Camber .... -0.44-degree(s)
   Toe-in .... -3/32-inch
  Rear
   Camber .... -0.63-degree(s)
   Toe-in 1/16-inch
1988
 Except wagon
  Front
   Caster .... 3.90-degree(s)
   Camber .... -0.44-degree(s)
   Toe-in.. -3/32-inch

Rear
    Camber .... -0.88-degree(s)
    Toe-in.. 1/16-inch

Wagon
  Front
    Caster .... 3.80-degree(s)
    Camber .... -0.88-degree(s)
    Toe-in.. -3/32-inch
  Rear
    Camber .... -0.88-degree(s)
    Toe-in.. 1/16-inch

1989
  Except wagon
    Front
      Caster .... 3.80-degree(s)
      Camber .... -0.50-degree(s)
      Toe-in.. -3/32-inch
    Rear
      Camber .... -0.88-degree(s)
      Toe-in.. 1/16-inch

Wagon
  Front
    Caster .... 3.70-degree(s)
    Camber .... -0.88-degree(s)
    Toe-in.. -3/32-inch
  Rear
    Camber .... -0.88-degree(s)
    Toe-in.. 1/16-inch

1990 thru 1996
  Except wagon
    Front
      Caster .... 3.80-degree(s)
      Camber .... -0.50-degree(s)
      Toe-in.. 7/64-inch
    Rear
      Camber .... -0.60-degree(s)
      Toe-in.. 1/16-inch

Wagon
  Front
    Caster .... 3.70-degree(s)
    Camber .... -0.40-degree(s)
    Toe-in.. -7/64-inch
  Rear
    Camber .... -0.90-degree(s)
    Toe-in.. 1/16-inch

Tempo
1984
  Front
    Caster... 1.30-degree(s)
    Toe-in.. -7/64-inch
  Left wheel
    Camber .... 1.88-degree(s)
  Right wheel
    Camber .... 1.50-degree(s)
  Rear
    Camber .... -0.25-degree(s)
    Toe-in.. 0-inch

1985
  Front
    Caster... 1.25-degree(s)
    Toe-in.. -7/64-inch
  Left wheel
    Camber .... 2.00-degree(s)
  Right wheel
    Camber .... 1.50-degree(s)
  Rear
    Camber .... -0.20-degree(s)
    Toe-in.. 0-inch

1986 thru 1994
  Front
    Caster... 2.45-degree(s)
    Camber
      Left wheel .... 1.40-degree(s)
      Right wheel .... 0.95-degree(s)
    Toe-in.. -7/64-inch
  Rear, 2WD (1986 thru 1992)
    Camber .... -0.16-degree(s)
    Toe-in.. 0-inch
  Rear, 2WD (1993 and 1994)
    Camber .... -0.55-degree(s)
    Toe-in.. 0-inch
  Rear, 4WD (1987 and 1988)
    Camber .... 0.40-degree(s)
    Toe-in.. 0-inch
  Rear, 4WD (1989 and 1990)
    Camber .... 0.34-degree(s)
    Toe-in.. 0-inch

Thunderbird
1983
  Front
    Caster... 1.25-degree(s)
    Camber .... 0.25-degree(s)
    Toe-in.. 3/16-inch
1984
  Front
    Caster... 1.00-degree(s)
    Camber .... 0.25-degree(s)
    Toe-in.. 3/16-inch
1985 and 1986
  Front
    Caster... 0.75-degree(s)
    Camber .... 0.25-degree(s)
    Toe-in.. 3/16-inch
1987
  Except turbo
    Front
      Caster .... 1.22-degree(s)
      Camber .... -0.66-degree(s)
      Toe-in.. 3/16-inch
  Turbo
    Front
      Caster .... 1.16-degree(s)
      Camber .... 0.22-degree(s)
      Toe-in.. 3/16-inch
1988
  Except turbo
    Front
      Caster .... 1.22-degree(s)
      Camber .... -0.66-degree(s)
      Toe-in.. 3/16-inch
  Turbo
    Front
      Caster .... 1.34-degree(s)
      Camber .... 0-degree(s)
      Toe-in.. 3/16-inch
1989 thru 1991
  Front
    Caster... 5.50-degree(s)
    Camber .... -0.50-degree(s)
    Toe-in 1/8-inch
  Rear
    Camber .... -0.50-degree(s)
    Toe-in 1/16-inch
1992 thru 1996
  Front
    Caster... 5.50-degree(s)

Camber .... -0.50-degree(s)
Toe-in 5/64-inch
Rear
  Camber .... -0.50-degree(s)
  Toe-in 1/16-inch

Windstar
1995 and 1996
  Front
    Caster
      Left Wheel .... 3.30-degree(s)
      Right Wheel .... 3.80-degree(s)
    Camber .... -0.25-degree(s)
    Toe-in .... -3/64-inch
  Rear
    Camber .... 0-degree(s)
    Toe-in 1/64-inch

FORD TRUCKS

Includes Mercury Mountaineer

Note: *Caster and camber adjustments vary with the ride height, which is measured between the bottom of the frame and the top of the I-beam axle housing. This measurement is refered to as Dimension "A". The vehicle must be parked on a smooth, level surface when the measurement is taken.*

Aerostar
1985 thru 1988
  Caster .... 4.00-degree(s)
  Camber .... 0.22.00-degree(s)
  Toe-in 1/32-inch
1989 thru 1991
  2WD
    Caster .... 3.50-degree(s)
    Camber .... 0-degree(s)
    Toe-in 1/32-inch
  4WD
    Caster .... 3.50-degree(s)
    Camber .... 0-degree(s)
    Toe-in .... -1/32-inch
1992
  2WD
    Caster .... 3.50-degree(s)
    Camber .... 0-degree(s)
    Toe-in 1/32-inch
  4WD
    Caster .... 4.00-degree(s)
    Camber .... 0-degree(s)
    Toe-in 1/32-inch
1993 thru 1996
  2WD
    Caster .... 4.00-degree(s)
    Camber .... 0.25-degree(s)
    Toe-in 1/32-inch
  4WD
    Caster .... 4.00-degree(s)
    Camber .... 0.25-degree(s)
    Toe-in .... -1/32-inch

Explorer and Mountaineer
1991 and 1992
  2WD
    Caster .... 5.00-degree(s)
    Camber .... 1.00-degree(s)
    Toe-in 1/32-inch
  4WD
    Caster .... 4.00-degree(s)

Camber .... 1.00-degree(s)
Toe-in 1/32-inch
1993 and 1994
  2WD
    Caster .... 4.00-degree(s)
    Camber .... 0.25-degree(s)
    Toe-in 1/32-inch
  4WD
    Caster .... 4.00-degree(s)
    Camber .... 0.25-degree(s)
    Toe-in 1/32-inch
1995 and 1996
  Without Automatic Ride Control
    Caster .... 4.00-degree(s)
    Camber .... -0.50-degree(s)
    Toe-in 5/32-inch
  With Automatic Ride Control
    Caster .... 4.00-degree(s)
    Camber .... -0.70-degree(s)
    Toe-in 5/32-inch

E100 and 150
1983 and 1984 (early)
  Dimension A - 4-inches
    Caster .... 5.50-degree(s)
    Camber .... 0-degree(s)
    Toe-in 1/32-inch
  Dimension A - 4 1/2-inches
    Caster .... 4.25-degree(s)
    Camber .... 1.00-degree(s)
    Toe-in 1/32-inch
  Dimension A - 5 3/8-inches
    Caster .... 3.00-degree(s)
    Camber .... 2.00-degree(s)
    Toe-in 1/32-inch
  Dimension A - 6 1/4-inches
    Caster .... 1.75-degree(s)
    Camber .... 3.00-degree(s)
    Toe-in 1/32-inch
1984 (late) thru 1988
  Dimension A - 4-inches
    Caster .... 8.50-degree(s)
    Camber .... -0.38-degree(s)
    Toe-in 1/32-inch
  Dimension A - 4 1/2-inches
    Caster .... 7.25-degree(s)
    Camber .... 0.63-degree(s)
    Toe-in 1/32-inch
  Dimension A - 5-inches
    Caster .... 6.00-degree(s)
    Camber .... 1.63-degree(s)
    Toe-in 1/32-inch
  Dimension A - 5 1/2-inches
    Caster .... 4.15-degree(s)
    Camber .... 2.63-degree(s)
    Toe-in 1/32-inch
  Dimension A - 6-inches
    Caster .... 3.50-degree(s)
    Camber .... 3.63-degree(s)
    Toe-in 1/32-inch
1989 thru 1991
  Caster .... 5.00-degree(s)
  Camber .... 0.25-degree(s)
  Toe-in 1/32-inch
1992 thru 1996
  Caster .... 4.00-degree(s)
  Camber .... 0.25-degree(s)
  Toe-in 1/32-inch

E250 and 350

1983 thru 1988
Dimension A - 3 1/2-inches
Caster .... 9.25-degree(s)
Camber .... -0.70-degree(s)
Toe-in 1/32-inch
Dimension A - 4-inches
Caster .... 8.20-degree(s)
Camber .... 0.25-degree(s)
Toe-in 1/32-inch
Dimension A - 4 1/2-inches
Caster .... 6.63-degree(s)
Camber .... 1.25-degree(s)
Toe-in 1/32-inch
Dimension A - 5 3/8-inches
Caster .... 5.50-degree(s)
Camber .... 2.25-degree(s)
Toe-in 1/32-inch
Dimension A - 6 1/4-inches
Caster .... 4.25-degree(s)
Camber .... 3.25-degree(s)
Toe-in 1/32-inch

1989 thru 1991
Caster .... 5.00-degree(s)
Camber .... 0.50-degree(s)
Toe-in 1/32-inch

1992 thru 1996
Caster .... 4.00-degree(s)
Camber .... 0.50-degree(s)
Toe-in 1/32-inch

F100, 150 and Bronco

1983 thru 1988
2WD
Dimension A - 2 3/4-inches
Caster .... 6.38-degree(s)
Camber .... -0.75-degree(s)
Toe-in 1/32-inch
Dimension A - 3 1/8-inches
Caster .... 6.00-degree(s)
Camber .... 0-degree(s)
Toe-in 1/32-inch
Dimension A - 3 1/2-inches
Caster .... 5.12-degree(s)
Camber .... 1.00-degree(s)
Toe-in 1/32-inch
Dimension A - 4-inches
Caster .... 4.38-degree(s)
Camber .... 1.75-degree(s)
Toe-in 1/32-inch
Dimension A - 4 5/16-inches
Caster .... 3.50-degree(s)
Camber .... 2.75-degree(s)
Toe-in 1/32-inch
Dimension A - 4 3/4-inches
Caster .... 2.75-degree(s)
Camber .... 4.00-degree(s)
Toe-in 1/32-inch
4WD
Dimension A - 3 1/8-inches
Caster .... 7.00-degree(s)
Camber .... -1.00 degree(s)
Toe-in 1/32-inch
Dimension A - 3 1/2-inches
Caster .... 6.00-degree(s)
Camber .... 0-degree(s)
Toe-in 1/32-inch

Dimension A - 4-inches
Caster .... 5.00-degree(s)
Camber .... 1.00-degree(s)
Toe-in 1/32-inch
Dimension A - 4 1/4-inches
Caster .... 4.00-degree(s)
Camber .... 2.00-degree(s)
Toe-in 1/32-inch
Dimension A - 4 3/4-inches
Caster .... 3.00-degree(s)
Camber .... 3.00-degree(s)
Toe-in 1/32-inch

1989 thru 1996
2WD
Caster .... 4.00-degree(s)
Camber .... 0.25-degree(s)
Toe-in 1/32-inch
4WD
Caster .... 4.00-degree(s)
Camber .... 0.25-degree(s)
Toe-in 1/32-inch

F250 and 350 (2WD)

1983 and 1984
Dimension A - 2 3/4-inches
Caster .... 6.00-degree(s)
Camber .... 0-degree(s)
Toe-in 1/32-inch
Dimension A - 3 1/8-inches
Caster .... 5.00-degree(s)
Camber .... 1.00-degree(s)
Toe-in 1/32-inch
Dimension A - 3 1/2-inches
Caster .... 4.12-degree(s)
Camber .... 1.75-degree(s)
Toe-in 1/32-inch
Dimension A - 4-inches
Caster .... 3.38-degree(s)
Camber .... 2.88-degree(s)
Toe-in 1/32-inch
Dimension A - 4 5/16-inches
Caster .... 2.50-degree(s)
Camber .... 3.88-degree(s)
Toe-in 1/32-inch

1985
Dimension A - 2 3/4-inches
Caster .... 6.00-degree(s)
Camber .... 0.12-degree(s)
Toe-in 3/32-inch
Dimension A - 3 1/2-inches
Caster .... 5.12-degree(s)
Camber .... 0.88-degree(s)
Toe-in 3/32-inch
Dimension A - 4 1/4-inches
Caster .... 4.38-degree(s)
Camber .... 1.75-degree(s)
Toe-in 3/32-inch
Dimension A - 4 3/4-inches
Caster .... 3.50-degree(s)
Camber .... 2.75-degree(s)
Toe-in 3/32-inch
Dimension A - 5 1/2-inches
Caster .... 2.63-degree(s)
Camber .... 3.75-degree(s)
Toe-in 3/32-inch

1986
Dimension A - 2 3/4-inches
Caster .... 7.00-degree(s)
Camber .... -1.00-degree(s)
Toe-in 1/32-inch

Dimension A - 3 1/8-inches
    Caster.. 6.00-degree(s)
    Camber .... 0-degree(s)
    Toe-in.. 1/32-inch

Dimension A - 3 1/2-inches
    Caster.. 5.00-degree(s)
    Camber .... 0.88-degree(s)
    Toe-in.. 1/32-inch

Dimension A - 4-inches
    Caster.. 4.00-degree(s)
    Camber .... 1.75-degree(s)
    Toe-in.. 1/32-inch

Dimension A - 4 5/16-inches
    Caster.. 3.00-degree(s)
    Camber .... 2.75-degree(s)
    Toe-in.. 1/32-inch

1987 and 1988

Dimension A - 3 1/8-inches
    Caster.. 9.50-degree(s)
    Camber .... -2.00-degree(s)
    Toe-in.. 1/32-inch

Dimension A - 3 1/2-inches
    Caster.. 8.75-degree(s)
    Camber .... -1.00-degree(s)
    Toe-in.. 1/32-inch

Dimension A - 4-inches
    Caster.. 7.50-degree(s)
    Camber .... 0-degree(s)
    Toe-in.. 1/32-inch

Dimension A - 4 5/16-inches
    Caster.. 6.50-degree(s)
    Camber .... 1.00-degree(s)
    Too in.. 1/32-inch

Dimension A - 4 3/4-inches
    Caster.. 5.75-degree(s)
    Camber .... 2.00-degree(s)
    Toe-in.. 1/32-inch

Dimension A - 5-inches
    Caster.. 4.50-degree(s)
    Camber .... 3.00-degree(s)
    Toe-in.. 1/32-inch

1989 thru 1992

F250
    Caster.. 4.00-degree(s)
    Camber .... 0.25-degree(s)
    Toe-in.. 1/32-inch

F350 (except Super duty)
    Caster.. 4.00-degree(s)
    Camber .... 0.50-degree(s)
    Toe-in.. 1/32-inch

Super Duty
    Caster.. 0.63-degree(s)
    Camber .... 0-degree(s)
    Toe-in.. 1/32-inch

1993 thru 1996

F250
    Caster.. 4.00-degree(s)
    Camber .... 0.25-degree(s)
    Toe-in.. 1/32-inch

F350
    Except with dual rear wheels or Super duty
        Caster.. 4.00-degree(s)
        Camber .... 0.50-degree(s)
        Toe-in.. 1/32-inch
    With dual rear wheels
        Caster.. 3.50-degree(s)
        Camber .... 0.50-degree(s)
        Toe-in.. 1/32-inch

Super duty
    Caster.. 4.00-degree(s)
    Camber .... 0-degree(s)
    Toe-in.. 1/32-inch

F250 and 350 4WD

1983 thru 1988

Except with Monobeam front axle
    Dimension A - 5-inches
        Caster.. 4.12-degree(s)
        Camber .... -1.00-degree(s)
        Toe-in.. 1/32-inch
    Dimension A - 5 3/8-inches
        Caster.. 4.25-degree(s)
        Camber .... 0-degree(s)
        Toe-in.. 1/32-inch
    Dimension A - 5 7/8-inches
        Caster.. 4.38-degree(s)
        Camber .... 1.12-degree(s)
        Toe-in.. 1/32-inch
    Dimension A - 6 1/4-inches
        Caster.. 4.50-degree(s)
        Camber .... 2.12-degree(s)
        Toe-in.. 1/32-inch
    Dimension A - 6 3/4-inches
        Caster.. 4.63-degree(s)
        Camber .... 125-degree(s)
        Toe-in.. 1/32-inch
With Monobearn front axle
    Dimension A - 4 3/4-inches
        Caster.. 5.00-degree(s)
        Camber .... 1.50-degree(s)
        Too in.. 1/32-inch

1989 thru 1992

F250
    Caster.. 4.00-degree(s)
    Camber .... 0.25-degree(s)
    Toe-in.. 1/32-inch

F350
    Except with dual rear wheels or Super Duty
        Caster.. 3.50-degree(s)
        Camber .... 0-degree(s)
        Toe-in.. 1/32-inch
    With dual rear wheels
        Caster.. 3.50-degree(s)
        Camber .... 0.50-degree(s)
        Toe-in.. 1/32-inch
    Super duty
        Caster.. 3.50-degree(s)
        Camber .... 0-degree(s)
        Toe-in.. 1/32-inch

1993 thru 1996

F250
    Caster.. 4.00-degree(s)
    Camber .... 0.25-degree(s)
    Toe-in.. 1/32-inch

F350
    Caster.. 4.00-degree(s)
    Camber .... 0-degree(s)
    Toe-in.. 1/32-inch

Ranger and Bronco II

1983 thru 1988

2WD (with stamped axle)
    Dimension A - 3 1/8-inches
        Caster .... 6.75-degree(s)
        Camber .... -1.00-degree(s)
        Toe-in.. 1/32-inch

Dimension A - 3 1/4-inches
- Caster .... 6.00-degree(s)
- Camber .... -0.50-degree(s)
- Toe-in 1/32-inch

Dimension A - 3 5/8-inches
- Caster .... 5.50-degree(s)
- Camber .... 0-degree(s)
- Toe-in 1/32-inch

Dimension A - 3 1/2-inches
- Caster .... 5.00-degree(s)
- Camber .... 0.50-degree(s)
- Toe-in 1/32-inch

Dimension A - 3 3/4-inches
- Caster .... 4.25-degree(s)
- Camber .... 1.00-degree(s)
- Toe-in 1/32-inch

Dimension A - 4-inches
- Caster .... 3.88-degree(s)
- Camber .... 1.50-degree(s)
- Toe-in 1/32-inch

Dimension A - 4 1/8-inches
- Caster .... 3.25-degree(s)
- Camber .... 2.00-degree(s)
- Toe-in 1/32-inch

Dimension A - 4 5/16-inches
- Caster .... 2.75-degree(s)
- Camber .... 2.50-degree(s)
- Toe-in 1/32-inch

2WD (with forged axle)

Dimension A - 3 1/2-inches
- Caster .... 5.75-degree(s)
- Camber .... -0.25-degree(s)
- Toe-in 1/32-inch

Dimension A - 3 3/4-inches
- Caster .... 5.25-degree(s)
- Camber .... 0.25-degree(s)
- Toe-in 1/32-inch

Dimension A - 4-inches
- Caster .... 4.75-degree(s)
- Camber .... 0.75-degree(s)
- Toe-in 1/32-inch

Dimension A - 4 1/8-inches
- Caster .... 4.25-degree(s)
- Camber .... 1.25-degree(s)
- Toe-in 1/32-inch

Dimension A - 4 3/8-inches
- Caster .... 3.75-degree(s)
- Camber .... 1.75-degree(s)
- Toe-in 1/32-inch

Dimension A - 4 1/2-inches
- Caster .... 3.25-degree(s)
- Camber .... 2.25-degree(s)
- Toe-in 1/32-inch

4WD (except Ranger STX)

Dimension A - 2 1/2-inches
- Caster .... 7.25-degree(s)
- Camber .... -2.00-degree(s)
- Toe-in 1/32-inch

Dimension A - 2 3/4-inches
- Caster .... 6.75-degree(s)
- Camber .... -1.00-degree(s)
- Toe-in 1/32-inch

Dimension A - 3 1/8-inches
- Caster .... 5.50-degree(s)
- Camber .... 0-degree(s)
- Toe-in 1/32-inch

Dimension A - 3 1/2-inches
- Caster .... 4.50-degree(s)

- Camber .... 1.00-degree(s)
- Toe-in 1/32-inch

Dimension A - 4-inches
- Caster .... 3.50-degree(s)
- Camber .... 2.00-degree(s)
- Toe-in 1/32-inch

Dimension A - 4 5/16-inches
- Caster .... 2.50-degree(s)
- Camber .... 3.00-degree(s)
- Toe-in 1/32-inch

4WD (Ranger STX)

Dimension A - 4 1/8-inches
- Caster .... 7.00-degree(s)
- Camber .... -1.50-degree(s)
- Toe-in 1/32-inch

Dimension A - 4 3/8-inches
- Caster .... 6.00-degree(s)
- Camber .... -0.38-degree(s)
- Toe-in 1/32-inch

Dimension A - 4 5/8-inches
- Caster .... 4.88-degree(s)
- Camber .... 0.50-degree(s)
- Toe-in 1/32-inch

Dimension A - 4 7/8-inches
- Caster .... 3.88-degree(s)
- Camber .... 1.50-degree(s)
- Toe-in 1/32-inch

Dimension A - 5 1/8-inches
- Caster .... 2.88-degree(s)
- Camber .... 2.50-degree(s)
- Toe-in 1/32-inch

1989 thru 1991

2WD
- Caster .... 4.50-degree(s)
- Camber .... 0.50-degree(s)
- Toe-in 0-inch

4WD (except STX)

With Dana 28 axle
- Caster .... 3.50-degree(s)
- Camber .... 0.50-degree(s)
- Toe-in 1/32-inch

With Dana 35 axle
- Caster .... 4.00-degree(s)
- Camber .... 0.50-degree(s)
- Toe-in 1/32-inch

4WD (STX)

With Dana 28 axle
- Caster .... 3.50-degree(s)
- Camber .... 0.50-degree(s)
- Toe-in .... .03-inch

With Dana 35 axle
- Caster .... 4.00-degree(s)
- Camber .... 0.50-degree(s)
- Toe-in 1/32-inch

1992

2WD
- Caster .... 5.00-degree(s)
- Camber .... 1.00-degree(s)
- Toe-in 1/32-inch

4WD
- Caster .... 4.00-degree(s)
- Camber .... 1.00-degree(s)
- Toe-in 1/32-inch

1993 thru 1996

2WD
- Caster .... 4.00-degree(s)
- Camber .... 0.25-degree(s)
- Toe-in 1/32-inch

4WD
- Caster .... 4.00-degree(s)
- Camber .... 0.25-degree(s)
- Toe-in 1/32-inch

GEO

Metro
1989 thru 1996
  Front
- Caster .... 3.00-degree(s)
- Camber .... 0-degree(s)
- Toe-in 5/64-inch
  Rear
- Camber .... 0-degree(s)
- Toe-in 5/64-inch

Prizm
1989 thru 1992
  Except GS1
    Front
- Caster .... 1.45-degree(s)
- Camber .... -0.15-degree(s)
- Toe-in 1/32-inch
    Rear
- Camber .... -0.56-degree(s)
- Toe-in 5/32-inch
  GS1
    Front
- Caster .... 1.45-degree(s)
- Camber .... 0-degree(s)
- Toe-in 1/32-inch
    Rear
- Camber .... -0.70-degree(s)
- Toe-in 5/32-inch
1993 thru 1996
  Front
- Caster .... 1.33-degree(s)
- Camber .... -0.17-degree(s)
- Toe-in 3/64-inch
  Rear (1993 thru 1995)
- Camber .... -0.90-degree(s)
- Toe-in 5/32-inch
  Rear (1996)
- Camber .... -0.90-degree(s)
- Toe-in 13/64-inch

Spectrum
1989
  Front
- Caster .... 2.25-degree(s)
- Camber .... 0.30-degree(s)
- Toe-in 0-inch
  Rear
- Camber .... 0-degree(s)
- Toe-in 7/32-inch

Storm
1989 thru 1995
  Front
- Caster .... 4.50-degree(s)
- Camber .... -0.50-degree(s)
- Toe-in 0-inch
  Rear
- Camber .... -0.50-degree(s)
- Toe-in 5/64-inch

1990 and 1991
  Front
- Caster .... 3.00-degree(s)
- Camber .... -0.50-degree(s)
- Toe-in 0-inch
  Rear
- Camber .... -0.50-degree(s)
- Toe-in 5/32-inch

Tracker
1989 thru 1995
  Front
- Caster .... 1.50-degree(s)
- Camber .... 0.50-degree(s)
- Toe-in 5/32-inch

HONDA

Accord
1983 and 1984
  Front
- Caster .... 1.44-degree(s)
- Camber .... 0-degree(s)
- Toe-in 0-inch
  Rear
- Camber .... 0-degree(s)
- Toe-in 0-inch
1985
  Front
- Caster .... 1.50-degree(s)
- Camber .... 0-degree(s)
- Toe-in 0-inch
  Rear
- Camber .... 0-degree(s)
- Toe-in 3/32-inch
1986 thru 1989
  Front
- Caster .... 0.50-degree(s)
- Camber .... 0-degree(s)
- Toe-in 0-inch
  Rear
- Camber .... 0-degree(s)
- Toe-in 3/32-inch
1990 thru 1993
  Front
- Caster .... 3.00-degree(s)
- Camber .... 0-degree(s)
- Toe-in 0-inch
  Rear (except wagon)
- Camber .... -0.50-degree(s)
- Toe-in 5/64-inch
  Rear (wagon)
- Camber .... -0.50-degree(s)
- Toe-in 5/32-inch
1994 thru 1996
  Front
- Caster .... 3.00-degree(s)
- Camber .... 0-degree(s)
- Toe-in 0-inch
  Rear (1994)
- Camber .... -0.42-degree(s)
- Toe-in 5/64-inch
  Rear (1995)
- Camber .... -0.50-degree(s)
- Toe-in 5/64-inch
  Rear (1996)
- Camber .... -0.25-degree(s)
- Toe-in 5/64-inch

Civic and CRX

1983
Hatchback and sedan
Front
Caster .... 2.00-degree(s)
Camber .... 0-degree(s)
Toe-in 0-inch
Rear
Camber .... -0.25-degree(s)
Toe-in 5/64-inch
Wagon
Front
Caster .... 1.30-degree(s)
Camber .... 0-degree(s)
Toe-in 0-inch

1984
Except wagon and CRX
Front
Caster .... 2.34-degree(s)
Camber .... 0-degree(s)
Toe-in 0-inch
Rear
Camber .... -0.75-degree(s)
Toe-in 3/32-inch
Wagon
Front
Caster .... 2.00-degree(s)
Camber .... 0-degree(s)
Toe-in 0-inch
Rear
Camber .... -0.75-degree(s)
Toe-in 3/32-inch
CRX
Front
Caster .... 2.44-degree(s)
Camber .... 0-degree(s)
Toe-in 0-inch
Rear
Camber .... -0.75-degree(s)
Toe-in 3/32-inch

1985 thru 1987
Except wagon and CRX
Front
Caster
  w/o power steering .... 2.50-degree(s)
  w/power steering .... 3.00-degree(s)
Camber .... 0-degree(s)
Toe-in 0-inch
Rear
Camber .... -0.75-degree(s)
Toe-in 3/32-inch
Wagon
2WD
Front
Caster .... 2.00-degree(s)
Camber .... 0-degree(s)
Toe-in 0-inch
4WD
Front
Caster .... 1.90-degree(s)
Camber .... 0.70-degree(s)
Toe-in 0-inch
Rear
Camber .... 0-degree(s)
Toe-in 0-inch

CRX
Except 1987
Front
Caster .... 2.44-degree(s)
Camber .... -0.20-degree(s)
Toe-in 0-inch
Rear
Camber .... -0.75-degree(s)
Toe-in 3/32-inch
1987
Front
Caster .... 2.50-degree(s)
Camber .... 0-degree(s)
Toe-in 0-inch
Rear
Camber .... -0.75-degree(s)
Toe-in 3/32-inch

1988 thru 1991
Except wagon and CRX
Front
Caster .... 3.00-degree(s)
Camber .... 0-degree(s)
Toe-in 0-inch
Rear
Camber .... -0.40-degree(s)
Toe-in 3/32-inch
Wagon
2WD
Front
Caster .... 3.00-degree(s)
Camber .... 0.30-degree(s)
Toe-in 0-inch
Rear
Camber .... -0.38-degree(s)
Toe-in 3/32-inch
4WD
Front
Caster .... 2.90-degree(s)
Camber .... 0.56-degree(s)
Toe-in 0-inch
Rear
Camber .... 0-degree(s)
Toe-in 3/32-inch
CRX
Front
Caster .... 3.00-degree(s)
Camber .... 0-degree(s)
Toe-in 0-inch
Rear
Camber .... -0.40-degree(s)
Toe-in 3/32-inch

1992 thru 1995
Front
Caster .... 1.17-degree(s)
Camber .... 0-degree(s)
Toe-in 0-inch
Rear
Camber .... -0.33-degree(s)
Toe-in 5/64-inch

1996
Front
Caster .... 1.66-degree(s)
Camber .... 0-degree(s)
Toe-in 3/64-inch
Rear
Camber .... -1.00-degree(s)
Toe-in 5/64-inch

Del Sol

1993 thru 1995

- Front
  - Caster.................................... 1.17-degree(s)
  - Camber
    - Except VTEC engine......................... -0.25-degree(s)
    - VTEC engine .... -0.33-degree(s)
  - Toe-in..................................... 0-inch
- Rear
  - Camber .... -0.50-degree(s)
  - Toe-in..................................... 5/64-inch

Odyssey

1995 and 1996

- Front
  - Caster.................................... 3.00-degree(s)
  - Camber .... 0-degree(s)
  - Toe-in..................................... 0-inch
- Rear (with 13-inch wheels)
  - Camber .... 0-degree(s)
  - Toe-in..................................... 0-inch
- Rear (with 14-inch or 15-inch wheels)
  - Camber .... -0.50-degree(s)
  - Toe-in..................................... 0-inch

Prelude

1983 thru 1987

- Front
  - Caster.................................... 0-degree(s)
  - Camber .... 0-degree(s)
  - Toe-in..................................... 0-inch
- Rear
  - Camber .... 0-degree(s)
  - Toe-in..................................... 3/16-inch

1988 thru 1991

- 2-wheel steering
  - Front
    - Caster .... 2.30-degree(s)
    - Camber .... 0-degree(s)
    - Toe-in.................................... 3/32-inch
  - Rear
    - Camber .... -0.30-degree(s)
    - Toe-in.................................... 3/32-inch
- 4-wheel steering
  - Front
    - Caster .... 2.30-degree(s)
    - Camber .... 0-degree(s)
    - Toe-in.................................... 0-inch
  - Rear
    - Camber .... -0.30-degree(s)
    - Toe-in.................................... 1/8-inch

1992 thru 1996

- Front
  - Caster.................................... 2.66-degree(s)
  - Camber .... 0-degree(s)
  - Toe-in..................................... 0-inch
- Rear (2-wheel steering)
  - Camber .... -0.75-degree(s)
  - Toe-in..................................... 5/64-inch
- Rear (4-wheel steering)
  - Camber .... -0.75-degree(s)
  - Toe-in..................................... 5/64-inch

Passport

1994 thru 1996

- Front
  - Caster.................................... 2.33-degree(s)
  - Camber .... 0.50-degree(s)
  - Toe-in..................................... 5/64-inch

HYUNDAI

Accent

1995 and 1996

- Front
  - Caster.................................... 2.00-degree(s)
  - Camber .... 0-degree(s)
  - Toe-in..................................... 0-inch
- Rear
  - Camber .... -0.68-degree(s)
  - Toe-in..................................... 5/64-inch

Elantra

1992

- Front
  - Caster.................................... 2.35-degree(s)
  - Camber .... 0-degree(s)
  - Toe-in..................................... 0-inch
- Rear
  - Camber .... -0.66-degree(s)
  - Toe-in..................................... 0-inch

1993 thru 1995

- Front
  - Caster.................................... 2.56-degree(s)
  - Camber .... 0-degree(s)
  - Toe-in..................................... 0-inch
- Rear
  - Camber .... -0.66-degree(s)
  - Toe-in..................................... 0-inch

Excel

1986 thru 1989

- Front
  - Caster.................................... 0.80-degree(s)
  - Camber .... 0.50-degree(s)
  - Toe-in..................................... 1/32-inch
- Rear
  - Camber .... -0.66-degree(s)
  - Toe-in..................................... 0-inch

1990 thru 1994

- Front
  - Caster
    - w/o power steering .... 1.00-degree(s)
    - w/power steering .... 1.66-degree(s)
  - Camber .... 0-degree(s)
  - Toe-in..................................... 1/32-inch
- Rear (1990 and 1991)
  - Camber .... -0.66-degree(s)
  - Toe-in..................................... 0-inch
- Rear (1992 thru 1994)
  - Camber .... -0.66-degree(s)
  - Toe-in..................................... 1/16-inch

Pony

1983 thru 1987

- Front
  - Caster.................................... 1.75-degree(s)
  - Camber .... 1.00-degree(s)
  - Toe-in..................................... 5/64-inch

Scoupe

1991 thru 1995

- Front
  - Caster
    - w/o power steering .... 1.00-degree(s)
    - w/power steering .... 1.75-degree(s)
  - Camber .... -0.15-degree(s)
  - Toe-in..................................... 1/32-inch

Haynes suspension, steering and driveline manual

Rear (1991)
  Camber .... -0.66-degree(s)
  Toe-in 0-inch
Rear (1992 thru 1995)
  Camber .... -0.66-degree(s)
  Toe-in 1/16-inch

Sonata
1989 thru 1994
  Front
    Caster .... 2.00-degree(s)
    Camber .... 0.50-degree(s)
    Toe-in 0-inch
  Rear
    Camber .... 0-degree(s)
    Toe-in 0-inch
1995 and 1996
  Front
    Caster .... 2.66-degree(s)
    Camber .... 0-degree(s)
    Toe-in 0-inch
  Rear
    Camber .... -0.50-degree(s)
    Toe-in 0-inch

Stellar
1984 thru 1986
  Except vehicles manufactured after 7-1-86
    Front
      Caster .... 2.50-degree(s)
      Camber .... 0-degree(s)
      Toe-in 1/16-inch
  Vehicles manufactured after 7-1-86
    Front
      Caster .... 2.75-degree(s)
      Camber .... 0.30-degree(s)
      Toe-in 1/32-inch
1987 and 1988
  Front
    Caster .... 2.75-degree(s)
    Camber .... 0.50-degree(s)
    Toe-in 5/64-inch

INFINITI

G20
1991 thru 1996
  Front
    Caster .... 1.80-degree(s)
    Camber .... 0-degree(s)
    Toe-in 3/64-inch
  Rear (1991 thru 1993)
    Camber .... -1.25-degree(s)
    Toe-in 3/64-inch
  Rear (1993 thru 1995)
    Camber .... -1.00-degree(s)
    Toe-in 0-inch
  Rear (1996)
    Camber .... -1.08-degree(s)
    Toe-in 3/64-inch

I30
1995 and 1996
  Front
    Caster .... 2.75-degree(s)
    Camber .... -0.25-degree(s)
    Toe-in 5/64-inch

Rear
  Camber .... -1.00-degree(s)
  Toe-in 3/64-inch

J30
1993 thru 1996
  Front
    Caster .... 6.58-degree(s)
    Camber .... -0.75-degree(s)
    Toe-in 3/64-inch
  Rear
    Camber .... -1.00-degree(s)
    Toe-in 7/64-inch

M30
1993-90
  Front
    Caster .... 4.70-degree(s)
    Camber .... 0.20-degree(s)
    Toe-in 0-inch
  Rear
    Camber .... -0.30-degree(s)
    Toe-in .... -3/32-inch

Q45
1990 thru 1996
  w/o Active Suspension
    Front
      Caster .... 6.50-degree(s)
      Camber .... -0.80-degree(s)
      Toe-in 3/64-inch
    Rear
      Camber .... -1.10-degree(s)
      Toe-in 5/64-inch
  w/Active Suspension
    Front
      Caster .... 6.90-degree(s)
      Camber .... -0.90-degree(s)
      Toe-in 0-inch
    Rear
      Camber .... -1.50-degree(s)
      Toe-in 5/64-inch

ISUZU

I-Mark
1983
  Front
    Caster .... 5.12-degree(s)
    Camber .... 0.12-degree(s)
    Toe-in 1/8-inch
1984 and 1985 (rear-wheel drive models)
  Front
    Caster .... 4.90-degree(s)
    Camber .... -0.12-degree(s)
    Toe-in 1/8-inch
1985 thru 1989 (front-wheel drive models)
  Front
    Caster .... 2.25-degree(s)
    Camber .... 0.30-degree(s)
    Toe-in 0-inch
  Rear (1985 and 1986)
    Camber .... 0-degree(s)
    Toe-in 1/16-inch
  Rear (1987 thru 1989)
    Camber .... -0.15-degree(s)
    Toe-in 5/64-inch

Impulse
1983 thru 1987
  Front
    Caster .... 4.75-degree(s)
    Camber .... -0.25-degree(s)
    Toe-in 1/16-inch
1988 and 1989
  Front
    Caster .... 4.50-degree(s)
    Camber .... 0-degree(s)
    Toe-in 1/16-inch
1990 thru 1992
  Front
    Caster .... 3.00-degree(s)
    Camber .... -0.50-degree(s)
    Toe-in 0-inch
  Rear
    Camber .... -0.50-degree(s)
    Toe-in 5/32-inch

Oasis
1996
  Front
    Caster .... 3.00-degree(s)
    Camber .... 0-degree(s)
    Toe-in 0-inch
  Rear
    Camber .... -0.50-degree(s)
    Toe-in 0-inch

Stylus
1991 thru 1993
  Front
    Caster .... 3.00-degree(s)
    Camber .... -0.50-degree(s)
    Toe-in 0-inch
  Rear
    Camber .... -0.50-degree(s)
    Toe-in 5/32-inch

Amigo
1989 thru 1994
  Front
    Caster .... 2.50-degree(s)
    Camber .... 0.50-degree(s)
    Toe-in 5/64-inch

Hombre
1996
  Front
    Caster .... 2.00-degree(s)
    Camber .... 0.80-degree(s)
    Toe-in 5/32-inch

Pickup
1983
  Except 4WD
    Front
      Caster .... 0.50-degree(s)
      Camber .... 0.50-degree(s)
      Toe-in 1/16-inch
  4WD
    Front
      Caster .... 0.30-degree(s)
      Camber .... 0.56-degree(s)
      Toe-in 0-inch
1984 thru 1987
  Except 4WD
    Front
      Caster .... 0.50-degree(s)

    Camber .... 0.50-degree(s)
    Toe-in 5/64-inch
  4WD
    Front
      Caster .... 0.30-degree(s)
      Camber .... 0.56-degree(s)
      Toe-in 0-inch
1988 thru 1996
  Shortbed 2WD
    Front
      Caster .... 1.56-degree(s)
      Camber .... 0.50-degree(s)
      Toe-in 5/64-inch
  Shortbed 4WD
    Front
      Caster .... 1.90-degree(s)
      Camber .... 0.50-degree(s)
      Toe-in 5/64-inch
  Longbed 2WD
    Front
      Caster .... 1.88-degree(s)
      Camber .... 0.50-degree(s)
      Toe-in 5/64-inch
  Longbed 4WD
    Front
      Caster .... 2.12-degree(s)
      Camber .... 0.50-degree(s)
      Toe-in 5/64-inch

Rodeo
1991 thru 1996
  Front
    Caster .... 2.33-degree(s)
    Camber .... 0.50-degree(s)
    Toe-in 5/64-inch

Trooper
1984 thru 1986
  Front
    Caster .... 0.50-degree(s)
    Camber .... 0.56-degree(s)
    Toe-in 0-inch
1987 thru 1991
  Front
    Caster .... 2.50-degree(s)
    Camber .... 0.50-degree(s)
    Toe-in 5/64-inch
1992 thru 1996
  Short wheelbase
    Front
      Caster .... 2.00-degree(s)
      Camber .... 0-degree(s)
      Toe-in 0-inch
  Long wheelbase
    Front
      Caster .... 2.15-degree(s)
      Camber .... 0-degree(s)
      Toe-in 0-inch

JAGUAR

XJ6
1983
  Front
    Caster .... 2.25-degree(s)
    Camber .... 0.50-degree(s)
    Toe-in 3/32-inch

Rear
  Camber .... -0.75-degree(s)
  Toe-in 0-inch

1984 thru 1987
  Front
    Caster .... 2.25-degree(s)
    Camber .... -0.50-degree(s)
    Toe-in 1/16-inch
  Rear
    Camber .... -0.75-degree(s)
    Toe-in 0-inch

1988 thru 1993
  Front
    Caster .... 4.00-degree(s)
    Camber .... -0.25-degree(s)
    Toe-in 1/32-inch
  Rear
    Camber .... -0.75-degree(s)
    Toe-in 1/32-inch

1994
  Front
    Caster .... 4.20-degree(s)
    Camber .... -0.15-degree(s)
    Toe-in 3/64-inch
  Rear
    Camber .... -0.75-degree(s)
    Toe-in 1/32-inch

1995 and 1996
  Front
    Caster .... 4.50-degree(s)
    Camber .... -0.25-degree(s)
    Toe-in 3/64-inch
  Rear
    Camber .... -0.75-degree(s)
    Toe-in 1/8-inch

XJ12 and XJR

1994
  Front
    Caster .... 4.20-degree(s)
    Camber .... -0.15-degree(s)
    Toe-in 3/64-inch
  Rear
    Camber .... -0.75-degree(s)
    Toe-in 1/32-inch

1995 and 1996
  Front
    Caster .... 4.50-degree(s)
    Camber .... -0.25-degree(s)
    Toe-in 3/64-inch
  Rear (XJ12)
    Camber .... -0.75-degree(s)
    Toe-in 1/8-inch
  Rear (XJR)
    Camber .... -1.60-degree(s)
    Toe-in 1/8-inch

XJS

1983 thru 1993
  Front
    Caster .... 3.50-degree(s)
    Camber .... -0.50-degree(s)
    Toe-in 1/16-inch
  Rear
    Camber .... -0.75-degree(s)
    Toe-in 0-inch

1994 thru 1996
  Front
    Caster .... 4.00-degree(s)

Camber .... -0.50-degree(s)
Toe-in 5/32-inch
Rear
  Camber .... -0.75-degree(s)
  Toe-in 0-inch

JEEP

Cherokee

1983
  Caster .... 4.00-degree(s)
  Camber .... 0-degree(s)
  Toe-in 1/16-inch

CJ-5 and CJ-7

1983 thru 1986
  Caster .... 6.00-degree(s)
  Camber .... 0-degree(s)
  Toe-in 1/16-inch

Comanche, Cherokee, and Wagoneer

1984 thru 1989
  Caster .... 7.50-degree(s)
  Camber .... 0-degree(s)
  Toe-in 0-inch

1990 thru 1994
  Caster .... 6.00-degree(s)
  Camber .... 0-degree(s)
  Toe-in 0-inch

1995 and 1996
  Caster .... 7.00-degree(s)
  Camber .... -0.25-degree(s)
  Toe-in 0-inch

Grand Cherokee

1993 thru 1996
  Caster .... 7.00-degree(s)
  Camber .... -0.25-degree(s)
  Toe-in 1/8-inch

Grand Wagoneer and J Series Truck

1983 thru 1989
  Caster .... 4.00-degree(s)
  Camber .... 0-degree(s)
  Toe-in 1/16-inch

1990
  Caster .... 4.00-degree(s)
  Camber .... 0-degree(s)
  Toe-in 5/32-inch

1991
  Caster .... 4.00-degree(s)
  Camber .... 0-degree(s)
  Toe-in 0-inch

Wrangler

1987 and 1988
  Caster .... 8.00-degree(s)
  Camber .... 0-degree(s)
  Toe-in 0-inch

1989 thru 1995
  with M/T
    Caster .... 6.50-degree(s)
    Camber .... 0-degree(s)
    Toe-in 0-inch
  with A/T
    Caster .... 8.00-degree(s)
    Camber .... 0-degree(s)
    Toe-in 0-inch

LEXUS

ES250

1990
Front
  Caster.. 1.70-degree(s)
  Camber .... 0.50-degree(s)
  Toe-in.. 1/32-inch
Rear
  Camber .... -0.70-degree(s)
  Toe-in.. 5/32-inch

1991
Front
  Caster.. 1.56-degree(s)
  Camber .... 0.44-degree(s)
  Toe-in.. 1/32-inch
Rear
  Camber .... -0.70-degree(s)
  Toe-in.. 5/32-inch

ES300

1992 thru 1996
Front
  Caster.. 1.25-degree(s)
  Camber .... -0.66-degree(s)
  Toe-in.. 0-inch
Rear
  Camber .... -0.50-degree(s)
  Toe-in.. 5/32-inch

GS300

1993 thru 1996
Front
  Caster.. 7.10-degree(s)
  Camber .... -0.25-degree(s)
  Toe-in.. 3/64-inch
Rear
  Camber .... -7.10-degree(s)
  Toe-in.. 5/32-inch

LS400

1990
  w/o air suspension
    Front
      Caster .... 9.25-degree(s)
      Camber .... 0.10-degree(s)
      Toe-in...................................... 3/32-inch
    Rear
      Camber .... 0-degree(s)
      Toe-in...................................... 3/32-inch
  w/air suspension
    Front
      Caster .... 9.88-degree(s)
      Camber .... -0.10-degree(s)
      Toe-in...................................... 1/32-inch
    Rear
      Camber .... -0.75-degree(s)
      Toe-in...................................... 1/8-inch

1991 and 1992
  w/o air suspension
    Front
      Caster .... 9.15-degree(s)
      Camber .... 0-degree(s)
      Toe-in...................................... 3/32-inch
    Rear
      Camber .... 0-degree(s)
      Toe-in...................................... 3/32-inch

  w/air suspension
    Front
      Caster .... 9.88-degree(s)
      Camber .... -0.10-degree(s)
      Toe-in...................................... 1/32-inch
    Rear
      Camber .... -0.75-degree(s)
      Toe-in...................................... 1/8-inch

1993 and 1994
  w/o air suspension
    Front
      Caster .... 9.33-degree(s)
      Camber .... 0.10-degree(s)
      Toe-in...................................... 1/8-inch
    Rear
      Camber .... -0.16-degree(s)
      Toe-in...................................... 5/64-inch
  w/air suspension
    Front
      Caster .... 9.80-degree(s)
      Camber .... -0.10-degree(s)
      Toe-in...................................... 3/64-inch
    Rear
      Camber .... -0.75-degree(s)
      Toe-in...................................... 1/8-inch

1995 and 1996
  w/o air suspension
    Front
      Caster .... 7.00-degree(s)
      Camber .... 0.33-degree(s)
      Toe-in...................................... 1/8-inch
    Rear
      Camber .... -0.80-degree(s)
      Toe-in...................................... 5/64-inch
  w/air suspension
    Front
      Caster .... 7.42-degree(s)
      Camber .... 0.08-degree(s)
      Toe-in...................................... 3/64-inch
    Rear
      Camber .... -1.42-degree(s)
      Toe-in...................................... 1/8-inch

SC300 and SC400

1992 thru 1996
Front
  Caster.. 2.90-degree(s)
  Camber .... 0-degree(s)
  Toe-in.. 3/64-inch
Rear
  Camber .... -0.88-degree(s)
  Toe-in.. 13/64-inch

LINCOLN

Continental

1983
Front
  Caster.. 1.25-degree(s)
  Camber .... 0.38-degree(s)
  Toe-in.. 1/8-inch
1984
Front
  Caster.. 1.75-degree(s)
  Camber .... 0-degree(s)
  Toe-in.. 1/8-inch

1985 thru 1987
  Front
    Caster .... 1.50-degree(s)
    Camber .... 0-degree(s)
    Toe-in 1/8-inch
1988
  Front
    Caster .... 4.80-degree(s)
    Camber .... -0.88-degree(s)
    Toe-in .... -7/64-inch
  Rear
    Camber .... -1.30-degree(s)
    Toe-in 7/64-inch
1989 thru 1991
  Front
    Caster .... 4.38-degree(s)
    Camber .... -1.12-degree(s)
    Toe-in .... -7/64-inch
  Rear
    Camber .... -1.30-degree(s)
    Toe-in 7/64-inch
1992 thru 1994
  Front
    Caster .... 4.50-degree(s)
    Camber .... -1.10-degree(s)
    Toe-in .... -7/64-inch
  Rear
    Camber .... -1.30-degree(s)
    Toe-in 7/64-inch
1995 and 1996
  Front
    Caster .... 4.50-degree(s)
    Camber .... -0.70-degree(s)
    Toe-in .... -7/64-inch
  Rear
    Camber .... -0.70-degree(s)
    Toe-in 7/64-inch

Mark VI
1983
  Front
    Caster .... 3.00-degree(s)
    Camber .... 0.50-degree(s)
    Toe-in 1/16-inch

Mark VII
1984
  Front
    Caster .... 1.75-degree(s)
    Camber .... 0-degree(s)
    Toe-in 1/8-inch
1985 thru 1992
  Front
    Caster .... 1.50-degree(s)
    Camber .... 0-degree(s)
    Toe-in 1/8-inch

Mark VIII
1993 thru 1996
  Front
    Caster .... 5.50-degree(s)
    Camber .... -0.50-degree(s)
    Toe-in 1/8-inch
  Rear
    Camber .... -0.50-degree(s)
    Toe-in 1/16-inch

Town Car
1983 thru 1986
  Front
    Caster .... 3.00-degree(s)
    Camber .... 0.50-degree(s)
    Toe-in 1/16-inch
1987
  Front
    Caster .... 3.25-degree(s)
    Camber .... -0.50-degree(s)
    Toe-in 1/16-inch
1988
  Front
    Caster .... 4.00-degree(s)
    Camber .... -0.50-degree(s)
    Toe-in 1/16-inch
1989 and 1990
  Front
    Caster .... 3.50-degree(s)
    Camber .... -0.50-degree(s)
    Toe-in 1/16-inch
1991
  Front
    Caster .... 5.50-degree(s)
    Camber .... 0.50-degree(s)
    Toe-in 1/16-inch
1992 thru 1996
  Front
    Caster .... 6.00-degree(s)
    Camber .... -0.50-degree(s)
    Toe-in .... -1/16-inch

MAZDA

323 and Protege
1986 and 1987
  Front
    Caster .... 1.63-degree(s)
    Camber .... 0.80-degree(s)
    Toe-in 5/64-inch
  Rear
    Camber .... 0-degree(s)
    Toe-in 0-inch
1988 and 1989
  2WD
    Front
      Caster .... 1.56-degree(s)
      Camber .... 0.80-degree(s)
      Toe-in 5/64-inch
    Rear
      Camber .... 0-degree(s)
      Toe-in 5/64-inch
  4WD
    Front
      Caster .... 1.80-degree(s)
      Camber .... 1.00-degree(s)
      Toe-in 5/64-inch
    Rear
      Camber .... -0.44-degree(s)
      Toe-in 5/64-inch
1990 thru 1994
  Front
    2WD
      Caster .... 1.90-degree(s)
      Camber .... -0.10-degree(s)
      Toe-in 3/32-inch

4WD
  Caster .... 2.56-degree(s)
  Camber .... -0.75-degree(s)
  Toe-in 3/32-inch
  Rear
    Camber .... -0.33-degree(s)
    Toe-in 3/32-inch
1995 and 1996
  Sedan
    Front
      Caster .... 1.90-degree(s)
      Camber .... -0.75-degree(s)
      Toe-in 5/64-inch
    Rear
      Camber .... -0.80-degree(s)
      Toe-in 5/64-inch
  Hatchback
    Front
      Caster .... 1.88-degree(s)
      Camber .... -0.75-degree(s)
      Toe-in 5/64-inch

626

1983 thru 1987
  Front
    Caster .... 1.70-degree(s)
    Camber .... 0.30-degree(s)
    Toe-in 1/8-inch
  Rear
    Camber .... 0.06-degree(s)
    Toe-in 0-inch
1988 and 1989
  Front
    Caster .... 1.22-degree(s)
    Camber .... 0.28-degree(s)
    Toe-in 0-inch
  Rear
    Camber .... -0.50-degree(s)
    Toe-in 1/8-inch
1990 thru 1992
  Front
    Caster .... 1.70-degree(s)
    Camber .... 0.28-degree(s)
    Toe-in 0-inch
  Rear
    Camber .... -0.50-degree(s)
    Toe-in 1/8-inch
1993 thru 1996
  Front
    Caster .... 2.62-degree(s)
    Camber .... -0.60-degree(s)
    Toe-in 1/8-inch
  Rear
    Camber .... -0.15-degree(s)
    Toe-in 1/8-inch

929

1988 thru 1991
  Front
    Caster .... 4.50-degree(s)
    Camber .... 1.00-degree(s)
    Toe-in 5/32-inch
  Rear
    Camber .... -0.25-degree(s)
    Toe-in 3/32-inch
1992
  Front
    Caster .... 5.44-degree(s)
    Camber .... 0.10-degree(s)
    Toe-in 3/32-inch

Rear
  Camber .... -0.30-degree(s)
  Toe-in 1/8-inch
1993 and 1994
  Front
    Caster .... 5.25-degree(s)
    Camber .... 0.16-degree(s)
    Toe-in 3/32-inch
  Rear
    Camber .... -0.42-degree(s)
    Toe-in 1/8-inch
1995
  Front
    Caster .... 5.35-degree(s)
    Camber .... 0.25-degree(s)
    Toe in 3/32-inch
  Rear
    Camber .... -0.25-degree(s)
    Toe-in 1/16-inch

GLC

1983 and 1984
  Except wagon
    Front
      Caster .... 1.90-degree(s)
      Camber .... 0.90-degree(s)
      Toe-in 0-inch
    Rear
      Toe-in 0-inch
  Wagon
    Front
      Caster .... 1.56-degree(s)
      Camber .... 0.75-degree(s)
      Toe-in 1/8-inch
1985
  Front
    Caster .... 1.90-degree(s)
    Camber .... 0.90-degree(s)
    Toe-in 1/8-inch
  Rear
    Toe-in 0-inch

Miata

1990 thru 1996
  Front
    Caster .... 4.44-degree(s)
    Camber .... 0.40-degree(s)
    Toe-in 5/32-inch
  Rear
    Camber .... -0.70-degree(s)
    Toe-in 5/32-inch

Millenia

1995 and 1996
  Front
    Caster .... 2.58-degree(s)
    Camber .... -0.18-degree(s)
    Toe-in 1/8-inch
  Rear
    Camber .... -0.30-degree(s)
    Toe-in 1/8-inch

MX-3

1992 thru 1995
  Front
    Caster .... 2.75-degree(s)
    Camber .... -0.80-degree(s)
    Toe-in 1/8-inch

Rear
  Camber .... -0.95-degree(s)
  Toe-in 5/64-inch

MX-6

1988 and 1989
Front
  Caster .... 1.22-degree(s)
  Camber .... 0.28-degree(s)
  Toe-in 0-inch
Rear
  2-wheel steering
    Camber .... -0.50-degree(s)
    Toe-in 1/8-inch
  4-wheel steering
    Camber .... 0-degree(s)
    Toe-in 1/8-inch

1990 thru 1992
Front
  Caster .... 1.70-degree(s)
  Camber .... 0.28-degree(s)
  Toe-in 0-inch
Rear
  2-wheel steering
    Camber .... -0.50-degree(s)
    Toe-in 1/8-inch
  4-wheel steering
    Camber .... 0-degree(s)
    Toe-in 1/8-inch

1993 thru 1996
Front
  Caster .... 3.00-degree(s)
  Camber .... -0.70-degree(s)
  Toe-in 1/8-inch
Rear
  Camber .... -0.36-degree(s)
  Toe-in 1/8-inch

RX-7

1983 thru 1985
Front
  Caster
    Right side .... 4.20-degree(s)
    Left side .... 3.70-degree(s)
  Camber .... 1.00-degree(s)
  Toe-in 1/8-inch

1986 thru 1992
Front
  Caster .... 4.70-degree(s)
  Camber .... 0.30-degree(s)
  Toe-in 1/8-inch
Rear
  Camber .... -0.75-degree(s)
  Toe-in 1/8-inch

1993 and 1994
Front
  Caster .... 6.08-degree(s)
  Camber .... 0.10-degree(s)
  Toe-in 5/64-inch
Rear
  Camber .... -1.22-degree(s)
  Toe-in 5/64-inch

1995 and 1996
Front
  Caster .... 6.66-degree(s)
  Camber .... 0.10-degree(s)
  Toe-in 3/64-inch

Rear
  Camber .... -1.22-degree(s)
  Toe-in 5/64-inch

MPV

1989 thru 1991
2WD
  Front
    Caster .... 5.44-degree(s)
    Camber .... 0.38-degree(s)
    Toe-in 5/32-inch
4WD
  Front
    Caster .... 5.50-degree(s)
    Camber .... 0.15-degree(s)
    Toe-in 5/32-inch

1992 and 1993
2WD
  Front
    Caster .... 4.80-degree(s)
    Camber .... 0.38-degree(s)
    Toe-in 5/32-inch
4WD
  Front
    Caster .... 5.00-degree(s)
    Camber .... 0.15-degree(s)
    Toe-in 5/32-inch

1994
2WD
  Front
    Caster
      Right side .... 5.95-degree(s)
      Left side .... 5.45-degree(s)
    Camber .... 0.38-degree(s)
    Toe-in 7/64-inch
4WD
  Front
    Caster
      Right side .... 6.00-degree(s)
      Left side .... 5.50-degree(s)
    Camber .... 0.16-degree(s)
    Toe-in 7/64-inch

1995 and 1996
2WD
  Front
    Caster
      Right side .... 5.95-degree(s)
      Left side .... 5.45-degree(s)
    Camber .... 0.38-degree(s)
    Toe-in 13/64-inch
4WD
  Front
    Caster
      Right side .... 6.00-degree(s)
      Left side .... 5.50-degree(s)
    Camber .... 0.16-degree(s)
    Toe-in 13/64-inch

Navajo

1991 and 1992
Front
  Caster .... 5.00-degree(s)
  Camber .... 0-degree(s)
  Toe-in 0-inch

1993 and 1994
Front
  Caster .... 4.50-degree(s)
  Camber .... 0.25-degree(s)
  Toe-in 1/32-inch

B2000, B2200 and B2600

1983 and 1984
  Front
    Caster.. 1.00-degree(s)
    Camber .... 0.75-degree(s)
    Toe-in.. 1/8-inch
1985 thru 1993
  2WD
    Front
      Caster
        w/o power steering .... 0.80-degree(s)
        w/power steering .... 1.80-degree(s)
      Camber .... 0.75-degree(s)
      Toe-in.. 1/8-inch
  4WD
    Front
      Caster .... 2.00-degree(s)
      Camber .... 1.00-degree(s)
      Toe-in.. 1/8-inch

B2300, B3000 and B4000

1994 thru 1996
  Front
    Caster.. 4.00-degree(s)
    Camber .... 0.25-degree(s)
    Toe-in.. 1/32-inch

MERCEDES-BENZ

190 Series

1984 and 1985
  Front
    Caster.. 10.16-degree(s)
    Camber .... 0.33-degree(s)
    Toe-in.. 5/32-inch
  Rear
    Toe-in.. 13/64-inch
1986 thru 1993
  Front
    Except 2.3L 16V
      Caster .... 10.16-degree(s)
      Camber .... 0-degree(s)
      Toe-in.. 5/32-inch
    2.3 16V
      Caster .... 10.50-degree(s)
      Camber .... -0.33-degree(s)
      Toe-in.. 5/32-inch
  Rear
    Toe-in.. 13/64-inch

240D, 260E, 300D, 300CD, 300TD, 300TE and 300SD

1983 thru 1985
  All except 300SD
    Front
      Caster .... 8.75-degree(s)
      Camber .... 0-degree(s)
      Toe-in.. 13/64-inch
  300SD
    Front
      Caster .... 9.75-degree(s)
      Camber .... 0-degree(s)
      Toe-in.. 21/64-inch
1986 thru 1993
  Except 300E, TE with 4MATIC
    Front
      Caster .... 10.16-degree(s)
      Camber .... 0-degree(s)
      Toe-in.. 5/32-inch

300E, TE with 4MATIC

  Front
    Caster .... 10.33-degree(s)
    Camber .... -0.25-degree(s)
    Toe-in.. 5/32-inch
  Rear
    Toe-in.. 13/64-inch

380SEC, 380SF, 380SL, 380SEL, 500SEL and 500SEC

1983 thru 1985
  All except 380SL and Hydropneumatic suspension
    Front
      Caster .... 9.75-degree(s)
      Camber .... 0-degree(s)
      Toe-in.. 21/64-inch
  380SL
    Front
      Caster .... 3.25-degree(s)
      Camber .... 0-degree(s)
      Toe-in.. 1/8-inch
    Rear
      Toe-in.. 13/64-inch
  All w/Hydropneumatic suspension
    Front
      Caster.. 10.50-degree(s)
      Camber .... -0.33-degree(s)
      Toe-in.. 1/8-inch

300SDL, 300SE, 300SEL, 350SD, 350SDL, 420SEL, 560SEL and 560SEC

1986 thru 1991
  All except Hydropneumatic suspension
    Front
      Caster .... 9.75-degree(s)
      Camber .... 0-degree(s)
      Toe-in.. 21/64-inch
  All w/Hydropneumatic suspension
    Front
      Caster .... 10.50-degree(s)
      Camber .... -0.33-degree(s)
      Toe-in.. 1/8-inch

SL320, SL500, SL600, 300SL, 500SL, 560SL and 600SL

1986 thru 1989
  Front
    Caster.. 10.50-degree(s)
    Camber .... -0.16-degree(s)
    Toe-in.. 5/32-inch
  Rear
    Toe-in.. 13/64-inch
1990 thru 1996
  Front
    Caster.. 10.50-degree(s)
    Camber .... -0.80-degree(s)
    Toe-in.. 5/32-inch
  Rear
    Toe-in.. 13/64-inch

300CE, E300 and E320

1986 thru 1996
  Except 4MATIC, Sportline and convertible
    Front
      Caster .... 10.16-degree(s)
      Camber .... 0-degree(s)
      Toe-in.. 5/32-inch
  Sportline
    Front
      Caster .... 10.66-degree(s)
      Camber .... -0.80-degree(s)
      Toe-in.. 5/32-inch

Rear, all
  Toe-in 13/64-inch

4MATIC
  Front
    Caster .... 10.33-degree(s)
    Camber .... -0.25-degree(s)
    Toe-in 5/32-inch
  Rear
    Toe-in 13/64-inch

Convertible
  Front
    Caster .... 10.50-degree(s)
    Camber .... -0.66-degree(s)
    Toe-in 5/32-inch
  Rear
    Toe-in 13/64-inch

400SE, 400SEL,500SEL, 500SEC, 600SEC, 600SEL, S320, S350, S420, S500 and S600

1992 thru 1996
  Front
    Caster .... 10.00-degree(s)
    Camber .... -0.10-degree(s)
    Toe-in 17/64-inch
  Rear
    Toe-in 3/8-inch

400E and E420

1992 thru 1996
  Front
    Caster .... 10.50-degree(s)
    Camber .... -0.66-degree(s)
    Toe-in 5/32-inch
  Rear
    Toe-in 13/64-inch

500E and E500

1992 thru 1996
  Front
    Caster .... 10.80-degree(s)
    Camber .... -1.00-degree(s)
    Toe-in 5/32-inch
  Rear
    Toe-in 13/64-inch

C220 and C280

1994 thru 1996
  Front
    Caster .... 4.66-degree(s)
    Camber .... -0.58-degree(s)
    Toe-in 5/32-inch
  Rear
    Toe-in 17/64-inch

C36 AMG

1995 and 1996
  Front
    Caster .... 4.90-degree(s)
    Camber .... -0.58-degree(s)
    Toe-in 13/64-inch
  Rear
    Toe-in 17/64-inch

MERCURY

Capri

1983 and 1984
  Front
    Caster .... 1.25-degree(s)
    Camber .... 0-degree(s)
    Toe-in 3/16-inch

1985 and 1986
  Front
    Caster .... 1.00-degree(s)
    Camber .... 0-degree(s)
    Toe-in 3/16-inch

1991 thru 1995
  Front
    Caster .... 1.58-degree(s)
    Camber .... 0.80-degree(s)
    Toe-in 5/32-inch
  Rear
    Camber .... 0-degree(s)
    Toe-in 5/32-inch

Cougar

1983
  Front
    Caster .... 1.25-degree(s)
    Camber .... 0.25-degree(s)
    Toe-in 3/16-inch

1984
  Front
    Caster .... 1.00-degree(s)
    Camber .... 0.25-degree(s)
    Toe-in 3/16-inch

1985 and 1986
  Front
    Caster .... 0.75-degree(s)
    Camber .... 0.25-degree(s)
    Toe-in 3/16-inch

1987 and 1988
  Front
    Caster .... 1.22-degree(s)
    Camber .... -0.64-degree(s)
    Toe-in 3/16-inch

1989 thru 1991
  Front
    Caster .... 5.50-degree(s)
    Camber .... -0.50-degree(s)
    Toe-in 1/8-inch
  Rear
    Camber .... -0.50-degree(s)
    Toe-in 1/16-inch

1992 thru 1996
  Front
    Caster .... 5.50-degree(s)
    Camber .... -0.50-degree(s)
    Toe-in 5/64-inch
  Rear
    Camber .... -0.50-degree(s)
    Toe-in 1/16-inch

Grand Marquis and Colony Park

1983 thru 1986
  Front
    Caster .... 3.00-degree(s)
    Camber .... 0.50-degree(s)
    Toe-in 1/16-inch

1987 thru 1991
    Front
        Caster.. 3.50-degree(s)
        Camber .... -0.50-degree(s)
        Toe-in 1/16-inch
1992 thru 1996
    Front
        Caster.. 5.50-degree(s)
        Camber .... -0.50-degree(s)
        Toe-in .... -1/16-inch

Lynx
1983
    Front
        Caster.. 1.30-degree(s)
        Camber
            Left wheel .... 2.20-degree(s)
            Right wheel .... 1.70-degree(s)
        Toe-in .... -7/64-inch
    Rear
        Camber .... -1.25-degree(s)
        Toe-in 3/16-inch
1984
    Front
        Caster.. 1.38-degree(s)
        Camber
            Left wheel .... 2.12-degree(s)
            Right wheel .... 1.70-degree(s)
        Toe-in .... -7/64-inch
    Rear
        Camber .... -1.25-degree(s)
        Toe-in 3/16-inch
1085
    Front
        Caster.. 1.30-degree(s)
        Camber
            Left wheel .... 1.88-degree(s)
            Right wheel .... 1.44-degree(s)
        Toe-in .... -7/64-inch
    Rear
        Camber .... -1.20-degree(s)
        Toe-In 3/16-inch
1986 and 1987
    Front
        Caster.. 2.34-degree(s)
        Camber
            Left wheel .... 1.00-degree(s)
            Right wheel .... 0.75-degree(s)
        Toe-in .... -3/32-inch
    Rear
        Camber .... -0.44-degree(s)
        Toe-in 3/16-inch

Marquis
1983
    Except wagon
        Front
            Caster .... 1.12-degree(s)
            Camber .... 0.44-degree(s)
            Toe-in 3/16-inch
    Wagon
        Front
            Caster .... 0.88-degree(s)
            Camber .... 0.50-degree(s)
            Toe-in 3/16-inch
1984
    Except wagon
        Front
            Caster .... 1.00-degree(s)

Camber .... 0.44-degree(s)
        Toe-in 3/16-inch
    Wagon
        Front
            Caster .... 1.00-degree(s)
            Camber .... 0.50-degree(s)
            Toe-in 3/16-inch
1985 and 1986
    Except wagon
        Front
            Caster .... 0.88-degree(s)
            Camber .... 0.38-degree(s)
            Toe-in 3/16-inch
    Wagon
        Front
            Caster .... 0.75-degree(s)
            Camber .... 0.44-degree(s)
            Toe-in 3/16-inch

Mystique
1995 and 1996
    Front
        Caster.. 2.30-degree(s)
        Camber .... -0.40-degree(s)
        Toe-in .... -11/64-inch
    Rear
        Camber .... -0.40-degree(s)
        Toe-in 5/32-inch

Sable
1986 and 1987
    Except wagon
        Front
            Caster .... 3.80-degree(s)
            Camber .... -0.50-degree(s)
            Toe-in .... -3/32-inch
        Rear
            Camber .... -0.88-degree(s)
            Toe-in 1/16-inch
    Wagon
        Front
            Caster .... 3.80-degree(s)
            Camber .... -0.44-degree(s)
            Toe-in .... -3/32-inch
        Rear
            Camber .... -0.63-degree(s)
            Toe-in 1/16-inch
1988
    Except wagon
        Front
            Caster .... 3.90-degree(s)
            Camber .... -0.44-degree(s)
            Toe-in .... -3/32-inch
        Rear
            Camber .... -0.88-degree(s)
            Toe-in 1/16-inch
    Wagon
        Front
            Caster .... 3.80-degree(s)
            Camber .... -0.88-degree(s)
            Toe-in .... -3/32-inch
        Rear
            Camber .... -0.88-degree(s)
            Toe-in 1/16-inch
1989
    Except wagon
        Front
            Caster .... 3.70-degree(s)
            Camber .... -0.50-degree(s)
            Toe-in .... -3/32-inch

Rear
Camber .... -0.88-degree(s)
Toe-in 1/16-inch
Wagon
Front
Caster .... 3.70-degree(s)
Camber .... -0.88-degree(s)
Toe-in .... -3/32-inch
Rear
Camber .... -0.88-degree(s)
Toe-in 1/16-inch
1990 thru 1992
Except wagon
Front
Caster .... 3.70-degree(s)
Camber .... -0.50-degree(s)
Toe-in .... -7/64-inch
Rear
Camber .... -0.90-degree(s)
Toe-in 1/16-inch
Wagon
Front
Caster .... 3.63-degree(s)
Camber .... -0.44-degree(s)
Toe-in .... -3/32-inch
Rear
Camber .... -0.88-degree(s)
Toe-in 1/16-inch
1993 thru 1995
Except wagon
Front
Caster .... 3.80-degree(s)
Camber .... -0.50-degree(s)
Toe-in .... -7/64-inch
Rear
Camber .... -0.90-degree(s)
Toe-in 1/16-inch
Wagon
Front
Caster .... 3.70-degree(s)
Camber .... -0.40-degree(s)
Toe-in .... -7/64-inch
Rear
Camber .... -0.90-degree(s)
Toe-in 1/16-inch
1996
Front
Caster .... 3.80-degree(s)
Camber .... -0.50-degree(s)
Toe-in .... -7/64-inch
Rear
Camber .... -0.60-degree(s)
Toe-in 1/16-inch

Topaz
1984
Front
Caster .... 1.30-degree(s)
Camber
Left wheel .... 1.88-degree(s)
Right wheel .... 1.50-degree(s)
Toe-in .... -7/64-inch
Rear
Camber .... -0.25-degree(s)
Toe-in 0-inch

1985
Front
Caster .... 1.25-degree(s)
Camber
Left wheel .... 2.00-degree(s)
Right wheel .... 1.50-degree(s)
Toe-in .... -7/64-inch
Rear
Camber .... -0.20-degree(s)
Toe-in 0-inch
1986 thru 1994
Front
Caster .... 2.45-degree(s)
Camber
Left wheel .... 1.40-degree(s)
Right wheel .... 0.95-degree(s)
Toe-in .... -7/64-inch
Rear (2WD)
1986 thru 1992
Camber .... -0.15-degree(s)
Toe-in 0-inch
1993 and 1994
Camber .... -0.55-degree(s)
Toe-in 0-inch
Rear (4WD)
1987 and 1988
Camber .... 0.40-degree(s)
Toe-in 0-inch
1989 thru 1991
Camber .... 0.34-degree(s)
Toe-in 0-inch

Tracer
1987 thru 1989
Front
Caster .... 1.56-degree(s)
Camber .... 0.80-degree(s)
Toe-in 3/32-inch
Rear
Camber .... 0-degree(s)
Toe-in .... -3/32-inch
1990 thru 1996
Front
Caster .... 1.90-degree(s)
Camber .... -0.06-degree(s)
Toe-in 3/32-inch
Rear
Camber .... -0.33-degree(s)
Toe-in 3/32-inch

Villager
1993 thru 1996
Front
Caster .... 0.80-degree(s)
Camber .... 0.30-degree(s)
Toe-in 1/8-inch
Rear
Camber .... 0-degree(s)
Toe-in 0-inch

Zephyr
1983
Front
Caster .... 1.12-degree(s)
Camber .... 0.44-degree(s)
Toe-in 3/16-inch

MITSUBISHI

3000GT
1991 thru 1996
- Front
  - Caster.. 3.90-degree(s)
  - Camber.. 0-degree(s)
  - Toe-in... 0-inch
- Rear (2WD)
  - Camber.. 0-degree(s)
  - Toe-in... 1/64-inch
- Rear (4WD)
  - Camber.. -0.15-degree(s)
  - Toe-in... 1/64-inch

Diamante
1992 thru 1995
- Front
  - Caster.. 2.75-degree(s)
  - Camber.. 0-degree(s)
  - Toe-in... 0-inch
- Rear
  - Camber.. 0-degree(s)
  - Toe-in... 0-inch

Eclipse
1990 thru 1994
- Front
  - 1.8L
    - Caster.. 2.30-degree(s)
    - Camber.. 0.25-degree(s)
    - Toe-in... 0-inch
  - 2.0L 2WD
    - Caster.. 2.40-degree(s)
    - Camber.. 0.10-degree(s)
    - Toe-in 0-inch
  - 2.0L 4WD
    - Caster.. 2.30-degree(s)
    - Camber.. 0.15-degree(s)
    - Toe-in... 0-inch
- Rear (2WD)
  - Camber.. -0.75-degree(s)
  - Toe-in... 0-inch
- Rear (4WD)
  - Camber.. -1.56-degree(s)
  - Toe-in... 5/32-inch

1995
- Front
  - RS
    - Caster.. 4.66-degree(s)
    - Camber.. -0.08-degree(s)
    - Toe-in... 0-inch
  - GS, GST, GSX
    - Caster.. 4.66-degree(s)
    - Camber.. -0.33-degree(s)
    - Toe-in... 0-inch
- Rear
  - 2WD with 14-inch wheels or 4WD with A/T
    - Camber.. -1.33-degree(s)
    - Toe-in 1/8-inch
  - 2WD with 16-inch wheels or 4WD with M/T
    - Camber.. -1.66-degree(s)
    - Toe-in... 1/8-inch

1996
- Front
  - With 14-inch wheels
    - Caster.. 4.33-degree(s)
    - Camber.. 0-degree(s)
    - Toe-in 0-inch

- With 16-inch wheels
  - Caster.. 4.66-degree(s)
  - Camber.. -0.08-degree(s)
  - Toe-in... 0-inch
- Rear
  - Camber.. -1.33-degree(s)
  - Toe-in... 1/8-inch

Expo
1992 thru 1996
- Front
  - FWD
    - Caster.. 2.16-degree(s)
    - Camber.. 0.33-degree(s)
    - Toe-in... 0-inch
  - 4WD
    - Caster.. 2.00-degree(s)
    - Camber.. 0.66-degree(s)
    - Toe-in... 0-inch
- Rear
  - Camber.. -0.50-degree(s)
  - Toe-in... 3/32-inch

Galant and Sigma
1985 thru 1987
- Front
  - Caster.. 0.70-degree(s)
  - Camber.. 0.50-degree(s)
  - Toe-in... 0-inch
- Rear
  - Camber.. 0-degree(s)
  - Toe-in... 0-inch

1988 thru 1990
- Front
  - Caster.. 0.70-degree(s)
  - Camber.. 0.50-degree(s)
  - Toe-in... 0-inch
- Rear
  - Camber.. -0.75-degree(s)
  - Toe-in... 0-inch

1991 thru 1993
- Front
  - Caster.. 2.00-degree(s)
  - Camber.. 0.38-degree(s)
  - Toe-in... 0-inch
- Rear
  - Camber.. -0.75-degree(s)
  - Toe-in... 0-inch

1994 thru 1996
- Front
  - Caster.. 4.33-degree(s)
  - Camber.. 0-degree(s)
  - Toe-in... 0-inch
- Rear
  - Camber.. -1.33-degree(s)
  - Toe-in... 1/8-inch

Mirage
1985 and 1986
- Front
  - Caster.. 0.75-degree(s)
  - Camber.. 0-degree(s)
  - Toe-in... 0-inch
- Rear
  - Camber.. -1.00-degree(s)
  - Toe-in... 0-inch

1987 and 1988
- Front
  - Caster.. 1.00-degree(s)

Camber
w/o power steering .... 0-degree(s)
w/power steering .... 1.70-degree(s)
Toe-in 0-inch
Rear
Camber .... -0.70-degree(s)
Toe-in 1/8-inch
1989 thru 1992
Front
Caster .... 2.30-degree(s)
Camber .... 0-degree(s)
Toe-in 0-inch
Rear
Camber .... -0.70-degree(s)
Toe-in 0-inch
1993 thru 1996
Front
Caster .... 2.25-degree(s)
Camber .... 0-degree(s)
Toe-in 0-inch
Rear
Camber .... -0.66-degree(s)
Toe-in 1/8-inch

Precis
1987 thru 1989
Front
Caster .... 0.80-degree(s)
Camber .... 0.50-degree(s)
Toe-in 1/32-inch
Rear
Camber .... -0.66-degree(s)
Toe-in 0-inch
1990 thru 1994
Front
Caster
w/o power steering .... 1.00-degree(s)
w/Power Steering .... 1.66-degree(s)
Camber .... 0-degree(s)
Toe-in 1/32-inch
Rear (1990 and 1991)
Camber .... -0.70-degree(s)
Toe-in 0-inch
Rear (1992 thru 1994)
Camber .... -0.66-degree(s)
Toe-in 1/16-inch

Starion
1983 thru 1985
Front
Caster .... 5.30-degree(s)
Camber .... 0-degree(s)
Toe-in 0-inch
Rear
Camber .... 0-degree(s)
Toe-in 0-inch
1986 thru 1989
Front
Caster .... 5.88-degree(s)
Camber .... -0.50-degree(s)
Toe-in 0-inch
Rear (1986 and 1987)
Camber .... 0-degree(s)
Toe-in 0-inch
Rear (1988 and 1989)
Camber .... -0.25-degree(s)
Toe-in 0-inch

Tredia, Cordia
1983 thru 1988
Front
Caster .... 0.75-degree(s)
Camber .... 0.44-degree(s)
Toe-in 0-inch
Rear (1983 thru 1985)
Camber .... -0.56-degree(s)
Toe-in .... 0-degree(s)
Rear (1986 thru 1988)
Camber .... 0.70-degree(s)
Toe-in 0-inch

Montero
1983 thru 1986
Front
Caster .... 2.90-degree(s)
Camber .... 1.00-degree(s)
Toe-in 7/32-inch
1987 and 1988
Front
Caster .... 3.00-degree(s)
Camber .... 1.00-degree(s)
Toe-in 7/32-inch
1989 thru 1991
Front
Caster .... 2.90-degree(s)
Camber .... 1.00-degree(s)
Toe-in 7/32-inch
1992 thru 1996
Front
Caster .... 3.00-degree(s)
Camber .... 0.66-degree(s)
Toe-in 9/64-inch

Pickup
1983 thru 1986
2WD
Front
Caster .... 2.50-degree(s)
Camber .... 1.00-degree(s)
Toe-in 7/32-inch
4WD
Front
Caster .... 2.00-degree(s)
Camber .... 1.00-degree(s)
Toe-in 7/32-inch
1987 thru 1996
2WD
Front
Caster .... 2.50-degree(s)
Camber .... 0.66-degree(s)
Toe-in 7/32-inch
4WD
Front
Caster .... 2.00-degree(s)
Camber .... 1.00-degree(s)
Toe-in 7/32-inch

Van
1987 thru 1989
Front
Caster .... 3.12-degree(s)
Camber .... 0.50-degree(s)
Toe-in 1/32-inch
1990 and 1991
Front
Caster .... 3.00-degree(s)
Camber .... 0.50-degree(s)
Toe-in 1/32-inch

NISSAN

200SX
1983
  Front
    Caster.. 2.50-degree(s)
    Camber...................................... 0.06-degree(s)
    Toe-in.. 1/32-inch
1984 thru 1989
  Front
    Caster.. 3.50-degree(s)
    Camber...................................... 0.30-degree(s)
    Toe-in.. 1/32-inch
  Rear
    Camber...................................... -0.50-degree(s)
    Toe-in.. -3/64-inch
1995 and 1996
  Front
    Caster.. 1.42-degree(s)
    Camber...................................... -0.58-degree(s)
    Toe-in.. 5/64-inch
  Rear
    Camber...................................... -1.00-degree(s)
    Toe-in.. 3/64-inch

240SX
1989 thru 1996
  Front
    Caster.. 6.75-degree(s)
    Camber...................................... -0.75-degree(s)
    Toe-in.. 1/16-inch
  Rear
    Camber...................................... -1.17-degree(s)
    Toe-in.. 3/32-inch

280ZX
1983
  Front
    Caster.. 4.90-degree(s)
    Camber...................................... 0.20-degree(s)
    Toe-in.. 3/32-inch
  Rear
    Camber...................................... 0.70-degree(s)
    Toe-in.. 1/16-inch

300ZX
1984 thru 1989
  Front
    Caster.. 6.56-degree(s)
    Camber...................................... 0.15-degree(s)
    Toe-in.. 3/32-inch
  Rear
    Camber...................................... -1.15-degree(s)
    Toe-in.. 1/32-inch
1990 and 1991
  Front
    Caster.. 9.75-degree(s)
    Camber...................................... -0.80-degree(s)
    Toe-in.. 3/64-inch
  Rear
    Camber...................................... -1.15-degree(s)
    Toe-in.. 3/32-inch
1992 thru 1996
  Front
    Caster.. 9.66-degree(s)
    Camber...................................... -0.80-degree(s)
    Toe-in.. 3/64-inch
  Rear
    Camber...................................... -1.10-degree(s)
    Toe-in.. 7/64-inch

Altima
1993 thru 1996
  Front
    Caster.. 2.66-degree(s)
    Camber...................................... -0.08-degree(s)
    Toe-in.. 3/64-inch
  Rear
    Camber...................................... -1.25-degree(s)
    Toe-in.. 5/64-inch

Maxima and 810
1983
  Front
    Caster.. 3.70-degree(s)
    Camber...................................... 0.44-degree(s)
    Toe-in.. 0-inch
  Rear
    Camber...................................... 1.70-degree(s)
    Toe-in.. 3/32-inch
1984
  Front
    Caster.. 3.70-degree(s)
    Camber...................................... 0.44-degree(s)
    Toe-in.. 0-inch
  Rear
    Camber...................................... 2.00-degree(s)
    Toe-in.. -1/32-inch
1985 and 1986
  Front
    Caster.. 2.00-degree(s)
    Camber...................................... 0.30-degree(s)
    Toe-in.. 3/32-inch
  Rear (except wagon)
    Camber...................................... 0.25-degree(s)
    Toe-in.. -5/32-inch
  Rear (wagon)
    Camber...................................... 0.44-degree(s)
    Toe-in.. -7/32-inch
1987 and 1988
  Front
    Caster.. 2.00-degree(s)
    Camber...................................... 0.30-degree(s)
    Toe-in.. 3/32-inch
  Rear
    Camber...................................... -0.44-degree(s)
    Toe-in.. -3/16-inch
1989 thru 1991
  Front
    Caster.. 1.25-degree(s)
    Camber...................................... -0.25-degree(s)
    Toe-in.. 3/32-inch
  Rear
    Camber...................................... -0.56-degree(s)
    Toe-in.. -3/32-inch
1992 thru 1994
  Front
    Caster.. 1.25-degree(s)
    Camber...................................... -0.25-degree(s)
    Toe-in.. 3/32-inch
  Rear
    Camber...................................... -0.56-degree(s)
    Toe-in.. -5/64-inch
1995 and 1996
  Front
    Caster.. 2.75-degree(s)
    Camber...................................... -0.25-degree(s)
    Toe-in.. 5/64-inch

Rear
    Camber .... -1.00-degree(s)
    Toe-in 3/64-inch

NX Coupe

1991 thru 1994
  Front
    Caster .... 1.80-degree(s)
    Camber .... -0.25-degree(s)
    Toe-in 3/32-inch
  Rear
    Camber .... -1.15-degree(s)
    Toe-in 1/32-inch

Pulsar

1983 and 1984
  Except turbo
    Front
      Caster .... 1.50-degree(s)
      Camber .... 0.30-degree(s)
      Toe-in 7/32-inch
    Rear
      Camber .... 0-degree(s)
      Toe-in 0-inch
  Turbo
    Front
      Caster .... 1.50-degree(s)
      Camber .... 0.30-degree(s)
      Toe-in 1/16-inch
    Rear
      Camber .... -1.00-degree(s)
      Toe-in 0-inch
1985 and 1986
  Front
    Caster .... 1.50-degree(s)
    Camber .... 0.30-degree(s)
    Toe-in 7/32-inch
  Rear
    Camber .... -1.00-degree(s)
    Toe-in 0-inch
1987 thru 1990
  Front
    Caster .... 1.90-degree(s)
    Camber .... -0.50-degree(s)
    Toe-in 0-inch
  Rear
    Camber .... -1.25-degree(s)
    Toe-in 1/32-inch

Quest

1993 thru 1996
  Front
    Caster .... 0.80-degree(s)
    Camber .... 0.30-degree(s)
    Toe-in 1/8-inch
  Rear (1993 thru 1995)
    Camber .... 0-degree(s)
    Toe-in 0-inch
  Rear (1996)
    Camber .... -0.70-degree(s)
    Toe-in 0-inch

Sentra

1983
  Front
    Caster .... 1.50-degree(s)
    Camber .... 0.25-degree(s)
    Toe-in 3/16-inch
  Rear
    Camber .... 0-degree(s)

    Toe-in 0-inch
1984 thru 1986
  Front
    Caster .... 1.50-degree(s)
    Camber .... 0.30-degree(s)
    Toe-in 7/32-inch
  Rear
    Camber .... -1.00-degree(s)
    Toe-in 0-inch
1987 thru 1990
  2WD (except coupe)
    Front
      Caster .... 1.50-degree(s)
      Camber .... -0.15-degree(s)
      Toe-in 1/32-inch
    Rear
      Camber .... -1.00-degree(s)
      Toe-in .... -3/32-inch
  2WD coupe
    Front
      Caster .... 1.64-degree(s)
      Camber .... -0.30-degree(s)
      Toe-in 1/64-inch
    Rear
      Camber .... -1.15-degree(s)
      Toe-in .... -5/64-inch
  4WD wagon
    Front
      Caster .... 0.90-degree(s)
      Camber .... -0.06-degree(s)
      Toe-in 1/32-inch
    Rear
      Camber .... -0.15-degree(s)
      Toe-in .... -3/32-inch
1991 thru 1994
  Front
    Caster .... 1.44-degree(s)
    Camber .... -0.25-degree(s)
    Toe-in 3/32-inch
  Rear
    Camber .... -0.90-degree(s)
    Toe-in 1/32-inch
1995 and 1996
  Front
    Caster .... 1.42-degree(s)
    Camber .... -0.58-degree(s)
    Toe-in 5/64-inch
  Rear
    Camber .... -1.00-degree(s)
    Toe-in 3/64-inch

Stanza (except wagon)

1983 and 1984
  Front
    Caster .... 1.44-degree(s)
    Camber .... 0-degree(s)
    Toe-in 1/16-inch
  Rear (1983)
    Camber .... 0.75-degree(s)
    Toe-in 3/32-inch
  Rear (1984)
    Camber .... 0.75-degree(s)
    Toe-in .... -1/32-inch
1985 and 1986
  Front
    Caster .... 1.44-degree(s)
    Camber .... 0.30-degree(s)
    Toe-in 1/16-inch

Rear
- Camber .... 0.75-degree(s)
- Toe-in .... -1/32-inch

1987 thru 1989
- Front
  - Caster .... 2.00-degree(s)
  - Camber .... 0.30-degree(s)
  - Toe-in 3/32-inch
- Rear
  - Camber .... -0.44-degree(s)
  - Toe-in .... -3/16-inch

1990 thru 1992
- Front
  - Caster .... 1.30-degree(s)
  - Camber .... 0.25-degree(s)
  - Toe-in 3/32-inch
- Rear
  - Camber .... -0.56-degree(s)
  - Toe-in .... -3/32-inch

Stanza Wagon

1985 thru 1988
- 2WD
  - Front
    - Caster .... 1.50-degree(s)
    - Camber .... 0.50-degree(s)
    - Toe-in 3/32-inch
  - Rear
    - Camber .... 0-degree(s)
    - Toe-in 5/32-inch
- 4WD
  - Front
    - Caster .... 1.30-degree(s)
    - Camber .... 0.25-degree(s)
    - Toe-in 1/64-inch
  - Rear
    - Camber .... 0.75-degree(s)
    - Toe-in .... -3/32-inch

Pickup and Pathfinder

1983 (Datsun)
- 2WD
  - Front
    - Caster .... 1.30-degree(s)
    - Camber .... 0.50-degree(s)
    - Toe-in 1/4-inch
- 4WD
  - Front
    - Caster .... 1.70-degree(s)
    - Camber .... 0.50-degree(s)
    - Toe-in 7/32-inch

1983 and 1984 (Nissan)
- 2WD
  - Front
    - Caster .... 1.70-degree(s)
    - Camber .... 0.50-degree(s)
    - Toe-in 1/4-inch
- 4WD
  - Front
    - Caster .... 1.44-degree(s)
    - Camber .... 0.70-degree(s)
    - Toe-in 1/4-inch

1985 and 1986 (except Hardbody)
- 2WD
  - Front
    - Caster .... 1.30-degree(s)
    - Camber .... 0.50-degree(s)
    - Toe-in 1/4-inch

- 4WD
  - Front
    - Caster .... 1.44-degree(s)
    - Camber .... 0.70-degree(s)
    - Toe-in 3/32-inch

1986 thru 1996 (Hardbody)
- 2WD Pickup
  - Front
    - Caster .... 0.38-degree(s)
    - Camber .... 0.40-degree(s)
    - Toe-in 5/32-inch
- 4WD Pickup and Pathfinder
  - Front
    - Caster .... 1.30-degree(s)
    - Camber .... 0.64-degree(s)
    - Toe-in 5/32-inch

Van

1987 and 1988
- Front
  - Caster .... 1.50-degree(s)
  - Camber .... 0.25-degree(s)
  - Toe-in 3/64-inch

OLDSMOBILE

Bravada and Silhouette - See *Chevrolet/GMC Trucks*

88, 98 and Custom Cruiser (RWD)

1983 thru 1986
- Front
  - Caster .... 3.00-degree(s)
  - Camber .... 0.80-degree(s)
  - Toe-in 5/32-inch

1987 thru 1990
- Front
  - Caster .... 2.80-degree(s)
  - Camber .... 0.80-degree(s)
  - Toe-in 3/64-inch

1991 thru 1993
- Front
  - Caster .... 3.50-degree(s)
  - Camber .... 0-degree(s)
  - Toe-in 5/64-inch

88, 98 and LSS (FWD)

1985 and 1986
- Front
  - Caster .... 2.30-degree(s)
  - Camber
    - Left wheel .... -0.50-degree(s)
    - Right wheel .... 0.50-degree(s)
  - Toe-in 0-inch
- Rear
  - Camber .... -0.30-degree(s)
  - Toe-in 3/32-inch

1987 thru 1996
- Front
  - Caster .... 3.00-degree(s)
  - Camber .... 0.20-degree(s)
  - Toe-in 0-inch
- Rear
  - Camber .... -0.30-degree(s)
  - Toe-in 3/64-inch

Haynes suspension, steering and driveline manual

Achieva

1992

Front
Caster.. 1.70-degree(s)
Camber .... 0-degree(s)
Toe-in 0-inch

Rear
Camber .... -0.25-degree(s)
Toe-in 1/8-inch

1993 and 1994

Front
Caster.. 1.45-degree(s)
Camber .... 0-degree(s)
Toe-in 0-inch

Rear
Camber .... -0.25-degree(s)
Toe-in 0-inch

1995 and 1996

Front
Caster.. 1.45-degree(s)
Camber .... 0-degree(s)
Toe-in 0-inch

Rear
Camber .... -0.40-degree(s)
Toe-in 7/64-inch

Aurora

1995

Front
Caster.. 5.50-degree(s)
Camber .... 0.20-degree(s)
Toe-in 0-inch

Rear
Camber .... -0.30-degree(s)
Toe-in 7/64-inch

1996

Front
Caster.. 6.00-degree(s)
Camber .... 0.20-degree(s)
Toe-in 0-inch

Rear
Camber .... -0.30-degree(s)
Toe-in 7/64-inch

Calais

1985 and 1986

Front
Caster.. 1.70-degree(s)
Camber .... 0.80-degree(s)
Toe-in .... -1/16-inch

Rear
Camber .... -0.25-degree(s)
Toe-in 1/8-inch

1987 thru 1989

Front
Caster.. 1.70-degree(s)
Camber .... 0.80-degree(s)
Toe-in 0-inch

Rear
Camber .... -0.25-degree(s)
Toe-in 1/8-inch

1990 and 1991

Front
Caster.. 1.70-degree(s)
Camber .... 0-degree(s)
Toe-in 0-inch

Rear
Camber .... -0.25-degree(s)
Toe-in 1/8-inch

Ciera, Cutlass Cruiser (FWD) and Omega

1983

Front
Caster.. 2.00-degree(s)
Camber .... 0-degree(s)
Toe-in 0-inch

Rear
Camber .... 0-degree(s)
Toe-in 1/32-inch

1984 thru 1986

Front
Caster.. 1.88-degree(s)
Camber .... 0-degree(s)
Toe-in 0-inch

Rear
Camber .... 0-degree(s)
Toe-in 0-inch

1987 thru 1992

Front
Caster.. 1.70-degree(s)
Camber .... 0-degree(s)
Toe-in 0-inch

Rear
Camber .... 0-degree(s)
Toe-in 0-inch

1993 thru 1996

Front
Caster.. 3.00-degree(s)
Camber .... 0-degree(s)
Toe-in 0-inch

Rear
Camber .... 0-degree(s)
Toe-in 0-inch

Cutlass (FWD)

1988 and 1989

Front
Caster.. 1.80-degree(s)
Camber .... 0.70-degree(s)
Toe-in 0-inch

Rear (1988)
Camber .... 0.25-degree(s)
Toe-in .... -3/64-inch

Rear (1989)
Camber .... 0.25-degree(s)
Toe-in 0-inch

1990

Front
Caster.. 2.00-degree(s)
Camber .... 0.70-degree(s)
Toe-in 0-inch

Rear (w/o FE3 suspension)
Camber .... 0.10-degree(s)
Toe-in 0-inch

Rear (w/FE3 suspension)
Camber .... -0.15-degree(s)
Toe-in 0-inch

1991

Front
Caster.. 1.80-degree(s)
Camber .... 0.70-degree(s)
Toe-in 0-inch

Rear
Camber .... -0.15-degree(s)
Toe-in .... -3/64-inch

1992 thru 1995

Front
Caster.. 2.00-degree(s)
Camber .... 0.70-degree(s)
Toe-in 0-inch

Rear
    Camber .... 0.10-degree(s)
    Toe-in .... -3/64-inch

1996
  Front
    Caster .... 1.80-degree(s)
    Camber .... 0.70-degree(s)
    Toe-in 0-inch
  Rear (15-inch wheels)
    Camber .... -0.35-degree(s)
    Toe-in 0-inch
  Rear (16-inch wheels)
    Camber .... -0.45-degree(s)
    Toe-in 0-inch

Cutlass (RWD)

1983 thru 1986
  Front
    Caster .... 3.00-degree(s)
    Camber .... 0.50-degree(s)
    Toe-in 5/32-inch

1987 and 1988
  Front
    Caster .... 2.80-degree(s)
    Camber .... 0.50-degree(s)
    Toe-in 3/64-inch

Firenza

1983 and 1984
  Front
    Caster .... 1.70-degree(s)
    Camber .... 0.80-degree(s)
    Toe-in .... -1/16-inch
  Hear
    Camber .... -0.25-degree(s)
    Toe-in 1/16-inch

1985 and 1986
  Front
    Caster .... 1.70-degree(s)
    Camber .... 0.80-degree(s)
    Toe-in .... -1/16-inch
  Rear
    Camber .... -0.25-degree(s)
    Toe-in 5/32-inch

1987 thru 1989
  Front
    Caster .... 1.70-degree(s)
    Camber .... 0.80-degree(s)
    Toe-in 0-inch
  Rear
    Camber .... -0.25-degree(s)
    Toe-in 1/8-inch

Toronado and Trofeo

1983 thru 1985
  Front
    Caster .... 2.50-degree(s)
    Camber .... 0-degree(s)
    Toe-in 0-inch
  Rear
    Camber .... 0-degree(s)
    Toe-in 3/32-inch

1986
  Front
    Caster .... 2.50-degree(s)
    Camber .... 0-degree(s)
    Toe-in 0-inch
  Rear
    Camber .... -0.70-degree(s)
    Toe-in 3/32-inch

1987 thru 1990
  Front
    Caster .... 2.30-degree(s)
    Camber .... 0-degree(s)
    Toe-in 0-inch
  Rear
    Camber .... -0.38-degree(s)
    Toe-in 3/32-inch

1991
  Front
    Caster .... 2.30-degree(s)
    Camber .... 0-degree(s)
    Toe-in 3/32-inch
  Rear
    Camber .... -0.20-degree(s)
    Toe-in 3/32-inch

1992
  Front
    Caster .... 2.30-degree(s)
    Camber
      Left wheel .... -0.50-degree(s)
      Right wheel .... 0.50-degree(s)
    Toe-in 3/32-inch
  Rear
    Camber .... -0.20-degree(s)
    Toe-in 3/32-inch

PEUGEOT

405

1989 thru 1992
  Front
    Caster .... 2.00-degree(s)
    Camber .... 0.15-degree(s)
    Toe-in 1/2-inch
  Rear
    Camber .... -1.30-degree(s)
    Toe-in .... -1/16-inch

504 Wagon

1983
  Front
    Caster .... 2.63-degree(s)
    Camber .... 0.63-degree(s)
    Toe-in 13/64-inch

505 Sedan

1983 and 1984
  Front
    Caster .... 3.50-degree(s)
    Camber .... 0.70-degree(s)
    Toe-in 9/32-inch
  Rear
    Camber .... -0.44-degree(s)
    Toe-in 1/4-inch

1985
  Front
    Caster .... 3.50-degree(s)
    Camber .... -1.00-degree(s)
    Toe-in 9/32-inch
  Rear
    Camber .... -1.00-degree(s)
    Toe-in 13/64-inch

1986 thru 1989
  4-cylinder (except Turbo)
    Front
      Caster .... 2.63-degree(s)
      Camber .... -0.75-degree(s)
      Toe-in 13/64-inch

Haynes suspension, steering and driveline manual

Rear
    Camber .... -1.00-degree(s)
    Toe-in 13/64-inch
1986 thru 1988
    Turbo (w/o ABS)
        Front
            Caster .... 2.63-degree(s)
            Camber .... -1.00-degree(s)
            Toe-in 5/32-inch
        Rear
            Camber .... -1.00-degree(s)
            Toe-in 13/64-inch
    Turbo (with ABS)
        Front
            Caster .... 3.00-degree(s)
            Camber .... -1.30-degree(s)
            Toe-in 13/64-inch
        Rear
            Camber .... -1.00-degree(s)
            Toe-in 13/64-inch
1989
    Front
        Caster .... 2.00-degree(s)
        Camber .... -0.88-degree(s)
        Toe-in 5/32-inch

505 Wagon

1984 and 1985
    Front
        Caster.. 2.50-degree(s)
        Camber .... 0.50-degree(s)
        Toe-in 1/4-inch
1986 and 1987
    Except turbo gas and turbo diesel
        Front
            Caster .... 2.00-degree(s)
            Camber .... -0.50-degree(s)
            Toe-in 1/4-inch
    Turbo gas
        Front
            Caster .... 2.00-degree(s)
            Camber .... -1.00-degree(s)
            Toe-in 13/64-inch
    Turbo diesel
        Front
            Caster .... 2.00-degree(s)
            Camber .... -0.88-degree(s)
            Toe-in 5/32-inch
1988 thru 1991
    Except Turbo
        Front
            Caster .... 2.70-degree(s)
            Camber .... -0.75-degree(s)
            Toe-in 13/32-inch
    Turbo
        Front
            Caster .... 2.00-degree(s)
            Camber .... -0.88-degree(s)
            Toe-in 5/32-inch

604

1983 and 1984
    Front
        Caster .... 3.50-degree(s)
        Camber .... -0.50-degree(s)
        Toe-in 13/64-inch
    Rear
        Camber .... -1.50-degree(s)
        Toe-in.. 5/32-inch

PLYMOUTH

Colt, Champ and Vista - see _Dodge_

Acclaim

1989 thru 1990
    Front
        Caster.. 1.20-degree(s)
        Camber .... 0.30-degree(s)
        Toe-in 1/16-inch
    Rear
        Camber .... -0.50-degree(s)
        Toe-in 0-inch
1991 thru 1995
    Front
        Caster.. 2.80-degree(s)
        Camber .... 0.30-degree(s)
        Toe-in 1/16-inch
    Rear
        Camber .... -0.50-degree(s)
        Toe-in 0-inch

Breeze

1996
    Front
        Caster.. 3.30-degree(s)
        Camber .... 0.30-degree(s)
        Toe-in 3/64-inch
    Rear
        Camber .... -0.20-degree(s)
        Toe-in 3/64-inch

Caravelle (FWD)

1984 thru 1988
    Front
        Caster.. 1.20-degree(s)
        Camber .... 0.30-degree(s)
        Toe-in 1/16-inch
    Rear
        Camber .... -0.50-degree(s)
        Toe-in 0-inch

Conquest

1984 and 1985
    Front
        Caster.. 5.30-degree(s)
        Camber .... 0-degree(s)
        Toe-in 0-inch
    Rear
        Camber .... 0-degree(s)
        Toe-in 0-inch
1986
    Front
        Caster.. 5.88-degree(s)
        Camber .... -0.50-degree(s)
        Toe-in 0-inch
    Rear
        Camber .... 0-degree(s)
        Toe-in 0-inch

Fury and Caravelle (RWD)

1983 thru 1989
    Front
        Caster.. 2.50-degree(s)
        Camber .... 0.50-degree(s)
        Toe-in 1/8-inch

Horizon
1983 thru 1990
  Front
    Caster.. 1.44-degree(s)
    Camber .... 0.30-degree(s)
    Toe-in 1/16-inch
  Rear
    Camber .... -0.75-degree(s)
    Toe-in 3/32-inch

Laser
1990 thru 1994
  Front
    1.8L
      Caster .... 2.30-degree(s)
      Camber .... 0.25-degree(s)
      Toe-in 0-inch
    2.0L (2WD)
      Caster .... 2.40-degree(s)
      Camber .... 0.10-degree(s)
      Toe-in 0-inch
    2.0L (4WD)
      Caster .... 2.30-degree(s)
      Camber .... 0.15-degree(s)
      Toe-in 0-inch
  Rear (2WD)
    Camber .... -0.75-degree(s)
    Toe-in 0-inch
  Rear (4WD)
    Camber .... -1.56-degree(s)
    Toe-in 5/32-inch

Neon
1995 and 1996
  Front
    Caster.. 2.80-degree(s)
    Camber .... 0-degree(s)
    Toe-in 3/64-inch
  Rear
    Camber .... -0.25-degree(s)
    Toe-in 3/64-inch

Reliant
1983
  Front
    Except wagon
      Caster .... 1.20-degree(s)
      Camber .... 0.30-degree(s)
      Toe-in 1/16-inch
    Wagon
      Caster .... 0.88-degree(s)
      Camber .... 0.30-degree(s)
      Toe-in 1/16-inch
  Rear
    Camber .... -0.50-degree(s)
    Toe-in 0-inch
1984 thru 1989
  Front
    Except wagon
      Caster .... 1.20-degree(s)
      Camber .... 0.30-degree(s)
      Toe-in 1/16-inch
    Wagon
      Caster .... 0.88-degree(s)
      Camber .... 0.30-degree(s)
      Toe-in 1/16-inch
  Rear
    Camber .... -0.50-degree(s)
    Toe-in 0-inch

Sapporo
1983
  Front
    Caster.. 2.70-degree(s)
    Camber .... 1.20-degree(s)
    Toe-in 5/32-inch

Scamp
1983
  Front
    Caster.. 1.88-degree(s)
    Camber .... 0.30-degree(s)
    Toe-in 1/16-inch
  Rear
    Camber .... -0.63-degree(s)
    Toe-in 3/32-inch

Sundance and Duster
1987 thru 1990
  Front
    Caster.. 1.20-degree(s)
    Camber .... 0.30-degree(s)
    Toe-in 1/16-inch
  Rear
    Camber .... -0.50-degree(s)
    Toe-in 0-inch
1991 thru 1994
  Front
    Caster.. 2.80-degree(s)
    Camber .... 0.30-degree(s)
    Toe-in 1/16-inch
  Rear
    Camber .... -0.50-degree(s)
    Toe-in 0-inch

Turismo and TC3
1983 thru 1987
  Front
    Caster.. 1.88-degree(s)
    Camber .... 0.30-degree(s)
    Toe-in 1/16-inch
  Rear
    Camber .... -0.75-degree(s)
    Toe-in 3/32-inch

Voyager
1984 thru 1990
  Front
    Caster.. 0.70-degree(s)
    Camber .... 0.30-degree(s)
    Toe-in 1/16-inch
  Rear (1984 thru 1989)
    Camber .... -0.63-degree(s)
    Toe-in 0-inch
  Rear (1990)
    Camber .... -0.20-degree(s)
    Toe-in 0-inch
1991
  Front
    Caster.. 1.70-degree(s)
    Camber .... 0.30-degree(s)
    Toe-in 1/16-inch
  Rear
    Camber .... -0.20-degree(s)
    Toe-in .... 0-degree(s)
1992 thru 1995
  Front
    Caster.. 1.30-degree(s)
    Camber .... 0.30-degree(s)
    Toe-in 1/16-inch

Rear
    Camber .... -0.20-degree(s)
    Toe-in 0-inch

1996
  Front
    Caster .... 1.40-degree(s)
    Camber .... 0.10-degree(s)
    Toe-in 3/64-inch
  Rear
    Camber .... 0-degree(s)
    Toe-in 0-inch

PONTIAC

Trans Sport - See *Chevrolet/GMC Trucks*

T1000
1983 thru 1986
  Front
    Caster .... 5.00-degree(s)
    Camber .... 0.20-degree(s)
    Toe-in 1/16-inch
1987
  Front
    Caster .... 4.70-degree(s)
    Camber .... 0.20-degree(s)
    Toe-in 1/16-inch

6000 and Phoenix
1983
  Front
    Caster .... 2.00-degree(s)
    Camber .... 0-degree(s)
    Toe-in 0-inch
  Rear
    Camber .... 0-degree(s)
    Toe-in 1/32-inch
1984 thru 1986
  Front
    Caster .... 1.88-degree(s)
    Camber .... 0-degree(s)
    Toe-in 0-inch
  Rear
    Camber .... 0-degree(s)
    Toe-in 0-inch
1987 thru 1990
  Front
    Caster .... 1.70-degree(s)
    Camber .... 0-degree(s)
    Toe-in 0-inch
  Rear (2WD)
    Camber .... 0-degree(s)
    Toe-in 0-inch
  Rear (4WD) (1988 and 1989)
    Camber .... -0.50-degree(s)
    Toe-in 5/32-inch
  Rear (4WD) (1990)
    Camber .... 0.38-degree(s)
    Toe-in 3/32-inch
1991 and 1992
  Front
    Caster .... 1.70-degree(s)
    Camber .... 0-degree(s)
    Toe-in 0-inch
  Rear
    Camber .... 0-degree(s)
    Toe-in 0-inch

Bonneville (FWD)
1987 thru 1996
  Front
    Caster .... 3.00-degree(s)
    Camber .... 0.20-degree(s)
    Toe-in 0-inch
  Rear
    Camber .... -0.30-degree(s)
    Toe-in 3/64-inch

Bonneville, Grand Prix and Grand LeMans (RWD)
1983 thru 1986
  Front
    Caster .... 3.00-degree(s)
    Camber .... 0.50-degree(s)
    Toe-in 5/32-inch
1987
  Front
    Caster .... 2.80-degree(s)
    Camber .... 0.50-degree(s)
    Toe-in 3/64-inch

Fiero
1984 thru 1986
  Front
    Caster .... 5.00-degree(s)
    Camber .... 0.50-degree(s)
    Toe-in 5/32-inch
  Rear
    Camber .... -1.00-degree(s)
    Toe-in 0-inch
1987
  Front
    Caster .... 5.00-degree(s)
    Camber .... 0.50-degree(s)
    Toe-in 0-inch
  Rear
    Camber .... -1.00-degree(s)
    Toe-in 0-inch
1988
  Front
    Caster .... 3.00-degree(s)
    Camber .... 0-degree(s)
    Toe-in 5/32-inch
  Rear
    Camber .... -1.00-degree(s)
    Toe-in 5/32-inch

Firebird
1983 thru 1986
  Front
    Caster .... 3.00-degree(s)
    Camber .... 1.00-degree(s)
    Toe-in 3/64-inch
1987 thru 1990
  Front
    Caster .... 4.70-degree(s)
    Camber .... 0.30-degree(s)
    Toe-in 0-inch
1991 and 1992
  Front
    Caster .... 4.80-degree(s)
    Camber .... 0.30-degree(s)
    Toe-in 0-inch
1993 thru 1995
  Front
    Caster .... 4.40-degree(s)
    Camber .... 0.40-degree(s)
    Toe-in 0-inch

1996
  Front
    Caster... 5.00-degree(s)
    Camber .... 0-degree(s)
    Toe-in.. 0-inch

Firefly
1985 and 1986
  Front
    Caster... 3.15-degree(s)
    Camber .... 1.00-degree(s)
    Toe-in.. 1/16-inch
  Rear
    Camber .... 0-degree(s)
    Toe-in.. 0-inch
1987 and 1988
  Front
    Caster... 3.15-degree(s)
    Camber .... 0.25-degree(s)
    Toe-in.. 0-inch
  Rear
    Camber .... 0-degree(s)
    Toe-in.. 0-inch
1989 thru 1991
  Front
    Caster... 3.00-degree(s)
    Camber .... 0-degree(s)
    Toe-in.. 0-inch
  Rear
    Camber .... 0-degree(s)
    Toe-in.. 3/32-inch

Grand Am
1985 and 1986
  Front
    Caster... 1.70-degree(s)
    Camber .... 0.80-degree(s)
    Toe-in.. -1/16-inch
  Rear
    Camber .... -0.25-degree(s)
    Toe-in.. 1/8-inch
1987 thru 1989
  Front
    Caster... 1.70-degree(s)
    Camber .... 0.80-degree(s)
    Toe-in.. 0-inch
  Rear
    Camber .... -0.25-degree(s)
    Toe-in.. 1/8-inch
1990 thru 1992
  Front
    Caster... 1.70-degree(s)
    Camber .... 0-degree(s)
    Toe-in.. 0-inch
  Rear (1990)
    Camber .... -0.25-degree(s)
    Toe-in.. 1/8-inch
  Rear (1991 and 1992)
    Camber .... -0.25-degree(s)
    Toe-in.. 1/8-inch
1993 and 1994
  Front
    Caster... 1.45-degree(s)
    Camber .... 0-degree(s)
    Toe-in.. 0-inch
  Rear
    Camber .... -0.25-degree(s)
    Toe-in.. 0-inch

1995 and 1996
  Front
    Caster... 1.45-degree(s)
    Camber .... 0-degree(s)
    Toe-in.. 0-inch
  Rear
    Camber .... -0.40-degree(s)
    Toe-in.. 7/64-inch

Grand Prix (FWD)
1988 and 1989
  Front
    Caster... 1.80-degree(s)
    Camber .... 0.70-degree(s)
    Toe-in.. 0-inch
  Rear (1988)
    Camber .... 0.25-degree(s)
    Toe-in.. -3/64-inch
  Rear (1989)
    Camber .... 0.25-degree(s)
    Toe-in.. 0-inch
1990 thru 1995
  Front
    Caster... 2.00-degree(s)
    Camber .... 0.70-degree(s)
    Toe-in.. 0-inch
  Rear, (1990)
    w/o FE3 suspension
      Camber .... 0.25-degree(s)
      Toe-in.. 0-inch
    w/FE3 suspension
      Caster .... 0.10-degree(s)
      Camber 0-inch
  Rear, (1991 thru 1995)
    Camber .... 0.10-degree(s)
    Toe-in.. -3/64-inch
1996
  Front
    Caster... 1.80-degree(s)
    Camber .... 0.70-degree(s)
    Toe-in.. 0-inch
  Rear (with 15-inch wheels)
    Camber .... -0.35-degree(s)
    Toe-in.. 0-inch
  Rear (with 16-inch wheels)
    Camber .... -0.45-degree(s)
    Toe-in.. 0-inch

LeMans
1988 and 1989
  Front
    Caster... 1.75-degree(s)
    Camber .... -0.30-degree(s)
    Toe-in.. 0-inch
  Rear (1988)
    Camber .... -0.50-degree(s)
    Toe-in.. 3/32-inch
  Rear (1989)
    Camber .... -0.50-degree(s)
    Toe-in.. 5/32-inch
1990 and 1991
  Front
    Caster... 1.75-degree(s)
    Camber .... -0.30-degree(s)
    Toe-in.. 0-inch
  Rear
    Camber .... -0.50-degree(s)
    Toe-in.. -7/32-inch

1992 thru 1994
Front
  Caster .... 1.75-degree(s)
  Camber .... -0.50-degree(s)
  Toe-in 0-inch
Rear
  Camber .... -0.50-degree(s)
  Toe-in 1/8-inch

Parisienne and Safari
1983 thru 1986
Front
  Caster .... 3.00-degree(s)
  Camber .... 0.80-degree(s)
  Toe-in 5/32-inch
1987 thru 1989
Front
  Caster .... 2.80-degree(s)
  Camber .... 0.80-degree(s)
  Toe-in 3/64-inch

Sunbird and J2000
1983 and 1984
Front
  Caster .... 1.70-degree(s)
  Camber .... 0.80-degree(s)
  Toe-in .... -1/16-inch
Rear
  Camber .... -0.25-degree(s)
  Toe-in 1/16-inch
1985 and 1986
Front
  Caster .... 1.70-degree(s)
  Camber .... 0.80-degree(s)
  Toe-in .... -1/16-inch
Rear
  Camber .... -0.25-degree(s)
  Toe-in 5/32-inch
1987 thru 1989
Front
  Caster .... 1.70-degree(s)
  Camber .... 0.80-degree(s)
  Toe-in 0-inch
Rear
  Camber .... -0.25-degree(s)
  Toe-in 1/8-inch
1990 thru 1992
Front
  Caster .... 1.70-degree(s)
  Camber .... 0-degree(s)
  Toe-in 0-inch
Rear (1990)
  Camber .... -0.25-degree(s)
  Toe-in 1/8-inch
Rear (1991 and 1992)
  Camber .... -0.25-degree(s)
  Toe-in 1/8-inch
1993 and 1994
Front
  Caster .... 1.30-degree(s)
  Camber .... -0.15-degree(s)
  Toe-in 0-inch
Rear
  Camber .... -0.25-degree(s)
  Toe-in 1/8-inch

Sunfire
1995 and 1996
Front
  Caster .... 4.30-degree(s)

Camber .... -0.20-degree(s)
Toe-in 3/64-inch
Rear
  Camber .... -0.25-degree(s)
  Toe-in 7/64-inch

Tempest
1987 and 1988
Front
  Caster .... 1.15-degree(s)
  Camber .... 0.50-degree(s)
  Toe-in 0-inch
Rear
  Camber .... -0.25-degree(s)
  Toe-in 1/8-inch
1989 thru 1991
Front
  Caster .... 1.15-degree(s)
  Camber .... 0.63-degree(s)
  Toe-in 0-inch
Rear
  Camber .... -0.25-degree(s)
  Toe-in 1/8-inch
1991 and 1992
Front
  Caster .... 1.15-degree(s)
  Camber .... 0.10-degree(s)
  Toe-in 0-inch
Rear
  Camber .... -0.25-degree(s)
  Toe-in 5/32-inch

PORSCHE

911 and 911 Turbo
1983
Front
  Caster .... 6.00-degree(s)
  Camber .... 0.50-degree(s)
  Toe-in 1/8-inch
Rear
  Camber .... 0-degree(s)
  Toe-in 3/32-inch
1984 thru 1989
911
  Front
    Caster .... 6.0-degree(s)
    Camber .... 0-degree(s)
    Toe-in 1/8-inch
  Rear
    Camber .... -1.00-degree(s)
    Toe-in 5/32-inch
911 Turbo
  Front
    Caster .... 6.00-degree(s)
    Camber .... 0-degree(s)
    Toe-in 1/8-inch
  Rear
    Camber .... -0.50-degree(s)
    Toe-in 5/32-inch
1990 thru 1993
Except RS and Speedster
  Front
    Caster .... 4.44-degree(s)
    Camber .... 0-degree(s)

Toe-in 7/32-inch
Rear
Camber .... -0.75-degree(s)
Toe-in 1/4-inch
RS America & Speedster
Front
Caster .... 4.44-degree(s)
Camber .... -0.25-degree(s)
Toe-in 1/8-inch
Rear
Camber .... -0.75-degree(s)
Toe-in 1/4-inch
1995 and 1996
Front
Caster .... 5.33-degree(s)
Camber .... -0.33-degree(s)
Toe-in 3/64-inch
Rear
Camber .... -1.16-degree(s)
Toe-in 5/64-inch

911 Carrera
1989 thru 1993
Front
Caster .... 4.44-degree(s)
Camber .... 0-degree(s)
Toe-in 7/32-inch
Rear
Camber .... -0.66-degree(s)
Toe-in 5/32-inch
1994
Front
Caster .... 5.33-degree(s)
Camber .... -0.33-degree(s)
Toe-in 3/64-inch
Rear
except Sport Suspension
Camber .... -0.90-degree(s)
Toe-in 1/8-inch
Sport Suspension
Camber .... -1.00-degree(s)
Toe-in 1/8-inch

924S
1987 and 1988
Front
Caster .... 2.50-degree(s)
Camber .... -0.34-degree(s)
Toe-in 3/32-inch
Rear
Camber .... -1.00-degree(s)
Toe-in 0-inch

928, 928S and 928 GT
1983 thru 1989
Front
Caster .... 3.50-degree(s)
Camber .... -0.50-degree(s)
Toe-in 1/8-inch
Rear
Camber .... -0.65-degree(s)
Toe-in 5/32-inch
1990
Front
Caster .... 3.50-degree(s)
Camber .... -0.50-degree(s)
Toe-in 1/8-inch
Rear
Camber .... -0.65-degree(s)
Toe-in 5/32-inch

1991 thru 1995
Front
Caster .... 4.00-degree(s)
Camber .... -0.50-degree(s)
Toe-in 1/8-inch
Rear
Camber .... -0.65-degree(s)
Toe-in 5/32-inch

944, 944S and 944 Turbo
1983 thru 1989
Front
Caster .... 2.50-degree(s)
Camber .... -0.34-degree(s)
Toe-in 3/32-inch
Rear
Camber .... -1.00-degree(s)
Toe-in 0-inch
1990 and 1991
Front
Caster .... 2.50-degree(s)
Camber .... 0-degree(s)
Toe-in 5/64-inch
Rear
Camber .... -1.00-degree(s)
Toe-in 5/64-inch

968
1992 thru 1995
Front
Caster .... 3.25-degree(s)
Camber .... 0-degree(s)
Toe-in 5/04-inch
Rear
Camber .... -0.75-degree(s)
Toe-in 11/64-inch

RENAULT

18i, Fuego and Sportwagon
1983 thru 1986
Front
Caster
w/o power steering .... 1.25-degree(s)
w/power steering .... 2.25-degree(s)
Camber .... 0-degree(s)
Toe-in .... -5/32-inch
Rear
Camber .... 0.25-degree(s)
Toe-in .... -1/16-inch

Le Car
1983 thru 1986
Front
Caster .... 2.50-degree(s)
Camber .... 0.50-degree(s)
Toe-in .... -1/16-inch
Rear
Camber .... 0.75-degree(s)
Toe-in .... -3/32-inch

Haynes suspension, steering and driveline manual

SAAB

900

1983 thru 1985
- Front
  - Caster
    - w/o power steering .... 1.00-degree(s)
    - w/power steering .... 2.00-degree(s)
  - Camber .... 0.50-degree(s)
  - Toe-in 1/8-inch
- Rear
  - Camber .... -0.50-degree(s)
  - Toe-in 9/32-inch

1986 thru 1992
- Front
  - Caster .... 2.00-degree(s)
  - Camber .... 0.50-degree(s)
  - Toe-in 1/8-inch
- Rear
  - Camber .... -0.50-degree(s)
  - Toe-in 9/32-inch

1993
- Standard chassis
  - Front
    - Caster .... 2.00-degree(s)
    - Camber .... -0.25-degree(s)
    - Toe-in 5/64-inch
- Sport chassis
  - Front
    - Caster .... 2.00-degree(s)
    - Camber .... 0.25-degree(s)
    - Toe-in 1/16-inch
- Rear (all)
  - Toe-in 5/32-inch

1994 thru 1996
- Front
  - Caster .... 2.10-degree(s)
  - Camber .... -0.50-degree(s)
  - Toe-in 1/16-inch
- Rear
  - Camber .... -1.66-degree(s)
  - Toe-in 3/64-inch

9000

1986 thru 1996
- Front
  - Caster .... 1.65-degree(s)
  - Camber .... -0.65-degree(s)
  - Toe-in 1/16-inch
- Rear
  - Camber .... -0.25-degree(s)
  - Toe-in 3/32-inch

SATURN

All models

1991 thru 1996
- Front
  - Caster .... 1.70-degree(s)
  - Camber .... -0.50-degree(s)
  - Toe-in 7/64-inch
- Rear
  - Camber .... -0.70-degree(s)
  - Toe-in 7/64-inch

SUBARU

Brat

1983 thru 1987
- Except Turbo
  - Front
    - Caster .... 2.44-degree(s)
    - Camber .... -0.70-degree(s)
    - Toe-in .... -3/16-inch
  - Rear
    - Camber .... 0.30-degree(s)
    - Toe-in 0-inch
- Turbo
  - Front
    - Caster .... -0.50-degree(s)
    - Camber .... 2.20-degree(s)
    - Toe-in .... -3/16-inch
  - Rear
    - Camber .... -0.25-degree(s)
    - Toe-in 0-inch

Hatchback, Sedan and Wagon

1983 and 1984
- 2WD Sedan
  - Front
    - Caster .... -0.44-degree(s)
    - Camber .... 1.50-degree(s)
    - Toe-in 1/32-inch
  - Rear
    - Camber .... 0-degree(s)
    - Toe-in 0-inch
- 2WD Hatchback
  - Front
    - Caster .... 1.50-degree(s)
    - Camber .... -0.44-degree(s)
    - Toe-in 1/32-inch
  - Rear
    - Camber .... 0-degree(s)
    - Toe-in 0-inch
- 2WD Wagon
  - Front
    - Caster .... -0.06-degree(s)
    - Camber .... 1.75-degree(s)
    - Toe-in 1/32-inch
  - Rear
    - Camber .... 0-degree(s)
    - Toe-in 0-inch
- 4WD Hatchback
  - Front
    - Caster .... -0.50-degree(s)
    - Camber .... 2.44-degree(s)
    - Toe-in .... -3/16-inch
  - Rear
    - Camber .... 0-degree(s)
    - Toe-in 0-inch
- 4WD Wagon (except Turbo)
  - Front
    - Caster .... -0.70-degree(s)
    - Camber .... 2.44-degree(s)
    - Toe-in .... -3/16-inch
  - Rear
    - Camber .... 0.30-degree(s)
    - Toe-in 0-inch
- 4WD Wagon (Turbo)
  - Front
    - Caster .... -0.50-degree(s)
    - Camber .... 2.20-degree(s)
    - Toe-in .... -3/16-inch

Rear
  Camber .... -0.25-degree(s)
  Toe-in 0-inch

1985 thru 1989
2WD Sedan
  Front
    Caster .... 2.50-degree(s)
    Camber .... 0.75-degree(s)
    Toe-in 3/32-inch
2WD Hatchback
  Front
    Caster .... 1.50-degree(s)
    Camber .... -0.44-degree(s)
    Toe-in 1/32-inch
2WD Wagon
  Front
    Caster .... 2.00-degree(s)
    Camber .... 1.00-degree(s)
    Toe-in 3/32-inch
  Rear (all)
    Camber .... 0-degree(s)
    Toe-in 0-inch
4WD Sedan
  Front
    Caster .... 1.80-degree(s)
    Camber .... 1.70-degree(s)
    Toe-in .... -3/16-inch
4WD Wagon
  Front (w/o air suspension)
    Caster .... 1.56-degree(s)
    Camber .... 1.75-degree(s)
    Toe-in .... -3/16-inch
  Front (w/air suspension)
    Caster .... 2.12-degree(s)
    Camber .... 1.12-degree(s)
    Toe-in .... -3/16-inch
  Rear (all)
    Camber .... 0-degree(s)
    Toe-in 0-inch

Impreza
1993 thru 1996
2WD Sedan
  Front
    Caster .... 3.00-degree(s)
    Camber .... 0-degree(s)
    Toe-in 0-inch
  Rear
    Camber .... -0.80-degree(s)
    Toe-in 0-inch
2WD Wagon
  Front
    Caster .... 3.00-degree(s)
    Camber .... 0-degree(s)
    Toe-in 0-inch
  Rear
    Camber .... -0.80-degree(s)
    Toe-in 0-inch
4WD Sedan
  Front
    Caster .... 3.00-degree(s)
    Camber .... 0-degree(s)
    Toe-in 0-inch
  Rear
    Camber .... -0.08-degree(s)
    Toe-in 0-inch

4WD Wagon
  Front
    Caster .... 3.00-degree(s)
    Camber .... 0-degree(s)
    Toe-in 0-inch
  Rear
    Camber .... -0.90-degree(s)
    Toe-in 0-inch

Justy
1987 thru 1994
  Front
    Caster .... 2.50-degree(s)
    Camber .... 0.63-degree(s)
    Toe-in 3/32-inch
  Rear
    Camber .... 0-degree(s)
    Toe-in 0-inch

Legacy
1990 and 1991
2WD Sedan (except Turbo)
  Front
    Caster .... 3.10-degree(s)
    Camber .... -0.25-degree(s)
    Toe-in 0-inch
  Rear
    Camber .... -0.50-degree(s)
    Toe-in 0-inch
2WD Sedan (Turbo)
  Front
    Caster .... 2.80-degree(s)
    Camber .... -0.25-degree(s)
    Toe-in 0-inch
  Rear
    Camber .... -1.00-degree(s)
    Toe-in 0-inch
2WD Wagon
  Front
    Caster .... 2.80-degree(s)
    Camber .... -0.25-degree(s)
    Toe-in 0-inch
  Rear
    Camber .... -0.30-degree(s)
    Toe-in 0-inch
4WD Sedan (except Turbo)
  Front
    Caster .... 3.00-degree(s)
    Camber .... 0-degree(s)
    Toe-in 0-inch
  Rear
    Camber .... -0.50-degree(s)
    Toe-in 0-inch
4WD Sedan (Turbo)
  Front
    Caster .... 2.80-degree(s)
    Camber .... -0.25-degree(s)
    Toe-in 0-inch
  Rear
    Camber .... -1.00-degree(s)
    Toe-in 0-inch
4WD Wagon (w/o air suspension)
  Front
    Caster .... 2.75-degree(s)
    Camber .... 0-degree(s)
    Toe-in 0-inch
  Rear
    Camber .... -0.30-degree(s)
    Toe-in 0-inch

Haynes suspension, steering and driveline manual

4WD Wagon (w/air suspension)
- Front
  - Caster .... 3.00-degree(s)
  - Camber .... 0-degree(s)
  - Toe-in 0-inch
- Rear
  - Camber .... -0.50-degree(s)
  - Toe-in 0-inch

1992 thru 1996
- 2WD Sedan (except Turbo)
  - Front
    - Caster .... 3.10-degree(s)
    - Camber .... -0.25-degree(s)
    - Toe-in 0-inch
  - Rear
    - Camber .... -1.00-degree(s)
    - Toe-in 0-inch
- 2WD Sedan (Turbo)
  - Front
    - Caster .... 2.80-degree(s)
    - Camber .... -0.25-degree(s)
    - Toe-in 0-inch
  - Rear
    - Camber .... -1.00-degree(s)
    - Toe-in 0-inch
- 2WD Wagon
  - Front
    - Caster .... 2.80-degree(s)
    - Camber .... -0.25-degree(s)
    - Toe-in 0-inch
  - Rear
    - Camber .... -0.80-degree(s)
    - Toe-in 0-inch
- 4WD Sedan (except Turbo)
  - Front
    - Caster .... 3.00-degree(s)
    - Camber .... 0-degree(s)
    - Toe-in 0-inch
  - Rear
    - Camber .... -1.00-degree(s)
    - Toe-in 0-inch
- 4WD Sedan (Turbo)
  - Front
    - Caster .... 2.80-degree(s)
    - Camber .... -0.25-degree(s)
    - Toe-in 0-inch
  - Rear
    - Camber .... -1.00-degree(s)
    - Toe-in 0-inch
- 4WD Wagon (w/o air suspension)
  - Front
    - Caster .... 2.75-degree(s)
    - Camber .... 0-degree(s)
    - Toe-in 0-inch
  - Rear
    - Camber .... -0.80-degree(s)
    - Toe-in 0-inch
- 4WD Wagon (w/air suspension)
  - Front
    - Caster .... 3.00-degree(s)
    - Camber .... 0-degree(s)
    - Toe-in 0-inch
  - Rear
    - Camber .... -1.00-degree(s)
    - Toe-in 0-inch

Loyale

1990 thru 1993
- 2WD Sedan
  - Front
    - Caster .... 2.50-degree(s)
    - Camber .... 0.75-degree(s)
    - Toe-in 3/32-inch
- 2WD Wagon
  - Front
    - Caster .... 2.00-degree(s)
    - Camber .... 1.00-degree(s)
    - Toe-in 3/32-inch
  - Rear (all)
    - Camber .... 0-degree(s)
    - Toe-in 0-inch
- 4WD Sedan
  - Front
    - Caster .... 1.80-degree(s)
    - Camber .... 1.70-degree(s)
    - Toe-in .... -1/8-inch
- 4WD Wagon
  - Front
    - Caster .... 1.56-degree(s)
    - Camber .... 1.75-degree(s)
    - Toe-in .... -1/8-inch
  - Rear (all)
    - Camber .... 0-degree(s)
    - Toe-in 0-inch

1994 thru 1996
- 2WD Wagon
  - Front
    - Caster .... 2.00-degree(s)
    - Camber .... 1.00-degree(s)
    - Toe-in 3/32-inch
- 4WD Wagon
  - Front
    - Caster .... 1.56-degree(s)
    - Camber .... 1.75-degree(s)
    - Toe-in .... -1/8-inch
  - Rear (all)
    - Camber .... 0-degree(s)
    - Toe-in 0-inch

SVX

1992 thru 1996
- Front
  - Caster .... 4.88-degree(s)
  - Camber .... -0.44-degree(s)
  - Toe-in 0-inch
- Rear
  - Camber .... -0.70-degree(s)
  - Toe-in 0-inch

XT

1985 and 1986
- 2WD
  - Front
    - Caster .... 4.00-degree(s)
    - Camber .... 0-degree(s)
    - Toe-in 0-inch
- 4WD
  - Front
    - Caster .... 3.44-degree(s)
    - Camber .... 0.63-degree(s)
    - Toe-in 3/32-inch
  - Rear (all)
    - Camber .... 0-degree(s)
    - Toe-in 0-inch

1987 thru 1991
- 2WD
  - Front
    - Caster .... 4.00-degree(s)
    - Camber .... 0-degree(s)
    - Toe-in 0-inch
  - Rear
    - Camber .... 0-degree(s)
    - Toe-in 0-inch
- 4WD
  - Front
    - 4-cylinder
      - Caster .... 3.44-degree(s)
      - Camber .... 0.63-degree(s)
      - Toe-in .... -3/16-inch
    - 6-cylinder
      - Caster .... 3.50-degree(s)
      - Camber .... 0.80-degree(s)
      - Toe-in .... -3/16-inch
  - Rear
    - Camber .... -0.15-degree(s)
    - Toe-in 0-inch

SUZUKI

Esteem

1995 and 1996
- Front
  - Caster .... 2.70-degree(s)
  - Camber .... 0-degree(s)
  - Toe-in 0-inch
- Rear
  - Camber .... 0-degree(s)
  - Toe-in 0-inch

Samurai

1986 thru 1995
- Front
  - Caster .... 3.50-degree(s)
  - Camber .... 1.00-degree(s)
  - Toe-in 5/32-inch

Sidekick

1989 thru 1996
- Front
  - Caster .... 1.50-degree(s)
  - Camber .... 0.50-degree(s)
  - Toe-in 5/32-inch

Swift

1989 thru 1996
- Front
  - Caster .... 3.00-degree(s)
  - Camber .... 0-degree(s)
  - Toe-in 0-inch
- Rear
  - Toe-in 3/32-inch

TOYOTA

Avalon

1995 and 1996
- Front
  - Caster .... 1.16-degree(s)
  - Camber .... -0.58-degree(s)
  - Toe-in 0-inch

- Rear
  - Camber .... -0.75-degree(s)
  - Toe-in 5/32-inch

Camry

1983 and 1984
- Front
  - Caster
    - w/o power steering .... 1.00-degree(s)
    - w/power steering .... 2.50-degree(s)
  - Camber .... 0.56-degree(s)
  - Toe-in
    - w/o power steering 0-inch
    - w/power steering 3/32-inch
- Rear
  - Camber .... 0.50-degree(s)
  - Toe-in 0-inch

1985 and 1986
- Front
  - Caster .... 1.00-degree(s)
  - Camber .... 0.56-degree(s)
  - Toe-in 3/32-inch
- Rear
  - Camber .... 0.50-degree(s)
  - Toe-in 5/32-inch

1987 thru 1990
- Except wagon
  - Front
    - Caster .... 1.70-degree(s)
    - Camber .... 0.56-degree(s)
    - Toe-in 1/32-inch
  - Rear (2WD)
    - Camber .... 0.56-degree(s)
    - Toe-in 5/32-inch
  - Rear (4WD)
    - Camber .... -0.50-degree(s)
    - Toe-in 1/8-inch
- Wagon
  - Front
    - Caster .... 1.00-degree(s)
    - Camber .... 0.50-degree(s)
    - Toe-in 1/32-inch
  - Rear (4-cylinder)
    - Camber .... -0.56-degree(s)
    - Toe-in 5/32-inch
  - Rear (V6)
    - Camber .... -0.06-degree(s)
    - Toe-in 5/32-inch

1991
- Except wagon
  - Front
    - Caster .... 1.56-degree(s)
    - Camber .... 0.50-degree(s)
    - Toe-in 1/32-inch
  - Rear
    - Camber .... -0.56-degree(s)
    - Toe-in 5/32-inch
- Wagon
  - Front
    - Caster .... 0.90-degree(s)
    - Camber .... 0.44-degree(s)
    - Toe-in 1/32-inch
  - Rear (2WD)
    - Camber .... -0.06-degree(s)
    - Toe-in 5/32-inch
  - Rear (4WD)
    - Camber .... -0.50-degree(s)
    - Toe-in 1/8-inch

Haynes suspension, steering and driveline manual

1992 thru 1996
  Except wagon
    Front
      Caster .... 1.15-degree(s)
      Camber .... -0.58-degree(s)
      Toe-in 0-inch
    Rear (1992 and 1993)
      Camber .... -0.50-degree(s)
      Toe-in 5/32-inch
    Rear (1994 thru 1996)
      Camber .... -0.43-degree(s)
      Toe-in 5/32-inch
  Wagon
    Front
      Caster .... 1.00-degree(s)
      Camber .... -0.60-degree(s)
      Toe-in 0-inch
    Rear
      Camber .... -0.25-degree(s)
      Toe-in 5/32-inch

Celica

1983 thru 1985
  Front
    Caster .... 3.40-degree(s)
    Camber .... 0.90-degree(s)
    Toe-in 5/32-inch
  Rear
    Camber .... -0.15-degree(s)
    Toe-in 0-inch
1986 thru 1989
  Front
    Caster .... 1.15-degree(s)
    Camber .... -0.15-degree(s)
    Toe-in 0-inch
  Rear
    Camber .... -0.75-degree(s)
    Toe-in 3/16-inch
1990
  Front
    Caster .... 1.00-degree(s)
    Camber .... -0.15-degree(s)
    Toe-in 3/32-inch
  Rear
    Camber .... -0.75-degree(s)
    Toe-in 3/16-inch
1991 thru 1993
  2WD
    Front
      Caster .... 0.92-degree(s)
      Camber .... -0.15-degree(s)
      Toe-in 0-inch
  4WD
    Front
      Caster .... 0.80-degree(s)
      Camber .... -0.15-degree(s)
      Toe-in 0-inch
  Rear, all (1991)
    Camber .... -0.75-degree(s)
    Toe-in 3/16-inch
  Rear, all (1993-92)
    Camber .... -1.25-degree(s)
    Toe-in 3/16-inch
1994 thru 1996
  1.6L
    Front
      Caster .... 2.12-degree(s)
      Camber .... -0.85-degree(s)
      Toe-in 0-inch

    Rear
      Camber .... -1.25-degree(s)
      Toe-in 9/64-inch
  2.2L
    Front
      Caster .... 2.08-degree(s)
      Camber .... -0.75-degree(s)
      Toe-in 0-inch
    Rear
      Camber .... -1.17-degree(s)
      Toe-in 9/64-inch

Corolla

1983
  Except wagon
    Front
      Caster .... 1.75-degree(s)
      Camber .... 1.00-degree(s)
      Toe-in 1/32-inch
  Wagon
    Front
      Caster .... 1.56-degree(s)
      Camber .... 1.00-degree(s)
      Toe-in 1/32-inch
1984 and 1985
  FWD
    Front
      Caster .... 0.90-degree(s)
      Camber .... -0.50-degree(s)
      Toe-in 0-inch
    Rear
      Camber .... -0.50-degree(s)
      Toe-in 5/32-inch
  RWD
    Front
      Caster
        w/o power steering .... 2.75-degree(s)
        w/power steering .... 3.70-degree(s)
      Camber .... 0.25-degree(s)
      Toe-in 1/32-inch
1986 and 1987
  FWD
    Front
      Caster .... 0.88-degree(s)
      Camber .... -0.25-degree(s)
      Toe-in 1/32-inch
    Rear
      Camber .... -0.50-degree(s)
      Toe-in 9/64-inch
  RWD
    Front
      Caster
        w/o power steering .... 2.75-degree(s)
        w/power steering .... 3.70-degree(s)
      Camber .... 0.25-degree(s)
      Toe-in 1/32-inch
1988 thru 1990
  2WD
    4A-F engine (except Coupe)
      Front
        Caster .... 1.30-degree(s)
        Camber .... -0.15-degree(s)
        Toe-in 1/32-inch
      Rear
        Camber .... -0.56-degree(s)
        Toe-in 5/32-inch

Wheel alignment specifications

4A-F engine (Coupe)
Front
Caster .... 1.50-degree(s)
Camber .... -0.15-degree(s)
Toe-in 1/32-inch
Rear
Camber .... -0.70-degree(s)
Toe-in 5/32-inch
4A-GE engine
Front
Caster .... 1.30-degree(s)
Camber .... -0.25-degree(s)
Toe-in 1/32-inch
Rear
Camber .... -0.70-degree(s)
Toe-in 5/32-inch
4WD
Front
Caster .... 1.25-degree(s)
Camber .... 0.15-degree(s)
Toe-in 1/32-inch

1991
2WD
4A-FE engine (except Coupe)
Front
Caster .... 1.44-degree(s)
Camber .... -0.15-degree(s)
Toe-in 1/32-inch
Rear
Camber .... -0.56-degree(s)
Toe-in 5/32-inch
4A-FE engine (Coupe)
Front
Caster .... 1.50-degree(s)
Camber .... -0.25-degree(s)
Toe-in 1/32-inch
Rear
Camber .... -0.70-degree(s)
Toe-in 5/32-inch
4A-GE engine (M/T)
Front
Caster .... 1.30-degree(s)
Camber .... -0.30-degree(s)
Toe-in 1/32-inch
Rear
Camber .... -0.70-degree(s)
Toe-in 5/32-inch
4A-GE engine (A/T)
Front
Caster .... 1.50-degree(s)
Camber .... -0.25-degree(s)
Toe-in 1/32-inch
Rear
Camber .... -0.70-degree(s)
Toe-in 5/32-inch
4WD
Front
Caster .... 1.25-degree(s)
Camber .... 0.15-degree(s)
Toe-in 1/32-inch

1992
2WD
Front
Caster .... 1.44-degree(s)
Camber .... -0.15-degree(s)
Toe-in 1/32-inch
Rear
Camber .... -56-degree(s)
Toe-in 5/32-inch

4WD
Front
Caster .... 1.25-degree(s)
Camber .... 0.15-degree(s)
Toe-in 1/32-inch
1993 thru 1996
Front
Caster .... 1.33-degree(s)
Camber .... -0.15-degree(s)
Toe-in 3/64-inch
Rear (1993 thru 1995)
Camber .... -0.90-degree(s)
Toe-in 1/8-inch
Rear (1996)
Camber .... -0.90-degree(s)
Toe-in 5/32-inch

Cressida
1983 and 1984
Except wagon
Front
Caster .... 2.50-degree(s)
Camber .... 0.75-degree(s)
Toe-in 1/8-inch
Rear
Camber .... 0.38-degree(s)
Toe-in .... -3/32-inch
Wagon
Front
Caster .... 2.12-degree(s)
Camber .... 0.88-degree(s)
Toe-in 1/8-inch
1985
Front
Caster .... 4.80-degree(s)
Camber .... 0.44-degree(s)
Toe-in 3/32-inch
Rear
Camber .... -0.44-degree(s)
Toe-in 1/8-inch
1986 thru 1988
Except wagon
Front
Caster .... 4.88-degree(s)
Camber .... 0.44-degree(s)
Toe-in 3/32-inch
Rear
Camber .... 0.44-degree(s)
Toe-in 1/8-inch
Wagon
Front
Caster .... 4.25-degree(s)
Camber .... 0.44-degree(s)
Toe-in 3/32-inch
1989 thru 1992
Front
Caster .... 7.30-degree(s)
Camber .... 0.50-degree(s)
Toe-in 3/32-inch
Rear
Camber .... 0-degree(s)
Toe-in 3/32-inch

MR2
1985 and 1986
Front
Caster .... 5.38-degree(s)
Camber .... 0.25-degree(s)
Toe-in 1/32-inch

Haynes suspension, steering and driveline manual

Rear
 Camber .... -0.75-degree(s)
 Toe-in 5/16-inch
1987 thru 1989
 Front
  Caster .... 5.00-degree(s)
  Camber .... 0.25-degree(s)
  Toe-in 1/32-inch
 Rear
  Camber .... -0.88-degree(s)
  Toe-in 3/16-inch
1991 and 1992
 Front
  Caster .... 2.75-degree(s)
  Camber .... -0.90-degree(s)
  Toe-in 1/32-inch
 Rear
  Camber .... -1.30-degree(s)
  Toe-in 3/16-inch
1993
 Front
  Caster .... 2.80-degree(s)
  Camber .... -1.00-degree(s)
  Toe-in 3/64-inch
 Rear
  Camber .... -1.58-degree(s)
  Toe-in 15/64-inch
1994 thru 1996
 Front
  Caster .... 3.25-degree(s)
  Camber .... -1.00-degree(s)
  Toe-in 3/64-inch
 Rear
  Camber .... -1.58-degree(s)
  Toe-in 5/32-inch

Paseo
1992 thru 1996
 Front
  Caster .... 1.50-degree(s)
  Camber .... -0.42-degree(s)
  Toe-in 3/64-inch
 Rear
  Camber .... -0.50-degree(s)
  Toe-in 1/8-inch

Previa
1991 thru 1996
 2WD
  Front
   Caster .... 5.50-degree(s)
   Camber .... 0.08-degree(s)
   Toe-in 3/32-inch
 4WD
  Front
   Caster .... 5.33-degree(s)
   Camber .... 0.25-degree(s)
   Toe-in 1/8-inch

Starlet
1983
 Front
  Caster .... 2.00-degree(s)
  Camber .... 0.70-degree(s)
  Toe-in 5/64-inch
1984
 Front
  Caster .... 1.90-degree(s)
  Camber .... 0.70-degree(s)
  Toe-in 3/32-inch

Supra
1983 thru 1988
 2.8L
  Front
   Caster .... 4.20-degree(s)
   Camber .... 0.88-degree(s)
   Toe-in 1/8-inch
  Rear
   Camber .... -0.15-degree(s)
   Toe-in 0-inch
 3.0L
  Front
   Caster .... 7.50-degree(s)
   Camber .... -0.06-degree(s)
   Toe-in 0-inch
  Rear
   Camber .... -025-degree(s)
   Toe-in 1/8-inch
1989 and 1990
 Front
  Caster .... 7.63-degree(s)
  Camber .... -0.15-degree(s)
  Toe-in 0-inch
 Rear
  Camber .... -0.75-degree(s)
  Toe-in 5/32-inch
1991 and 1992
 Front
  Caster .... 7.56-degree(s)
  Camber .... -0.15-degree(s)
  Toe-in 0-inch
 Rear
  Camber .... -0.80-degree(s)
  Toe-in 5/32-inch
1993 thru 1996
 Except turbo
  Front
   Caster .... 3.33-degree(s)
   Camber .... -0.33-degree(s)
   Toe-in 0-inch
  Rear
   Camber .... -1.58-degree(s)
   Toe-in 1/8-inch
 Turbo
  Front
   Caster .... 3.50-degree(s)
   Camber .... -0.50-degree(s)
   Toe-in 0-inch
  Rear
   Camber .... -1.50-degree(s)
   Toe-in 1/8-inch

Tercel
1983 and 1984
 Except wagon
  Front
   Caster
    w/o power steering .... 1.15-degree(s)
    w/power steering .... 2.70-degree(s)
   Camber .... 0.30-degree(s)
   Toe-in 0-inch
  Rear
   Camber .... -0.06-degree(s)
   Toe-in 0-inch
 Wagon (2WD)
  Front
   Caster
    w/o power steering .... 0.75-degree(s)
    w/power steering .... 2.25-degree(s)

Camber .... 0.25-degree(s)
Toe-in 0-inch
Rear
  Camber .... -0.15-degree(s)
  Toe-in 0-inch
Wagon (4WD)
  Front
    Caster .... 2.44-degree(s)
    Camber .... 0.80-degree(s)
    Toe-in 0-inch
1985
  Except wagon
    Front
      Caster
        w/o power steering .... 1.15-degree(s)
        w/power steering .... 2.70-degree(s)
      Camber .... 0.20-degree(s)
      Toe-in 0-inch
    Rear
      Camber .... -0.06-degree(s)
      Toe-in 0-inch
  Wagon (2WD)
    Front
      Caster
        w/o power steering .... 0.75-degree(s)
        w/power steering .... 2.25-degree(s)
      Camber .... 0.10-degree(s)
      Toe-in 0-inch
    Rear
      Camber .... -0.15-degree(s)
      Toe-in 0-inch
  Wagon (4WD)
    Front
      Caster .... 2.44-degree(s)
      Camber .... 0.70-degree(s)
      Toe-in 0-inch
1986 thru 1988
  Except wagon
    Front
      Caster
        w/o power steering .... 1.15-degree(s)
        w/power steering .... 2.70-degree(s)
      Camber .... 0.10-degree(s)
      Toe-in .... -1/32-inch
    Rear
      Camber .... -0.06-degree(s)
      Toe-in 0-inch
  Wagon (2WD)
    Front
      Caster
        w/o power steering .... 0.75-degree(s)
        w/power steering .... 2.25-degree(s)
      Camber .... 0-degree(s)
      Toe-in .... -1/32-inch
    Rear
      Camber .... -0.15-degree(s)
      Toe-in 0-inch
  Wagon (4WD)
    Front
      Caster .... 2.44-degree(s)
      Camber .... 0.56-degree(s)
      Toe-in .... -1/32-inch
1987 thru 1990
  Front
    Caster
      w/o power steering .... 1.00-degree(s)
      w/power steering .... 2.50-degree(s)
    Camber .... 0-degree(s)
    Toe-in 0-inch

Rear
  Camber .... 0-degree(s)
  Toe-in 1/8-inch
1991
  Front
    Caster .... 1.25-degree(s)
    Camber .... -0.25-degree(s)
    Toe-in 1/32-inch
  Rear
    Camber .... -0.50-degree(s)
    Toe-in 1/8-inch
1992 and 1993
  Front
    Caster .... 1.75-degree(s)
    Camber .... -0.44-degree(s)
    Toe-in 3/64-inch
  Rear
    Camber .... -0.50-degree(s)
    Toe-in 1/8-inch
1994
  Front
    Caster .... 1.25-degree(s)
    Camber .... -0.25-degree(s)
    Toe-in 3/64-inch
  Rear
    Camber .... -0.50-degree(s)
    Toe-in 1/8-inch
1995 and 1996
  Front
    Caster .... 1.33-degree(s)
    Camber .... -0.33-degree(s)
    Toe-in 3/64-inch
  Rear
    Camber .... -0.50-degree(s)
    Toe-in 1/8-inch

4-Runner
1984
  Front
    Caster .... 2.25-degree(s)
    Camber .... 1.00-degree(s)
    Toe-in 1/32-inch
1985
  Front
    Caster .... 3.00-degree(s)
    Camber .... 1.00-degree(s)
    Toe-in 1/32-inch
1986 thru 1988
  Front
    Caster .... 2.00-degree(s)
    Camber .... 0.70-degree(s)
    Toe-in 1/8-inch
1989 and 1990
  4-Runner
    Front
      Caster .... 2.25-degree(s)
      Camber .... 0.78-degree(s)
      Toe-in 3/32-inch
1991 thru 1995
  Front
    Caster .... 1.75-degree(s)
    Camber .... 0.72-degree(s)
    Toe-in 5/64-inch

Land Cruiser
1983 thru 1987
  Front
    Caster .... 1.00-degree(s)
    Camber .... 1.00-degree(s)
    Toe-in 3/32-inch

1988 thru 1990
  Front
    Caster.. 0.88-degree(s)
    Camber... 1.00-degree(s)
    Toe-in... 1/32-inch
1991 thru 1996
  Front
    Caster.. 3.00-degree(s)
    Camber... 1.00-degree(s)
    Toe-in... 5/64-inch

Pickup (except T100 and Tacoma)

1983
  2WD
    Shortbed
      Front
        Caster.. 1.00-degree(s)
        Camber... 1.12-degree(s)
        Toe-in.. 3/32-inch
    Longbed
      Front
        Caster.. 1.00-degree(s)
        Camber... 1.12-degree(s)
        Toe-in.. 3/32-inch
  4WD
    Shortbed
      Front
        Caster.. 3.50-degree(s)
        Camber... 1.00-degree(s)
        Toe-in.. 1/32-inch
    Longbed
      Front
        Caster.. 3.50-degree(s)
        Camber... 1.00-degree(s)
        Toe-in.. 1/32-inch
1984
  2WD
    Shortbed
      Front
        Caster.. 0.70-degree(s)
        Camber... 0.50-degree(s)
        Toe-in.. 1/32-inch
    Longbed
      Front
        Caster.. 1.20-degree(s)
        Camber... 0.50-degree(s)
        Toe-in.. 1/32-inch
    3/4-Ton
      Front
        Caster.. 1.70-degree(s)
        Camber... 0.50-degree(s)
        Toe-in.. 1/32-inch
  4WD
    Front
      Caster.. 2.25-degree(s)
      Camber... 1.00-degree(s)
      Toe-in.. 1/32-inch
1985
  2WD
    Shortbed
      Front
        Caster.. 0.70-degree(s)
        Camber... 0.50-degree(s)
        Toe-in.. 1/32-inch
    Longbed
      Front
        Caster.. 1.15-degree(s)
        Camber... 0.50-degree(s)
        Toe-in.. 1/8-inch

1-Ton
  Front
    Caster.. 0.56-degree(s)
    Camber... 0.50-degree(s)
    Toe-in... 5/32-inch
Cab and chassis
  Front
    Caster.. 0.06-degree(s)
    Camber... 0.50-degree(s)
    Toe-in... 5/32-inch
4WD
  Front
    Caster.. 2.25-degree(s)
    Camber... 1.00-degree(s)
    Toe-in... 1/32-inch
1986 thru 1988
  2WD
    Short wheelbase
      Front
        Caster.. 0.70-degree(s)
        Camber... 0.50-degree(s)
        Toe-in.. 5/32-inch
    Cab and Chassis
      Front
        Caster.. 0.06-degree(s)
        Camber... 0.50-degree(s)
        Toe-in.. 5/32-inch
    w/dual rear wheels
      Front
        Caster.. 0.56-degree(s)
        Camber... 0.50-degree(s)
        Toe-in.. 5/32-inch
  4WD
    Short wheelbase
      Front
        Caster.. 1.25-degree(s)
        Camber... 0.70-degree(s)
        Toe-in.. 1/8-inch
    Long wheelbase
      Front
        Caster.. 1.50-degree(s)
        Camber... 0.70-degree(s)
        Toe-in.. 1/8-inch
1989 and 1990
  4-cylinder
    2WD (short wheelbase)
      Front
        Caster.. 0.70-degree(s)
        Camber... 0.50-degree(s)
        Toe-in.. 1/16-inch
    2WD (long wheelbase)
      Front
        Caster.. 1.00-degree(s)
        Camber... 0.44-degree(s)
        Toe-in.. 3/32-inch
    2WD (extra-long wheelbase)
      Front
        Caster.. 1.15-degree(s)
        Camber... 0.34-degree(s)
        Toe-in.. 5/32-inch
    4WD (standard cab)
      Front
        Caster.. 1.50-degree(s)
        Camber... 0.56-degree(s)
        Toe-in.. 3/32-inch
    4WD (Xtra cab)
      Front
        Caster.. 1.63-degree(s)
        Camber... 0.50-degree(s)
        Toe-in.. 1/8-inch

V6
  2WD (short wheelbase)
    Front
      Caster..................................... 1.12-degree(s)
      Camber.................................... 0.50-degree(s)
      Toe-in..................................... 3/32-inch
  2WD (long wheelbase, except 1-Ton)
  Front
      Caster..................................... 0.06-degree(s)
      Camber.................................... 0.56-degree(s)
      Toe-in..................................... 3/16-inch
  2WD (1-Ton)
  Front
      Caster..................................... 0.56-degree(s)
      Camber.................................... 0.56-degree(s)
      Toe-In..................................... 7/32-inch
  2WD (extra-long wheelbase)
    Front
      Caster..................................... 1.28-degree(s)
      Camber.................................... 0.40-degree(s)
      Toe-in..................................... 1/8-inch
  4WD (short wheelbase)
  Front
      Caster..................................... 1.50-degree(s)
      Camber.................................... 0.63-degree(s)
      Toe-in..................................... 3/32-inch
  4WD (long wheelbase)
    Front
      Caster..................................... 1.56-degree(s)
      Camber.................................... 0.56-degree(s)
      Toe-in..................................... 3/32-inch
  4WD (extra-long wheelbase)
    Front
      Caster..................................... 1.75-degree(s)
      Camber.................................... 0.56-degree(s)
      Toe-in..................................... 1/8-inch

1991 thru 1993
  4-cylinder
    2WD (short wheelbase)
      Front
        Caster..................................... 0.70-degree(s)
        Camber.................................... 0.50-degree(s)
        Toe-in..................................... 1/16-inch
    2WD (long wheelbase)
      Front
        Caster..................................... 0.97-degree(s)
        Camber.................................... 0.44-degree(s)
        Toe-in..................................... 3/32-inch
    2WD (extra-long wheelbase)
      Front
        Caster..................................... 1.25-degree(s)
        Camber.................................... 0.38-degree(s)
        Toe-in..................................... 1/8-inch
    4WD (short wheelbase)
      Front
        Caster..................................... 1.63-degree(s)
        Camber.................................... 0.70-degree(s)
        Toe-in..................................... 3/32-inch
    4WD (long wheelbase)
      Front
        Caster..................................... 1.70-degree(s)
        Camber.................................... 0.70-degree(s)
        Toe-in..................................... 3/32-inch
    4WD (extra-long wheelbase)
      Front
        Caster..................................... 1.80-degree(s)
        Camber.................................... 0.70-degree(s)
        Toe-in..................................... 7/64-inch

V6
  2WD (short wheelbase)
    Front
      Caster..................................... 0.56-degree(s)
      Camber.................................... 0.50-degree(s)
      Toe-in..................................... 7/32-inch
  2WD (long wheelbase)
    Front
      Caster..................................... 1.75-degree(s)
      Camber.................................... 0.50-degree(s)
      Toe-in..................................... 7/32-inch
  2WD (extra-long wheelbase)
    Front
      Caster..................................... 1.20-degree(s)
      Camber.................................... 0.40-degree(s)
      Toe-in..................................... 1/8-inch
  4WD (short wheelbase)
    Front
      Caster..................................... 1.70-degree(s)
      Camber.................................... 0.70-degree(s)
      Toe-in..................................... 3/32-inch
  4WD (long wheelbase)
    Front
      Caster..................................... 1.75-degree(s)
      Camber.................................... 0.70-degree(s)
      Toe-in..................................... 3/32-inch
  4WD (extra-long wheelbase)
    Front
      Caster..................................... 1.88-degree(s)
      Camber.................................... 0.70-degree(s)
      Toe-in..................................... 7/64-inch
1994 and 1995
  2WD
    Standard cab
      Front
        Caster..................................... 0.75-degree(s)
        Camber.................................... 0.50-degree(s)
        Toe-in..................................... 1/16-inch
    Xtra cab
      Front
        Caster..................................... 1.25-degree(s)
        Camber.................................... 0.33-degree(s)
        Toe-in..................................... 1/8-inch
    Cab and chassis
      Front
        Caster..................................... 1.25-degree(s)
        Camber.................................... 0.33-degree(s)
        Toe-in..................................... 5/32-inch
  4WD
    Standard cab
      Front
        Caster..................................... 1.58-degree(s)
        Camber.................................... 0.75-degree(s)
        Toe-in..................................... 5/64-inch
    Xtra-cab and cab and chassis
      Front
        Caster..................................... 1.85-degree(s)
        Camber.................................... 0.75-degree(s)
        Toe-in..................................... 1/16-inch

T100
1993 and 1994
  2WD
    1/2-Ton
      Front
        Caster..................................... 2.50-degree(s)
        Camber.................................... 0.50-degree(s)
        Toe-in..................................... 1/8-inch

1-Ton
  Front
    Caster .... 3.00-degree(s)
    Camber .... 0.50-degree(s)
    Toe-in 23/64-inch
4WD
  Front
    Caster .... 1.50-degree(s)
    Camber .... 0.75-degree(s)
    Toe-in 1/8-inch
1995
  2WD
    1/2-Ton
      Front
        Caster .... 2.66-degree(s)
        Camber .... 0.40-degree(s)
        Toe-in 1/8-inch
    1-Ton
      Front
        Caster .... 1.58-degree(s)
        Camber .... 0.40-degree(s)
        Toe-in 9/32-inch
  4WD
    Front
      Caster .... 1.58-degree(s)
      Camber .... 0.66-degree(s)
      Toe-in 1/8-inch
1996
  1/2-Ton
    Front
      Caster .... 2.46-degree(s)
      Camber .... 0.40-degree(s)
      Toe-in .... -3/32-inch
  1-Ton
    Front
      Caster .... 1.63-degree(s)
      Camber .... 0.36-degree(s)
      Toe-in .... -9/32-inch
  Extra Cab (1/2 and 1-Ton)
    Front
      Caster .... 2.00-degree(s)
      Camber .... 0.62-degree(s)
      Toe-in .... -15/64-inch
  4WD
    Front
      Caster .... 1.50-degree(s)
      Camber .... 0.75-degree(s)
      Toe-in 3/64-inch
  Extra Cab (4WD)
    Front
      Caster .... 1.66-degree(s)
      Camber .... 0.75-degree(s)
      Toe-in 5/64-inch

Tacoma
1995 and 1996
  2WD
    4-cylinder, Standard cab
      Front
        Caster .... 0.55-degree(s)
        Camber .... 0.06-degree(s)
        Toe-in 5/64-inch
    4-cylinder, Xtra-cab
      Front
        Caster .... 0.80-degree(s)
        Camber .... 0.06-degree(s)
        Toe-in 5/64-inch
    V6
      Front
        Caster .... 0.70-degree(s)

        Camber .... 0.06-degree(s)
        Toe-in 5/64-inch
  4WD
    4-cylinder, Standard cab
      Front
        Caster .... 1.95-degree(s)
        Camber .... 0.16-degree(s)
        Toe-in 1/16-inch
    4-cylinder, Xtra-cab
      Front
        Caster .... 2.22-degree(s)
        Camber .... 0.22-degree(s)
        Toe-in 1/16-inch
    V6, Standard cab
      Front
        Caster .... 2.00-degree(s)
        Camber .... 0.16-degree(s)
        Toe-in 1/16-inch
    V6, Xtra cab
      Front
        Caster .... 2.32-degree(s)
        Camber .... 0.22-degree(s)
        Toe-in 5/64-inch

Van
1984
  Front
    Caster .... 2.00-degree(s)
    Camber .... 0.50-degree(s)
    Toe-in 0-inch
1985
  Front
    Caster .... 2.70-degree(s)
    Camber .... 0.50-degree(s)
    Toe-in 0-inch
1986
  Front
    Caster .... 2.50-degree(s)
    Camber .... 0-degree(s)
    Toe-in 0-inch
1987 thru 1989
  2WD
    Front
      Caster .... 2.50-degree(s)
      Camber .... -0.06-degree(s)
      Toe-in 0-inch
  4WD
    Passenger Van
      Front
        Caster .... 2.88-degree(s)
        Camber .... 0.15-degree(s)
        Toe-in 0-inch
    Cargo Van
      Front
        Caster .... 2.15-degree(s)
        Camber .... 0.15-degree(s)
        Toe-in 0-inch

VOLKSWAGEN

Corrado
1990 thru 1995
  4-cylinder
    Front
      Caster .... 1.56-degree(s)
      Camber .... -0.66-degree(s)
      Toe-in 0-inch

Rear
  Camber .... -1.50-degree(s)
  Toe-in 5/32-inch
V6
  Front
    Caster .... 3.42-degree(s)
    Camber .... -1.33-degree(s)
    Toe-in 0-inch
  Rear
    Camber .... -1.50-degree(s)
    Toe-in 5/32-inch

EuroVan
1992 thru 1996
  Front
    Caster .... 1.66-degree(s)
    Camber .... 0.25-degree(s)
    Toe-in 5/32-inch
  Rear
    Camber .... -0.50-degree(s)
    Toe-in 5/32-inch

Fox
1987 thru 1993
  Except wagon
    Front
      Caster .... 2.00-degree(s)
      Camber .... -0.50-degree(s)
      Toe-in .... -5/64-inch
    Rear
      Camber .... -1.50-degree(s)
      Toe-in 7/32-inch
  Wagon
    Front
      Caster .... 1.75-degree(s)
      Camber .... -0.50-degree(s)
      Toe-in .... -5/64-inch
    Rear
      Camber .... -1.50-degree(s)
      Toe-in 3/16-inch

Golf, Jetta and GTI
1985 thru 1992
  Front
    Caster .... 1.50-degree(s)
    Camber .... -0.50-degree(s)
    Toe-in 0-inch
  Rear
    Camber .... -1.50-degree(s)
    Toe-in 5/32-inch
1993 thru 1996
  1.8L Gas and 1.9L Diesel
    Front
      Caster .... 1.75-degree(s)
      Camber .... -0.50-degree(s)
      Toe-in 0-inch
  2.0L
    Front
      Caster .... 1.80-degree(s)
      Camber .... -0.58-degree(s)
      Toe-in 0-inch
  2.8L
    Front
      Caster .... 3.25-degree(s)
      Camber .... -0.50-degree(s)
      Toe-in 0-inch
  Rear (all)
    Camber .... -1.50-inch
    Toe-in 5/32-inch

Passat
1990 thru 1996
  4-cylinder
    Front
      Caster .... 1.66-degree(s)
      Camber .... -1.33-degree(s)
      Toe-in 0-inch
  V6
    Front
      Caster .... 3.33-degree(s)
      Camber .... -1.16-degree(s)
      Toe-in 0-inch
  Rear (all)
    Camber .... -1.50-degree(s)
    Toe-in 5/32-inch

Pickup
1983
  Front
    Caster .... 1.33-degree(s)
    Camber .... 0.33-degree(s)
    Toe-in .... -1/8-inch
  Rear
    Camber .... 0-degree(s)
    Toe-in 0-inch

Quantum
1983 thru 1989
  Except Syncro
    Front
      Caster
        w/o power steering .... 0.50-degree(s)
        w/power steering .... 1.56-degree(s)
      Camber .... -0.66-degree(s)
      Toe-in 3/32-inch
    Rear
      Camber .... -1.63-degree(s)
      Toe-in 7/32-inch
  Syncro
    Front
      Caster .... 1.00-degree(s)
      Camber .... -0.33-degree(s)
      Toe-in 13/64-inch
    Rear
      Camber .... -0.56-degree(s)
      Toe-in 1/8-inch

Rabbit, Jetta and Scirocco
1983 and 1984
  Front
    Caster .... 1.80-degree(s)
    Camber .... 0.33-degree(s)
    Toe-in .... -1/8-inch
  Rear
    Camber .... -1.25-degree(s)
    Toe-in .... 0.15-degree(s)

Scirocco and Cabriolet
1985 thru 1993
  Front
    Caster .... 1.80-degree(s)
    Camber .... 0.33-degree(s)
    Toe-in .... -1/8-inch
  Rear
    Camber .... -1.25-degree(s)
    Toe-in 5/32-inch

Vanagon

1983 thru 1991
    Except Syncro
        Front
            Caster .... 7.25-degree(s)
            Camber .... 0-degree(s)
            Toe-in 5/32-inch
        Rear
            Camber .... -0.80-degree(s)
            Toe-in 0-inch
    Syncro
        Front
            Caster .... 4.66-degree(s)
            Camber .... 0.33-degree(s)
            Toe-in 5/64-inch
        Rear
            Camber .... -0.25-degree(s)
            Toe-in 5/64-inch

VOLVO

DL, GL, GLT and GLE

1983 and 1984
    Front
        Caster .... 3.50-degree(s)
        Camber .... 0.50-degree(s)
        Toe-in 1/8-inch

240 Series

1985 thru 1993
    Front
        Caster .... 3.50-degree(s)
        Camber .... 0.50-degree(s)
        Toe-in 1/8-inch

740 and 940

1984 thru 1995
    Front
        Caster .... 5.00-degree(s)
        Camber .... 0.10-degree(s)
        Toe-in 5/32-inch

760

1983 thru 1987
    Front
        Caster .... 5.00-degree(s)
        Camber .... 0.10-degree(s)
        Toe-in 5/32-inch

760, 780 and 960

1988 thru 1994
    Front
        Caster .... 5.00-degree(s)
        Camber .... 0.10-degree(s)
        Toe-in 5/32-inch
    Rear
        Camber .... 3.00-degree(s)
        Toe-in 1/64-inch

850

1993 thru 1996
    Front
        Caster .... 3.35-degree(s)
        Camber .... 0-degree(s)
        Toe-in 5/32-inch
    Rear
        Camber .... -1.00-degree(s)
        Toe-in 0-inch

960

1995 and 1996
    Front
        Caster .... 5.00-degree(s)
        Camber .... 0.10-degree(s)
        Toe-in 5/32-inch
    Rear
        Camber .... 0.70-degree(s)
        Toe-in 3/64-inch

Glossary

A

A-arm - A suspension locating link shaped like the letter "A." A typical contemporary front suspension setup consists of two unequal-length, non-parallel lateral locating arms for each front wheel. Often referred to as "unequal length A-arms," the two legs of the A are attached to the chassis by rubber bushings that allow the outer top of the A connecting the wheel to the assembly to move up and down freely. An A-arm can be used as an upper arm, a lower arm or for both locating links. Also referred to sometimes as *A-frame*.

Ackerman principle - Design having wheel spindles mounted on axle ends to permit spindles to be turned at an angle to axle for steering purposes.

Ackerman steering - Steering geometry designed so that, when the vehicle is turning a corner, the outside wheel is turned fewer degrees than the inside wheel to compensate for the larger circle the outside wheel is tracking along. The relationship between the angles of the inner and outer wheels isn't a fixed one; the tighter the turn, the greater the disparity between the desired turning angles of the wheels. Steering geometry that doesn't give the required amount of differential between the two front wheels in all types of turns results in tire scrub. Ackerman steering remains perfectly accurate only at very low speeds. At higher speeds, the nonlinear characteristics of the tires enter the picture. Ackerman steering minimizes tire scrub in the vast variety of conditions encountered by production vehicles, so most vehicles have this steering geometry, though it's usually a compromise design that corrects for normal driving conditions rather than for all possible turns.

Active suspension - A type of electronically controlled suspension system utilizing sensors at each wheel to anticipate changes in road surface and adjust spring and/or shock absorber resistance accordingly.

Adjustable shock absorbers - Shocks with adjustable *jounce* and *rebound* characteristics can be stiffened to compensate for wear or to fine tune a suspension for a particular application such as rough roads, heavy loads or racing.

Adjusting sleeve - A device on the steering linkage which changes *toe-in* or *toe-out*.

Aerodynamic drag (or air resistance) - The resistance of the air to forward movement of a body moving through it. Air resistance comes from three sources - (1) Drag resistance, a function of the aerodynamic shape of the body with respect to all outside surfaces. Protruding objects such as mirrors, mufflers and license plates can increase drag resistance considerably at higher speeds. The shape of the rear part of the body is of special importance because it determines the amount of turbulence in the wake of the vehicle. (2) Air friction on the outside surfaces (skin) of the body. For the more-or-less standardized quality of surface finish on passenger vehicles, this portion amounts to about 10 percent of the total air resistance. (3) Airflow through the vehicle for purposes of cooling or ventilation. This influence can be increase or decrease resistance, depending on the function, location and aerodynamic perfection of the channels. Air resistance increases as the square of the velocity of the vehicle and the power required to overcome this drag varies as the cube of the speed.

A-frame - See *A-arm*.

Air bag - The common term for a passive restraint system that uses an inflatable bag hidden in the steering wheel (driver's side) or the dash or glove box (passenger side). In a head-on collision, a sensor at the front of the vehicle activates the air bag inflater module, preventing the driver and/or front passenger from being thrown forward into the steering wheel, dash or windshield. Also, the inflatable bladder used in place of a spring in air suspension.

Air dam - A panel attached to the bodywork below the front bumper. An air dam is designed to reduce the air pressure beneath the vehicle and provide some downforce to the front tires. Air dams were originally used only on race cars, but are now used on a wide variety of sporty street vehicles as well.

Air drag - See *aerodynamic drag*.

Airfoil - An aerodynamic device used to improve traction by increasing the down force on either end of the vehicle. In cross-section, an airfoil is basically an inverted wing; instead of providing lift as in an airplane, the airfoil pushes the car closer to the ground. The use of airfoils (also called wings) increases the cornering capability of a vehicle and improves stability at speed, but at the expense of additional aerodynamic drag.

Air resistance - See *aerodynamic drag*.

Air shock - A shock absorber utilizing an air chamber, that when pressurized with compressed air, increases the shock absorber resistance.

Air spring - A bladder of compressed air which is used instead of a coil spring in an air suspension system. See *air suspension*.

Air suspension - A suspension system that uses air instead of metal springs to support the vehicle and control ride motions. Air springing results in a smoother ride because the natural frequency of vibration of an air spring doesn't vary with loading as it does with metal coil springs. Air springs can be made very soft for lightly loaded conditions and the pressure can be automatically increased to match any increase in load, thus maintaining a constant spring vibration period for any load.

Alignment - An adjustment to bring related components into a line. For example, the correct adjustment of a vehicle's front or rear suspension for camber, toe, caster and ride height.

Alloy wheel - A generic term used to describe any non-steel road wheel. The alloys are usually aluminum or magnesium (hence the term *mag wheel*, which refers to any non-steel wheel).

All-wheel-drive (AWD) - A vehicle drivetrain with all four wheels driven at all times. Generally utilizes a center differential with viscous-coupling. Sometimes referred to "full-time four-wheel drive".

Anti-dive, anti-lift, anti-squat - *Dive* is the nosing-down of the front end of a vehicle during braking, *lift* is the rising of the rear end during braking and *squat* is the settling of the rear end during hard acceleration. Resistance to these motions can be designed into the suspension. To get anti-dive reactions in an unequal length system, the upper arm is angled upward toward the front and the lower arm angled downward. This angling produces lift components in reaction to brake torque that "hold up" the front end. Trailing arms have strong resistance to lift and squat from braking and acceleration, respectively, and are often used at the rear to control these motions.

Anti-lock brake system (ABS) - A system, usually electronically controlled, that senses incipient wheel lockup during braking and relieves hydraulic pressure at wheels that are about to skid. A tire can be steered only if it's rolling, so a locked-up front wheel has no effect whatsoever on the car's direction of travel. Similarly, a locked rear wheel doesn't necessarily follow the same line as the front wheels. If the rear wheels lock, and there's some irregularity in the road that tends to throw the vehicle off course, they offer no resistance. An anti-lock brake system prevents wheel lock-up, improves steering control and reduces stopping distances, especially on wet or icy surfaces.

Anti-roll bar - See *stabilizer bar*.

Anti-seize compound - A coating that reduces the risk of seizing on fasteners that are subjected to high temperatures, such as exhaust manifold bolts and nuts.

Anti-skid - See anti-lock brake system.

Anti-sway bar - See *stabilizer bar*.

Aquaplaning - When a vehicle is driven at a high rate of speed on a wet surface the tire is unable to displace enough water to remain in contact with the pavement. When this condition occurs, the tire rides on top of the water. Also known as *hydroplaning*.

Articulated mounting - A term used where parts are connected by links and links are anchored to provide a double hinging action.

Aspect ratio - The ratio of section height to section width on a tire.

Asymmetrical tread - A tread grooved in an irregular pattern shape and size. Asymmetrical treads are designed to provide an optimum combination of braking, ride, handling and wet-dry road characteristics.

Automatic level control - A suspension system that compensates for variations in load at the front, rear or both ends of the vehicle, positioning the vehicle at a pre-designed level regardless of load.

Automatic steering effect - Built-in tendency of an automobile to resume travel in a straight line when released from a turn.

Automatic transmission - A mechanism in the drivetrain with gearsets that vary the power and torque delivered to the driven wheels as a function of engine load and speed, usually incorporating a fluid coupling or torque converter to allow changing gears and reversing direction without using a foot-operated clutch.

Auxiliary springs - Extra springs on a vehicle, usually at the rear, to support heavier loads.

Axle - A shaft on which a wheel revolves, or which revolves with a wheel. Also, a solid beam that connects the two wheels at one end of the vehicle. An axle is *live* if it transmits power, as in, for example, a front-engine, rear-wheel-drive vehicle. It's called a *beam axle* or a *dead axle* if it does nothing but support the wheels, as at the rear of a front-wheel-drive vehicle.

Axle bearing - A bearing which supports a rotating axle in an axle housing.

Axle boot - See *CV joint boot*.

Axle flange - Flange to which a road wheel attaches at the end of an axleshaft.

Axleshaft - A rotating shaft, splined to the differential, which delivers power from the final drive assembly to the drive wheels. Also called a *halfshaft*.

Axle wind-up - Also known as *axle tramp*, *axle hop* or *wheel tramp*. The tendency of an axle housing on a vehicle to rotate with the wheels as high torque is applied (as under acceleration), then quickly return to its original location.

B

Backbone frame - A frame with the cross-section of a rectangular box, runs along the center of the vehicle and occupies the space between the seats. This box generally divides at the front, running along each side of the transmission and engine up to a crossmember to which the front suspension pieces are attached. At the rear, a similar triangular frame encloses the final-drive housing and provides attachment points for the rear suspension.

Backing plate - The flat, round metal plate to which the brake shoes and wheel cylinders are attached.

Backlash - Clearance or freeplay between two parts. In gears, the amount of clearance between meshing teeth of two engaged gears that will allow rotation of the driven gear in a direction opposite to driving rotation, i.e. how much one gear can be moved back and forth without moving gear with which it's engaged or meshed.

Ball bearing - An anti-friction bearing consisting of a hardened inner and outer race with hardened steel balls interposed between two races.

Balljoints - The ball-and-socket connecting links used to attach the steering knuckles to the upper and/or lower control arms and the tie-rod ends to the steering arms. Balljoints act as pivots that allow turning of the front wheels and compensation for changes in the wheel and steering geometry's that occur during jounce, rebound and turning.

Beam axle - The rear axle on a front-wheel drive vehicle. It carries no drivetrain components. Also known as a *dead axle*. See *axle*.

Bearing - The curved surface on a shaft or in a bore, or the part assembled into either, that permits relative motion between them with minimum wear and friction.

Bearing caps - Retainers or caps, held in place by nuts and bolts, that hold bearings halves in place.

Bellhousing - The metal shroud, named for its shape, that surrounds the flywheel and clutch of a vehicle with a manual transmission or the driveplate (or flexplate) and torque converter of a vehicle with an automatic transmission.

Belt - A reinforcing band, usually fabric, fiberglass or steel, running around the circumference of a tire and strengthening the tread area.

Belted bias - Basic bias ply structure, plus two or more tread plies, or belts, around circumference of tire under the tread.

Bevel gear - A gear shaped like the wide end of a cone, used to transmit motion through an angle. Differential gears are bevel gears.

Bias - The acute angle at which the cords in the tire fabric intersect the circumferential centerline of the finished tire.

Bias ply - Pneumatic tire structure in which ply cords extend diagonally from bead to bead, laid at alternate angles.

Bias tire - A tire whose cords in its plies of structural fabric run at an angle called the *bias angle*. The carcass of a bias tire is constructed of adjacent layers of fabric that run continuously from bead to bead. A typical passenger-car bias tire has two plies of fabric with cords running at 30 to 40 degrees to the circumferential centerline.

Bias-belted tire - An interim tire design that bridged the gap between bias tires and radials. The bias-belted tire has two plies of conventional fabric with about the same bias angle as a pure bias tire. It also has a two-ply belt made of fiberglass, steel or fabric laid at an angle of 25 to 30 degrees, giving the bias-belted tire a main carcass or sidewall stiffness similar to that of the bias tire plus a belt that isn't as stiff as the radial's.

Bleeder valve - A valve on a wheel cylinder, caliper or other hydraulic component that is opened to purge the hydraulic system of air.

Brake - A device which converts a vehicle's forward motion (kinetic energy) into heat energy through the use of frictional force applied to the wheels, causing the vehicle to slow or stop.

Brake bleeding - Procedure for removing air from lines of a hydraulic brake system.

Brake booster - A mechanical device (usually air, vacuum or hydraulically actuated) that attaches to the brake system and multiplies the driver's force input to reduce the effort normally required to stop the vehicle. Such braking systems are commonly referred to as *power-assisted*, *vacuum-assisted*, *hydraulically-assisted* or simply *power brakes*. In most vehicles, the boost comes from intake vacuum.

Brake caliper - The component of a disc brake that converts hydraulic pressure into mechanical energy.

Brake cylinder - A cylinder in which a movable piston converts pressure to mechanical force to move brake shoes against the braking surface of the drum or rotor.

Brake disc - The component of a disc brake that rotates with the wheels and is squeezed by the brake caliper and pads, which creates friction and converts the energy of the moving vehicle into heat. Also referred to as a brake *rotor*.

Brake dive - See *anti-dive, anti-lift, anti-squat*.

Brake drum - The component of a drum brake that rotates with the wheel and is acted upon by the expanding brake shoes, which creates friction and converts the energy of the moving vehicle into heat.

Brake dust - The dust created as the brake linings wear down in normal use. Brake dust usually contains dangerous amounts of asbestos.

Brake "fade" - A condition in which repeated severe applications of brakes cause expansion of brake drum or loss of frictional ability or both, which results in impaired braking efficiency.

Brake fluid - A compounded liquid for use in hydraulic brake systems, which must meet exacting conditions (impervious to heat, freezing, thickening, bubbling, etc.).

Brake hose - A flexible conductor for transmission of fluid pressure in brake system.

Brake lines - The rigid metal tubing connecting the master cylinder to the brake calipers and/or wheel cylinders in a hydraulic brake system. Flexible brake hoses usually bridge the gap between the rigid metal lines and calipers/wheel cylinders to allow for the up-and-down motion of the suspension.

Brake linings - The replaceable friction material which contacts the brake drum to retard the vehicle's speed. The linings are bonded or riveted to the brake shoes. See *brake shoe*.

Brake pads - On disc-brake systems, the replaceable friction pads that pinch the brake disc when the brakes are applied. Brake pads consist of an organic or metallic friction material bonded or riveted to a rigid backing plate.

Brake shoe - The crescent-shaped carrier to which the friction linings are mounted and which force the lining against the rotating drum during braking.

Breaker bar - A socket wrench handle providing greater leverage. Also called a flex-handle.

Brinell hardness - A scale for designating degree of hardness possessed by a substance.

Bump - The upward movement of the wheels and suspension; also called *jounce*.

Bump steer - the slight turning or steering of a wheel away from its normal direction of travel as it moves through its suspension travel. At the front, bump steer is associated with the tie-rod/linkage arm relationship; the method of locating the rear suspension, the type of rear suspension and the geometry of the various linkages contribute to rear bump steer. In race cars, bump steer is normally designed out of the suspension to make the steering and handling as precise and predictable as possible; in street vehicles, front and rear bump steer are present to some extent because of design compromises dictated by economic and mass-production factors. Bump steer isn't always undesirable, anyway, because suspension engineers can use it to design some understeer or oversteer into the handling envelope of the vehicle.

Bump stop - A cushioning device, usually rubber, that limits the upward movement of the wheels and suspension to prevent metal-to-metal contact that could lead to suspension damage or failure. Also referred to as *jounce bumpers*.

Burr - A rough edge or area remaining on metal after it has been cast, cut or drilled.

Bushing - A one-piece sleeve placed in a bore to serve as a bearing surface for shaft, piston pin, etc. Usually replaceable.

C

Caliper - The non-rotating part of a disc-brake assembly that straddles the disc and contains the hydraulic components that pinch the disc when the brakes are applied. A caliper is also a measuring tool that can be set to measure inside or outside dimensions of an object; used for measuring things like the thickness of a block, the diameter of a shaft or the bore of a hole (inside caliper).

Caliper mounting plate - The component that connects a brake caliper to the steering knuckle, hub carrier or rear axle.

Camber - In wheel alignment, it is the outward or inward tilt of a wheel at it's top.

Camber thrust - The side force generated when a tire rolls with camber. Camber thrust can add to or subtract from the side force a tire generates.

Carcass - Tire structure except for sidewall and tread.

Cardan joint - A universal joint with corresponding yokes at a right angle with each other. It's named after a 16th century Italian who developed the concept. Over a century later, Robert Hooke of England developed and patented the conventional universal joint, or U-joint, which is sometimes referred to as a Cardan or Hooke universal.

Castellated - Resembling the parapets along the top of a castle wall. For example, a castellated balljoint stud nut.

Caster - In wheel alignment, the backward or forward tilt of the steering axis. The angle between the steering axis and the vertical plane, as viewed from the side. Caster is considered positive when the steering axis is inclined rearward at the top.

Center of gravity - Point of a body from which it could be suspended, or on which it could be supported, and be in balance. For example, the center of gravity of a wheel is the center of the wheel hub.

Centrifugal force - A force which tends to move a body away from its center of rotation. For example, a whirling weight attached to a string.

Chapman strut - A type of rear suspension utilizing a lower lateral link and a long spring-shock strut to determine wheel geometry. The basic principle is the same as that of the front MacPherson strut. It's named after Colin Chapman, who first used it on the original Lotus Elite, and later on Elan and Elite models. Datsun Z-cars have also used this design.

Chase - To repair or clean up damaged threads with a tap or die.

Chassis - A French word meaning framework of a vehicle without a body and fenders. Generally speaking, the suspension, steering, and braking components of a vehicle are all regarded as part of the chassis.

Clevis - A U-shaped metal piece with holes in each end through which a pin or bolt is run, used for attaching the brake pedal to the power brake booster pushrod, the clutch pedal to the clutch cable or master cylinder pushrod and for various other connections on an automobile.

Clutch - Any device which connects and disconnects a driven component from the driving component. For example, a clutch is used to disengage the engine from the transmission in vehicles with a manual transmission. Clutches function by using mechanical, magnetic or friction-type connections. In an air conditioning system, an electrically operated coupling device that connects or disconnects the compressor pulley and compressor shaft.

Clutch disc - The rotating circular metal plate splined to the transmission input shaft. The clutch disc has friction material on each face. It's located between the flywheel and the clutch pressure plate and is clamped tightly between these two members when the clutch is engaged, transmitting power from the flywheel through the clutch and into the transmission.

Clutch housing - See *bellhousing*.

Clutch interlock switch - A switch that prevents the vehicle from starting unless the clutch is pressed.

Clutch pressure plate - The member that, along with the clutch cover, is bolted to the flywheel and rotates with it. When the clutch is engaged, springs between the pressure plate and the cover force the clutch disc against the flywheel and pressure plate.

Clutch shaft - The shaft that takes power from the clutch into the transmission. Also called the *drive pinion*.

Coefficient of friction - The amount of friction developed between two surfaces pressed together and moved one on the other. Also, a numerical value indicating the amount of work required to slide one surface against another. The coefficient of friction equals the side force acting on the object divided by the weight of the object.

Coil spring - A spiral of elastic steel found in various sizes throughout a vehicle, most importantly as a springing medium in the suspension and in the valve train.

Concentric - Two circles having same center but different diameters.

Constant velocity (CV) joint - A double universal joint that cancels out vibrations caused by driving power being transmitted through an angle.

Contact patch - The area of a tire's tread in contact with the ground.

Cord - Textile or steel-wire strands which form the plies of a tire.

Counterclockwise rotation - Rotating in the direction opposite to the hands on a clock.

Countershaft - The shaft in a manual transmission that carries power by means of the gears from the clutch shaft to the driveshaft, turning in an opposite direction to them.

Countersink - To cut or form a depression to allow the head of a screw to go below surface.

Counterweight or counterbalancer - The weight added to a rotating shaft or wheel to balance the normal loads on the part and to offset vibration. Counterweights are used on the crankshaft and are often found on the flywheel and driveshaft.

Coupling - A connecting means for transferring movement from one part to another. May be mechanical, hydraulic or electrical.

CV joint boot - The flexible, accordion-plated rubber dust boot which encloses the constant velocity joint and the end of the axleshaft attached to it. The CV joint boot prevents dirt, dust, mud and moisture from entering the CV joint.

D

Damper - See *shock absorber*.

Dead axle - See *beam axle*.

Dead rear axle - A rear axle that does not turn. Example - rear axle of front wheel drive car.

Deceleration - Negative acceleration. The rate of change in velocity as a vehicle slows down during braking.

DeDion axle - A design that combines a fully-independent rear suspension with a live axle. It consists of a connecting tube located by some means - such as leaf springs or trailing arms - and a chassis-mounted differential driving the wheels through driveaxles with universal-joints.

Demountable rim - A rim for a tire that is readily removable from the wheel.

Dial caliper - A measuring device capable of measuring internal and external dimensions. Readings are read on a dial face as opposed to a vernier scale.

Dial gauge - A type of test instrument which indicates precise readings on a dial face.

Dial indicator - A precision measuring instrument that indicates movement to a thousandth of an inch with a needle sweeping around dial face.

Die - One of a pair of hardened metal blocks for forming metal into a desired shape, or a device for cutting external threads. See *thread die*.

Differential - A device with an arrangement of gears designed to permit the division of power to two shafts. For example, the differential permits one drive wheel to turn faster than the other.

Direct drive - In automobile transmissions, direct engagement between the engine and the driveshaft, where the engine crankshaft and the driveshaft turn at the same rpm.

Directional stability - The ability of a vehicle to be driven with safety and confidence in a straight line and at high speed without being unduly affected by pavement irregularities, crosswinds, aerodynamic lifting forces or other external influences that would tend to divert it from its intended direction of travel.

Disc - See *Brake disc*.

Disc brake - A brake design incorporating a flat, disc-like rotor onto which brake pads containing lining material are squeezed, generating friction and converting the energy of a moving vehicle into heat.

Distortion - A warpage or change in form from the original shape.

Dog clutch - Mating collars, flanges or lugs which can be moved as desired to engage or disengage similar collars, flanges or lugs in order to transmit rotary motion.

Double-reduction axle - A driveaxle construction in which two sets of reduction gears are used for extreme reduction of gear ratio.

Double-wishbone suspension - Another name for a double-A-arm suspension. A suspension system using two wishbones, or A-arms to connect the chassis to the spindle or knuckle.

Dowel pin - A steel pin pressed into matching shallow holes in two adjacent parts to provide proper alignment of the two parts.

Dozer - A portable frame straightening machine.

Drag - See *aerodynamic drag*.

Drag coefficient - A dimensionless number used in calculating the aerodynamic drag acting on a vehicle. The drag coefficient is a function of such factors as the shape of the vehicle and the airflow through the vehicle for cooling or ventilation. The drag coefficient is determined experimentally in wind-tunnel tests or by coasting tests performed on the actual vehicle. The higher the drag coefficient, the greater the aerodynamic drag force the engine must overcome at any given road speed.

Drag link - A connecting rod or link between steering gear Pitman arm and steering control linkage .

Drill - A tool for making a hole or to sink a hole with a pointed cutting tool rotated under pressure.

Drive - Any device which provides a fixed increase or decrease ratio of relative rotation between its input and output shafts. For example, a device located on the starter to allow for a method of engaging the starter to the flywheel.

Drive-fit - Term used when shaft is slightly larger than hole and must be forced in place.

Driveline - Universal joints, drive shaft and other parts connecting transmission with driving axles.

Driveshaft - The long hollow tube with universal joints at both ends that carries power from the transmission to the differential.

Drivetrain - the power-transmitting components in a vehicle. Usually consists of the clutch (on vehicles with a manual transmission), the (manual or automatic) transmission, the driveshaft, the universal joints, the differential and the driveaxle assemblies.

Drum brake - A type of brake using a drum-shaped metal cylinder that attaches to the inner surface of the wheel and rotates with it. When the brake pedal is pressed, curved brake shoes with friction linings press against the inner circumference of the drum to slow or stop the vehicle.

Dual braking system - A brake system designed to prevent complete loss of braking in the event of a hydraulic or some other sort of brake failure. A dual braking system usually divides the four wheels into either front and rear, or diagonal front-rear pairs with independent hydraulic circuits.

Dual master cylinder - Primary unit consisting of two actions for displacing fluid under pressure in a split hydraulic brake system.

Dual reduction axle - A drive axle construction with two sets of pinions and gears, either of which can be used.

Duo-servo drum brake - A type of self-energizing drum brake that has servo action in both forward and reverse.

Dust boot - A rubber diaphragm-like seal that fits over the end of a hydraulic component and around a pushrod or end of a piston, not used for sealing fluid in but keeping dust out.

Dynamic balance - The balance of an object when it's in motion. When the center line of weight mass of a revolving object is in the same plane as the center line of the object, that object is in dynamic balance. The opposite of *dynamic imbalance*.

Dynamic imbalance - An object that's out-of-balance when it's in motion. For example, a wheel that hops up and down when it's spinning because its weights have fallen off is in a state of dynamic imbalance.

E

Eccentric - One circle within another circle not having the same center. A disk, or offset section (of a shaft, for example) used to convert rotary motion to reciprocating motion. Sometimes called a cam.

Elliott steering knuckle - A type of axle in which the ends of the axle beam straddle the spindle.

Emergency brake - A braking system, independent of the main hydraulic system, that can be used to slow or stop the vehicle if the primary brakes fail, or to hold the vehicle stationary even though the brake pedal isn't depressed. It usually consists of a foot pedal or hand lever that actuates either front or rear brakes mechanically through a series of cables and linkages. Also known as a *parking brake*.

Endplay - The amount of lengthwise movement between two parts. As applied to a crankshaft, the distance that the crankshaft can move forward and back in the cylinder block.

F

Face - A machinist's term that refers to removing metal from the end of a shaft or the "face" of a larger part, such as a flywheel.

Fade - A condition brought about by repeated or protracted application of the brakes, resulting in a reduction or fading of brake effectiveness. Heat is the primary culprit, causing expansion of the brake drums/discs and lowering the friction coefficient of the brake shoe linings and/or disc brake pads.

Fatigue - A breakdown of material through a large number of loading and unloading cycles. The first signs are cracks followed shortly by breaks.

Feeler gauge - A thin strip or blade of hardened steel, ground to an exact thickness, used to check and/or measure clearances between parts. Feeler gauges are graduated in thickness by increments of .001 inch.

Final drive ratio - The ratio between the driveshaft or transmission output shaft rpm and the drive-wheel axleshaft rpm. It's determined by the ring and pinion gearing inside the differential. For example, if the ratio is 4:1, the driveshaft rotates four times for each rotation of the rear axle differential gear, the axleshafts and the wheels. The ratio is varied by changing the number of teeth on the ring and pinion gears.

Flange - A rib or rim used to provide strength to a panel or a means of attachment for another panel.

Flare-nut wrench - A wrench designed for loosening hydraulic fitting tube nuts (flare-nuts) without damaging them. Flare-nut wrenches are kind of like a six-point box-end wrench with one of the flats missing, which allows the wrench to pass over the tubing but still maintain a maximum amount of contact with the nut.

Floor shift - A type of transmission shift linkage in which the various gears are actuated by a lever attached to the floor rather than by a lever attached to the steering column.

Fluid - Any liquid or gas that is capable of flowing and that changes its shape at a steady rate when acted upon by a force tending to change its shape. A term used to refer to any substance having the above properties.

Fluid coupling - A hydraulic clutch used to transmit engine torque to the transmission gears. See *fluid drive*.

Fluid drive - A pair of vaned rotating elements held close to each other without touching. Rotation is imparted to driven member by driving member through resistance of a body of oil.

Flywheel - A heavy, usually metal, spinning wheel in which energy is absorbed and stored by means of momentum. On cars, this heavy metal wheel that's attached to the crankshaft to smooth out firing impulses. It provides inertia to keep the crankshaft turning smoothly during periods when no power is being applied. It also serves as part of the clutch and engine cranking systems.

Flywheel ring gear - A large gear pressed onto the circumference of the flywheel. When the starter gear engages the ring gear, the starter cranks the engine.

Foot-pound - A unit of measurement for work, equal to lifting one pound one foot.

Foot-pound (tightening) - A unit of measurement for torque, equal to one pound of pull one foot from the center of the object being tightened.

Four-wheel ABS - An anti-lock brake system that operates on all four wheels.

Four-wheel drive - A type of drive system in which both the front and rear wheels are connected through drivelines and driving axles to the transmission, usually via a transfer case. Four-wheel drive can be either full-time, with power delivered to both axles at all times, or part time, where the driver can select either rear-wheel drive only or four-wheel drive for marginal traction conditions.

Frame - The structural load-carrying members of a vehicle that support the engine and body and are in turn supported by the wheels.

Free height - The unloaded length or height of a spring.

Freeplay - The amount of travel before any action takes place. The "looseness" in a linkage, or an assembly of parts, between the initial application of force and actual movement. Usually perceived as "slop" or slight delay. For example, In a brake pedal it is the distance the pedal moves before the pistons in the master cylinder are actuated.

Freewheeling clutch - A mechanical device in which the driving member imparts motion to a driven member in one direction but not the other. Also known as an *overrunning clutch*.

Friction - Surface resistance to relative motion.

Friction drive - A method of power transmission used on early automobiles where power is transmitted from a driving to a driven wheel by means of pressing one wheel against another at a right angle.

Front-wheel drive - A drive system in which the transmission is connected by driving axles to the front wheels instead of the rear wheels.

Full-floating axle - A driveaxle design in which the axleshaft does not carry the vehicle's weight. Two roller bearings support the weight of the vehicle, so the axle can be removed without disturbing the wheel.

Fully-articulated rear suspension system - See *fully-independent rear suspension system*.

Fully-independent rear suspension - A suspension system that uses driveaxles and a series of links to allow each wheel to rise and fall independently of the other wheel. Also known as a *fully-articulated* independent rear suspension system.

Gearbox - See *manual transmission*.

Gear puller - A tool that's specially designed for removing press-fitted gears from their respective shafts.

Gear ratio - Number of revolutions made by a driving gear as compared to number of revolutions made by a driven gear of different size. For example, if one gear makes three revolutions while the other gear makes one revolution, the gear ratio is 3 to 1.

Gears - Wheel-like parts with teeth cut into the rim. Meshing of the teeth of two gears enables one to drive the other, thus transmitting power.

Gearshift - See *shift lever*.

Grind - To finish or polish a surface by means of an abrasive wheel.

Grinding - The process of resurfacing a brake disc or drum on a brake lathe using a power-driven abrasive stone.

Grommet - A round rubber seal which fits into a hole or recess, intended to seal or insulate the component passing through it. A donut-shaped rubber or plastic part used to protect wiring that passes through a panel, firewall or bulkhead.

Groove - The space between two adjacent tire tread ribs.

Guide pin - A caliper mounting bolt used for fastening a floating caliper to its mounting plate.

Halfshaft - A rotating shaft that transmits power from the final drive unit to the drive wheels, but usually refers to the two shafts that connect the wheels to the final drive with independent rear suspension or front-wheel drive rather than the axleshafts of a live rear axle.

Hat - The portion of a detachable brake disc that comes in contact with the wheel hub.

Heel - Outside or larger half of gear tooth. Also, end of brake shoe not against anchor.

Helical - Shaped like a coil of wire or a screw thread.

Helical gear - A gear design where gear teeth cut at an angle to shaft.

Heli-Coil - A rethreading device used when threads are worn or damaged. The device is installed in a retapped hole to reduce the thread size to the original size.

Herringbone gear - A pair of helical gears designed to operate together in form of a V.

Hold-down pin, spring and retainer - The most common method of retaining a brake shoe to the backing plate. The pin passes through the backing plate and brake shoe. The spring and retainer are fastened to the pin, which holds the shoe against the backing plate.

Hooke joint - See *Cardan joint*.

Hotchkiss drive - A live axle rear suspension design in which the axle is located by semi-elliptic leaf springs. The springs, mounted longitudinally, connect to the chassis at their ends and the axle is hung from them. Thus they not only spring the axle but also determine its freedom to move and transmit all cornering, braking and driving forces from the axle to the body.

Hydraulic - Any operation that uses the incompressibility of liquids, usually oil or water, and their ability to offer resistance when being forced into a small cylinder, thus transmitting an increase in applied force. Hydraulic brakes and clutches work on this principle.

Hydraulically operated power booster - A power booster that uses hydraulic pressure to assist the driver in the application of the brakes. This hydraulic pressure usually comes from the power steering pump or an electro-hydraulic pump.

Hydraulic brake system - System in which brake operation and control utilizes hydraulic brake fluid.

Hydraulic control unit - The portion of an anti-lock brake system that houses the solenoid valves and electro-hydraulic pump.

Hydropneumatic suspension - A suspension system using a gas and a liquid separated by a flexible bladder as a springing medium. Citroens have used a hydropneumatic suspension system for decades.

Hypoid gears - A design of pinion and ring gear in which the centerline of the pinion is offset from the centerline of the ring gear.

I

Idler arm - One of the connecting levers in a parallel, relay-type

steering linkage. The steering gearbox is attached to a Pitman arm which converts rotary motion to lateral motion. The Pitman arm usually connects to a transverse center link which in turn is connected to the idler arm attached to the frame rail on the opposite side of the vehicle. The ends of the center link connect to two adjustable tie-rods that transmit the lateral movement of the center link to the steering arms at each steering knuckle.

Idler gear - A gear interposed between two other gears to reverse the direction of rotation of the output gear.

Inboard brakes - A type of brake design in which the discs or drums and associated brake components are not located within the wheels. Inboard brakes are usually found only at the rear and are generally attached to the differential housing. This design offers a reduction in unsprung weight and better cooling.

Included angle - Combined angles of camber and steering axis inclination.

Independent suspension - A suspension design in which the wheel on one side of the vehicle may rise or fall independently of the wheel on the other side. Thus, a disturbance affecting one wheel has no effect on the other wheel.

Inertia - The tendency of a stationary body to remain at rest or resist motion, or a moving body's resistance to a change in its velocity. Effort (work) is required to start a mass moving or to retard it once it is in motion.

Input shaft - Transmission shaft which receives power from engine and transmits it to transmission gears.

Integral - Formed as a unit with another part.

Intermediate gear - In a transmission, the gear or gears between low and high. For example, in a 5-speed transmission, 2nd, 3rd and 4th gears are intermediate gears.

J

Jack - A hydraulic or mechanical device used to raise a vehicle to change a flat tire. The term also refers to a phenomenon which is characteristic off swing-axle rear suspensions: A cornering force acting on such a suspension tends to lift the body of the vehicle through the solidly mounted differential, forcing the outer wheel to tuck, or jack, under the vehicle. Carried to an extreme, jacking can turn the vehicle over.

Jam nut - A nut used to lock an adjustment nut, or other threaded component, in place. For example, a jam nut is employed to keep the adjusting nut on the rocker arm in position.

Jounce - See *bump*.

K

Key - A small piece of metal inserted into matching grooves machined into two parts fitted together - such as a gear pressed onto a shaft - which prevents slippage between the two parts.

Keyed - Prevented from rotating with a small metal device called a key.

Keyway - A slot cut in a shaft, pulley hub, etc. A square key is placed in the slot and engages a similar keyway in the mating piece.

Kinetic energy - The energy of a body in motion.

Kingpin - The shaft or journal around which the stub axle or steering spindle and wheel pivoted on some vehicles at one time. On modern front suspensions, the kingpin has been replaced by balljoints.

Kingpin inclination - The angle at which the kingpin is inclined inward from the true vertical center line. To determine the kingpin inclination, draw a line through the longitudinal axis of the kingpin and a vertical line through the centerline of the tire. With balljoints, the kingpin angle is between a line drawn through the balljoint centers and the vertical tire centerline. The choice of kingpin angle is a compromise between steering effort, returnability and wheel pull during braking. Also referred to as *steering axis inclination*.

Knurl - A roughened surface caused by a sharp wheel that displaces metal outward as its sharp edges push into the metal surface. To indent or roughen a finished surface.

L

Ladder frame - A type of frame construction consisting of two heavy-section longitudinal members connected together stepladder fashion by smaller transverse crossmembers. This design is no longer used much anymore because it's heavy and lacks torsional rigidity.

Lash - The amount of free motion in a gear train, between gears, or in a mechanical assembly, that occurs before movement can begin. Usually refers to the lash in a valve train.

Lateral acceleration - Sideways acceleration created when a vehicle corners. As a result of this lateral acceleration, centrifugal force acts on the vehicle and tries to pull it outward. The tires develop an equal and opposite force acting against the road to counteract this outward force. The side thrust generated by the tires in any corner is proportional to the mass and the velocity squared, inversely proportional to the radius of the corner. Steady-state lateral acceleration is usually measured on a *skidpad*, a flat smooth expanse of pavement with a circle marked on it.

Lateral runout - The amount of side-to-side movement of a rotating wheel, tire or brake disc from the vertical.

Lathe - Machine on which a piece of solid material is spun on a horizontal axis and shaped by a fixed cutting or abrading tool.

Leading/trailing drum brake - A drum brake design in which both brake shoes are attached to an anchor plate, and only one of the shoes is self-energized.

Leading arm - An independent suspension system having the wheel attached to the end of an arm that swings in a plane parallel to the longitudinal axis of the vehicle. The wheel is ahead of or leads the fixed pivot point of the arm. The Citroen 2CV used this type of front suspension.

Leaf spring - A slightly curved steel plate, usually mounted in multiples of two or more plates of varying lengths, installed on top of each other. When used with a live rear axle, the ends of the springs are attached to the vehicle frame and the center is fixed to the axle.

Level - The magnitude of a quantity considered in relation to an arbitrary reference value.

Limited slip differential - A differential that uses cone or disc clutches to lock the two independent axleshafts together, forcing both wheels to transmit their respective drive torque regardless of the available traction. It allows a limited amount of slip between the two axleshafts to accommodate the differential action. This design doubles the number of drive wheels in low-traction situation.

Link - Any series of rods, levers, etc. used to transmit motion from one component to another. Also, a general term used to indicate the existence of communication facilities between two points. A link can be electrical, electronic, mechanical or a combination of these.

Linkage - Any series of rods, yokes and levers, etc., used to transmit motion from one unit to another.

Live axle - A shaft through which power travels from the driveaxle gears to the driving wheels. See *axle*.

Load at installed height - The specified range of force required to compress a spring to its installed height, usually expressed in terms of so many pounds of force at so many inches.

Load range - Tire designation, with a letter (A, B, C, etc.), used to identify a given size tire with its load and inflation limits. Replaced the old term, *ply rating*.

Load Sensing Proportioning Valve (LSPV) - A hydraulic system control valve that works like a proportioning valve, but also takes into consideration the amount of weight carried by the rear axle.

Lock washer - A form of washer designed to prevent attaching nut from working loose.

Lower arm - The suspension arm which connects the vehicle chassis to the bottom of the steering knuckle.

Low speed - Gearing provided in an automobile which causes greatest number of revolutions of engine as compared to driving wheels.

Lubricant - Any substance, usually oil or grease, applied to moving parts to reduce friction between them.

Lug nuts - The nuts used to secure the wheels to a vehicle.

Haynes suspension, steering and driveline manual

M

Machining - The process of using a machine to remove metal from a metal part.

MacPherson strut - A type of front suspension system devised by Earle MacPherson at Ford of England. In its original form, this layout used a simple lateral link with the stabilizer bar to create the lower control arm. A long *strut* - an integral coil spring and shock absorber - was mounted between the body and the steering knuckle. Many modern MacPherson strut systems use a conventional lower A-arm and don't rely on the stabilizer bar for location.

Mag wheel - See *alloy wheel*.

Manual - Pertaining to or done with the hands. Also, requiring or using physical skill or energy.

Manual adjuster - A type of brake adjuster that must be adjusted from time-to-time, with the use of a hand tool.

Manual steering - A steering system without a power booster to reduce steering effort.

Manual transmission - A transmission with gearsets that vary the power and torque delivered to the driven wheels as a function of engine load and speed. It incorporates a steering-column or floor-mounted lever that the driver has to operate in conjunction with a clutch pedal to change from one gear to another.

Master cylinder - In brake systems, a cylinder containing a movable piston actuated by foot pressure, producing hydraulic pressure to push fluid through the lines and wheel cylinders and force the brake linings or pads against a drum or disc. In hydraulic clutch systems, a similar device is used to push hydraulic fluid into a slave cylinder which activates the clutch release arm with a rod.

Mechanical brakes - A brake system using cables or other mechanical linkages to actuate the shoes or pads rather than hydraulic pressure. The parking brake on most vehicles is a mechanical brake.

Moment of inertia - The rotational equivalent of inertia. The tendency of an object to resist being accelerated in rotation compared to pure inertia which is a tendency of objects to remain motionless if at rest or to keep moving in a straight line if already in motion. For a vehicle, the polar moment of inertia - the moment of inertia about a vertical axis through the center of rotation - has a considerable influence on steering and overall handling response.

Monocoque - A single skin or sheet with everything - suspension, engine, etc. - attached to it at a number of points by brackets designed to spread loads evenly into the shell. This is an especially efficient, albeit expensive, design for racing vehicles because it results in a very high strength-to-weight ratio. But because of the cost, monocoque design is rarely used on street vehicles.

Multiple disc clutch - A clutch having a number of driving and driven discs as compared to a single-plate clutch.

N

Needle bearing - An anti-friction bearing using a great number of rollers of small diameter in relation to their length.

Negative offset steering - A steering system in which a line drawn through the balljoint centers meets the ground outside the tire centerline when viewed from the front. The distance between the point where the line through the balljoints touches the pavement and the point in the center of the tire tread contact patch is the *scrub radius*. The scrub radius acts as a lever arm for braking forces developed by the tire and causes a torque which must be resisted by the steering linkage. With negative offset steering, the effect of a grabbing (or weak) brake or unequal road friction from side to side on the steering surface is reduced, the negative scrub radius actually helping the vehicle to stop in a straight line.

Neutral - A handling characteristic where the slip angles are equal at the front and at the rear. The vehicle exhibits neither oversteer or understeer characteristics and is described as "balanced."

Nonferrous metals - Metals which contain no iron or very little iron, and not subject to rusting.

Non-servo drum brake - A drum brake design in which the application of one shoe has no effect on the other.

O

O.D. - Outside diameter.

OEM (Original Equipment Manufacturer) - A designation used to describe the equipment and parts installed on a vehicle by the manufacturer, or those available from the vehicle manufacturer as replacement parts. See *aftermarket parts*.

One-way clutch - See *freewheeling clutch*.

Opposite lock - Turning the wheels in a direction opposite to the direction the vehicle is headed. Opposite lock is used to control rear-wheel skids.

Out-of-round - The condition of a brake drum when it has become distorted and is no longer perfectly round. In many cases an out-of-round brake drum can be salvaged by resurfacing on a brake lathe.

Output shaft - Shaft which receives power from transmission and transmits it to vehicle drive shaft.

Overdrive - Any arrangement of gearing which produces more revolutions of the driven shaft than the driving shaft. On some vehicles, this is achieved by adding a small gearbox behind the transmission; on others, the overdrive gear is integral with the main transmission. In either case, an overdrive gear provides a taller gear in addition to the three or four lower gears. The advantages of overdrive include reduced fuel consumption, lower engine noise and reduced engine wear.

Overhaul - To completely disassemble a unit, clean and inspect all parts, reassemble it with the original or new parts and make all adjustments necessary for proper operation.

Overrunning clutch - A device located on the starter to allow for a method of engaging the starter with the flywheel. The overrunning clutch uses a shift lever to actuate the drive pinion to provide for a positive meshing and de-meshing of the pinion with the flywheel ring gear. See *freewheeling clutch*.

Overrunning clutch, or coupling - See *free-wheeling*.

Oversteer - A handling characteristic where the slip angles are larger at the rear than at the front. During hard cornering, the rear tires slip while the front tires maintain traction. An oversteering vehicle breaks away at the rear, so the driver must countersteer with the front wheels and possibly apply opposite lock to keep the vehicle from spinning. When a vehicle exhibits oversteer it is described as "loose."

P

Pad - Disc brake friction material generally molded to metal backing, or shoe.

Panhard rod - A locating link running laterally across the vehicle, with its upper end attached to the body and its lower end attached to a live axle, beam axle or DeDion axle. The Panhard rod provides lateral location of the axle. Also known as a *track bar*.

Parking brake - The mechanically actuated portion of a drum brake or disc brake caliper, used to prevent the vehicle from rolling when it is parked, applied by a lever, pedal or rod. See *emergency brake*.

Park/Neutral - The selected nondrive modes of the transmission.

Pawl - A pivoted bar that slides over the teeth of a ratchet wheel to prevent or impart motion. For example, in an automatic transmission, the parking pawl locks the transmission in the Park position.

Peen - To stretch or cinch over by pounding with the rounded end of a hammer.

Penetrating oil - Special oil used to free rusted parts so they can be moved.

Periphery - Circumference of a circle. Example: tread of a tire.

Phillips screw or screwdriver - A type of screwhead having a cross instead of a slot for a corresponding type of screwdriver.

Pilot - A term which refers to any device used to center a cutting tool or the concentric installation of one part onto another. For example, the valve guide is used as a pilot for centering the grinding stone during a valve job; the pilot bearing is used to center a clutch alignment tool when bolting the clutch disc and pressure plate to the flywheel.

Pilot bearing - A small bearing installed in the center of the flywheel (or the rear end of the crankshaft) to support the front end of the input shaft of the transmission.

Pinion - A small gear which engages a larger geared wheel or rack. Pinions are used in rack-and-pinion steering gearboxes and in the differential ring-and-pinion set.

Pinion carrier - Mounting or bracket which retains bearings supporting a pinion shaft.

Pitch - A fore-and-aft rocking motion alternately compressing the front springs while the rear springs are extended, then reversing the spring movements. Also, the distance between two threads on a bolt or screw.

Pitman arm - The large lever, pressed onto the splined end of the steering gearbox Pitman shaft, which connects the steering gearbox with the center link of the steering linkage.

Pivot - A pin or short shaft upon which another part rests or turns, or about which another part rotates or oscillates.

Planetary gears - A system of gearing named after the solar system because of similarities between its function and the way the planets revolve around the sun. A pinion is surrounded by an internal ring gear with planet gears in mesh between the ring gear and the pinion.

Planet carrier - Carrier or bracket in a planetary system which contains shafts upon which pinions or planet gears turn.

Planet gears - Pinions or gears interposed between ring gear and sun gear and meshing with both in a planetary system.

Ply - The layer of rubber-coated parallel cords forming a tire body, or carcass, on a bias tire.

Ply rating - An index of tire strength. See *load range*.

Pneumatic - Pertaining to air. For example, any device operated by air pressure is a pneumatic device.

Pneumatic tire - A tire made of natural or synthetic rubber materials and containing compressed air to support the weight of the vehicle and cushion the road impacts.

Polyurethane - A chemical used in the production of resins.

Power brakes - Brakes with a vacuum or hydraulic assist to multiply the driver's effort so that braking effort is reduced. Also referred to as *vacuum-assisted* or *hydraulic-assisted brakes*.

Power steering - A system which provides additional force, usually hydraulic or vacuum-assisted, to the steering mechanism, reducing the driver's steering effort.

Powertrain - The components through which motive power is generated and transmitted to the driven axles. See *drivetrain*.

Preload - The amount of load placed on a bearing before actual operating loads are imposed. Proper preloading requires bearing adjustment and ensures alignment and minimum looseness in the system.

Press-fit - A tight fit between two parts that requires pressure to force the parts together. Also referred to as drive, or force, fit.

Pressure bleeder - A device that forces brake fluid under pressure into the master cylinder, so that by opening the bleeder screws all air will be purged from the hydraulic system.

Pressure plate - See *clutch pressure plate*.

Primary brake shoe - The brake shoe in a set which initiates the self-energizing action. Self-energization is determined by the location of the brake anchor pin and is caused by the tendency of the rotating drum to drag the lining along with it. When the drum is rotating in one direction, frictional force between the brake drum and the primary shoe tries to turn the shoe around the anchor pin. Because the drum itself prevents this, the shoe is forced even more strongly against the drum than the applying force is pushing it. If the drum is revolved in the opposite direction, there is no self-energization. But since there are two shoes, whichever way the drum is turned, one or the other shoe is effective. When self-energization is amplified to include both shoes by having the primary shoe push the *secondary shoe*, the amplified force is known as *servo action*.

Propeller shaft - Another name for the driveshaft connecting the transmission to the rear axle.

Proportioning valve - A hydraulic control valve located in the circuit to the rear wheels which limits the amount of pressure to the rear brakes during panic stops to prevent wheel lock-up.

Puller - A special tool designed to remove a bearing, bushing, hub, sleeve, etc. There are many, many types of pullers.

Q

Quarter panel - The sheet metal from the rear door opening to the taillights and from the rear wheel opening to the base of the roof and trunk opening.

R

Race (bearing) - The highly-finished inner or outer ring that provides a contact surface for ball or roller bearings.

Rack-and-pinion steering - A steering system with a pinion gear on the end of the steering shaft that mates with a rack (think of a geared wheel opened up and laid flat). When the steering wheel is turned, the pinion turns, moving the rack to the left or right. This movement is transmitted through the tie-rods to the steering arms at the wheels.

Radial ply - A pneumatic tire structure in which the ply cords extend from bead to bead, laid at right angles to center line of tire. Radial tires are, and have been for years, the favored tire design.

Radial runout - A variation in the diameter of a brake disc, wheel or tire from a specified amount.

Radius rods - Rods attached to the axle and to the frame to prevent fore-and-aft motion of the axle, yet permit vertical motion.

Ratio - Relation or proportion that one number bears to another.

Ream - To size, enlarge or smooth a hole by using a round cutting tool with fluted edges.

Recap - Used tire to which a top strip of synthetic or reclaimed rubber strip has been added.

Reciprocating - A back and forth movement. For example, the up and down motion of a piston in a cylinder is a reciprocating motion.

Recirculating-ball steering - A steering system in which the turning forces are transmitted through the ball bearings from a worm gear on the steering shaft to a sector gear on the Pitman arm shaft. In operation, a recirculating-ball steering assembly is filled with ball bearings which roll along grooves between the worm teeth and grooves inside the ball nut. When the steering wheel is turned, the worm gear on the end of the steering shaft rotates and the movement of the recirculating balls causes the ball nut to move up and down along the worm. Movement of the ball nut is carried by teeth to the sector gear which in turn moves with the ball nut to rotate the Pitman-arm shaft and activate the steering linkage. The balls recirculate from one end of the ball nut to the other through a pair of ball return guides. Also referred to as *recirculating-ball-and-nut steering*, *ball-and-nut steering* or *worm-and-recirculating-ball steering*.

Retread - Used tire with new rubber bonded to worn surface from shoulder to shoulder.

Reverse Elliot steering knuckle - Type of axle construction in which the steering spindle straddles the ends of the axle beam.

Revolutions per minute (rpm) - The speed the engine crankshaft is turning.

Ring and pinion - A term used to describe the differential drive pinion and ring gear. See *final drive ratio*.

Ring gear - The outer gear within which the other gears revolve in a planetary system. Term also refers to driven gear which mates with drive pinion in a differential assembly.

Rising rate suspension - A type of suspension system in which the spring rates increase as the wheels move further into jounce. This can be accomplished by using rising-rate geometry, by progressive-rate springs designed with varying wire diameter (or by varying the distances between the coils), by fitting two or more springs or through progressive compression of the rubber springs or rubber jounce stops. The aim of these designs is to maintain consistent ride and handling characteristics under maximum and minimum passenger, luggage or fuel loads and to maintain a soft ride on straight roads without sacrificing roll stiffness when cornering.

Rivet - To attach with rivets or to batter or upset end of a pin.

Riveted linings - Brake linings that are riveted to the pad backing plate or brake shoe.

Roll - A rotating motion about a longitudinal center line through the vehicle that causes the springs on one side of the vehicle to compress and those on the other side to extend.

Roll center - That point about which the body rolls when cornering.

Roller bearing - An inner and outer race upon which hardened steel rollers operate.

Rolling radius - The distance from the center of the tire's ground contact patch to the center of the wheel rim.

Rolling resistance - The motion-resisting forces present from the instant the wheels begin to turn. A large part of the power expended in a rolling wheel is converted into heat within the tire itself. The consequent temperature rise reduces both the abrasive resistance and the flex fatigue of the tire material and becomes the limiting factor on tire performance and longevity. On normal road surfaces, rolling resistance decreases with increasing tire pressures; it increases with vehicle weight.

Roll stiffness When the body rolls, the turning moment or torque the suspension exerts that tries to pull the body back to its normal upright position. Roll stiffness is expressed in pounds-feet per degree of roll and is basically a function of the spring rates and the perpendicular distance from the springs to the roll center. The stiffer the springs or the greater this distance, the larger the roll stiffness and the flatter the vehicle corners. Also known as *roll resistance*.

Rotor - Another name for a brake disc in a disc brake system. Also, in a distributor, the small rotating device mounted on the breaker cam inside the cap that connects the center electrode and the various outer spark plug terminals as it turns, distributing the high voltage from the coil secondary winding to the proper spark plug. Also, that part of an alternator which rotates inside the stator. Also, the rotating assembly of a turbocharger, including the compressor wheel, shaft and turbine wheel.

Rotor runout - Lateral movement of a rotor friction surface as it rotates past a fixed point.

Rubber - An elastic, vibration-absorbing material of either natural or synthetic origin. Rubber is used for engine mounts, dust covers, grommets, O-rings, seals, etc.

Runout - The amount of wobble (in-and-out movement) of a gear or wheel as it's rotated. The amount a shaft rotates "out-of-true." The out-of-round condition of a rotating part.

S

SAE - Society of Automotive Engineers.

SAE thread - Refers to a table of threads set up by Society of Automotive Engineers and determines number of threads per inch. Example: a quarter inch diameter rod with an SAE thread would have 28 threads per inch.

Safety factor - The degree of strength above normal requirements which serves as insurance against failure.

Safety rim - A type of wheel rim having a hump on the inner edge of the ledge on which the tire bead rides. the hump holds the tire on the rim in case of a blowout.

Schrader valve - A spring-loaded, one-way valve installed on all road wheels, for putting air into the tire. Also installed on the throttle body or the fuel rail on fuel injection systems, for attaching fuel pressure gauges and/or relieving system fuel pressure.

Also located inside the service valve fitting on air conditioning systems, to hold refrigerant in the system. Special adapters with built-in depressors must be used to attach service hoses to Schrader valves.

Score - A scratch, ridge or groove marring a finished surface, usually caused by dirt or some other foreign substance coming between the surface and a part that moves against it.

Scoring - Grooves or deep scratches on a friction surface caused by metal-to-metal contact (worn-out brake pads or shoes) or debris caught between the friction material and the friction surface.

Scuffing - A type of wear in which there's a transfer of material between parts moving against each other; shows up as pits or grooves in the mating surfaces.

Sealant (gasket) - A thick, tacky compound, usually spread with a brush, which may be used as a gasket or sealant, to seal small openings or surface irregularities.

Seat - The surface upon which another part rests or seats. For example, the valve seat is the matched surface upon which the valve face rests. Also used to refer to wearing into a good fit; for example, piston rings seat after a few miles of driving.

Secondary brake shoe or secondary shoe - The brake shoe which is energized by the primary shoe in a duo-servo drum brake assembly. The secondary shoe increases the servo, or self-energizing, action of brake. See *primary shoe*.

Section height - The height of an inflated tire from the bottom of the bead to the top of the tread.

Section width - The width between the exterior surfaces of the sidewalls of an inflated tire at its widest point.

Seize - When one surface sliding against another surface binds, then sticks, it is said to have seized. For example, a piston seizes in a cylinder because of a lack of lubrication, or because of over-expansion caused by excessive heat.

Selective transmission - An arrangement of gearing and shifting devices in which it's possible to go directly from neutral position into any desired pair of gears.

Self-leveling suspension - See *automatic level control*.

Semi-elliptic springs - A type of leaf spring that takes its name from the shape which is part of an ellipse. See *leaf springs*.

Semi-floating axle - A driveaxle construction in which the axleshafts support the weight of the vehicle. In this design, a single bearing is placed between the axleshaft and the axle bearing. To remove the axle, the wheel must be removed first.

Semi-independent rear suspension - A rear suspension system that allows up-and-down movement of the wheels.

Semi-trailing arm - A popular type of independent rear suspension in which the pivot axes are usually at about 25 degrees to a line running straight across the vehicle. This provides rear-wheel camber somewhere between that of a pure trailing arm (no camber change relative to the body) and a swing axle (large camber change).

Servo - In automatic transmissions, a hydraulic piston and cylinder assembly used to control the drum bands.

Servo action - A brake design in which a primary shoe pushes a secondary shoe to generate higher braking force.

Servo-action drum brake - See *Duo-servo drum brake.*

Shackle - A swinging support by which one end of a leaf spring is attached to the vehicle frame. The shackle is needed to take care of the changes in length of the spring as it moves up and down.

Shackle bolt - A link for connecting one end of a chassis spring to the frame, which allows the spring end to oscillate laterally.

Shear - To cut between two blades.

Shift lever - The mechanism that allows the driver to move the transmission gears into various drive positions. The shift lever is usually either a lever on the floor, or in a floor console between the seats, or on the steering column. A few vehicles, such as some Sixties era Chrysler vehicles, offered push-button control of their automatic transmissions.

Shift linkage - The rods, levers, etc. used to transmit motion of the shift lever into movement of the gears in the transmission.

Shim - Thin sheet used as a spacer between two parts. For example, alignment shims between the control arm pivot shaft and the frame serve to adjust castor and camber.

Shimming - Placing a shim or spacer under weak valve springs or springs with a short free height; the shim, which is placed between the spring and the cylinder head, compresses the spring a little more to restore the spring's installed and open loads.

Shimmy - A wobbling or shaking of the front wheels.

Shock absorber - A hydraulic and/or pneumatic device which provides mechanical or hydraulic friction to control the excessive deflection of the vehicle's springs. Also referred to as a *damper.*

Shrink fit - An exceptionally tight fit. For example, if a shaft or part is slightly larger than the hole in which it is to be inserted, the outer part is heated above its normal operating temperature or the inner part is chilled below its normal operating temperature, or both, and assembled in this condition. Upon cooling, a shrink fit is obtained.

Side clearance - The clearance between the sides of moving parts that don't serve as load-carrying surfaces.

Sidewall - The portion of the tire between the tread and the bead.

Slide hammer - A special puller that screws into or hooks onto the back of the a bearing; a heavy sliding handle on the shaft bottoms against the end of the shaft to knock the bearing free.

Sliding fit - Sufficient clearance between the shaft and the journal to allow free running without overheating.

Slip angle - In cornering, the angle formed by the direction the tire is pointed and the direction the vehicle is moving.

Slip-in bearing - A liner, made to extremely accurate measurements, which can be used for replacement purposes without additional fitting.

Slip joint - A variable-length connection that permits the driveshaft or axleshaft to change its length as the shaft moves up and down.

Snap-ring - A ring-shaped clip used to prevent endwise movement of the cylindrical parts and shafts. An internal snap-ring is installed in a groove in a housing; an external snap-ring fits into a groove cut out the outside of a cylindrical piece such as a shaft.

Solid height - The height of a coil spring when it's totally compressed to the point at which each coil touches the adjacent coil. See *coil binding.*

Solvent - A solution which dissolves some other material. A thinner used to dissolve or dilute another liquid or solid. For example, water is a solvent for sugar.

Spacer - Another name for a valve spring shim. See *shimming.*

Spacer washer - A sheet of metal or other material placed between two surfaces to reduce clearance or to provide a better thrust surface for a fastener.

Speed - The magnitude of velocity (regardless of direction).

Speedometer - A device for measuring and indicating the speed of a vehicle.

Spider gear - One of two to four small gears in the differential that mesh with the bevel gears on the ends of the axles. See *pinion gear.*

Spiral bevel gear - A ring gear and pinion in which the mating teeth are curved and placed at an angle to the pinion shaft.

Splayed spring - A design in which the leaf springs are placed at other than a 90 degree angle to axle.

Spline - A long keyway. One of several lengthwise grooves cut in a shaft or gear, either internally or externally. Meshing splines enforce mutual rotation of parts and at the same time permit freedom of lengthwise rotation.

Spline joint - Two mating parts each with a series of splines around its circumference, one inner and one outer, to provide a longitudinally movable joint without circumferential motion.

Split hydraulic brake system - Service brake system with two separate hydraulic circuits to provide braking action in one circuit if other one fails.

Spoiler - An aerodynamic wing, usually attached to a vehicle's decklid, designed to induce some downforce to the rear wheels.

Spring - An elastic device that yields under stress or pressure but returns to its original state or position when the stress or pressure is removed. A spring does this by first absorbing and then releasing a certain amount of energy. Basically, the function of the springs is to support the frame, engine, body, etc. of the vehicle and to absorb road shocks that result from the wheels hitting holes or bumps in the road. The three most common types of automotive springs are leaf springs, coil springs and torsion bars.

Sprung weight - A term used to describe all parts of an automobile that are supported by car springs. Example: frame, engine, body, etc.

Spur gear - A gear in which the teeth are cut parallel to the shaft.

Stabilizer bar - A transverse bar linking both sides (either front or rear) of the suspension, generally taking the form of a torsion bar with rubber bushing mounts on the chassis that allow it to turn freely. The ends of the bar are connected to or shaped as lever arms with attachments to the suspension linkages at each side via balljoint or rubber-bushed pivot links. The effect on a bump that both wheels take equally is to allow the wheels to move the same amount without deflecting the stabilizer bar. However, individual wheel movement or body roll will force the bar to twist as the lever arms are moved in different directions or amounts, thereby adding the bar's own spring rate to that of the vehicle's springs. The stabilizer bar's main function is to reduce roll, but it influences overall handling characteristics as well. Installing a front stabilizer bar increases understeer; installing one at the rear increases oversteer. Also referred to as an *anti-sway bar*, an *anti-roll bar* or simply a *roll bar*.

Standard thread - Refers to the U.S.S. table of the number of threads per inch. For example, a quarter-inch diameter standard thread is 20 threads per inch.

Static balance - The balance of an object while it's stationary.

Static imbalance - The lack of balance of an object while it's stationary.

Steering arm - The arm attached to the steering knuckle that turns the knuckle and wheel for steering.

Steering axis - See *kingpin inclination*.

Steering axis inclination - The angle formed by the centerline of the suspension balljoints and the true vertical centerline.

Steering column - A shaft connecting the steering wheel with the steering gear assembly. Also called the *steering shaft*.

Steering gear - A steering unit, whether it's recirculating ball, rack-and-pinion, etc.

Steering geometry - The various angles between the front wheels, the frame and the attachment points. Includes *camber*, *caster*, *steering axis inclination* and *toe-in* and *toe-out*. See *toe-out on turns*.

Steering knuckle - A steering and suspension component that is positioned by the control arm(s) and (on front-wheel-drive vehicles) the strut assembly. On a rear-wheel-drive vehicle, the steering knuckle is the component to which the wheel spindles are attached. On a front-wheel-drive vehicle, the knuckle serves as an attaching point for the outer end of the driveaxle and houses a hub and bearing assembly for the wheel.

Steering linkage - The rods, arms and other links that carry movement of the Pitman arm to the steering knuckles. See *idler arm*.

Steering lock - The number of degrees that the steering wheel can be turned from straight ahead before they are physically restrained from turning any further. Generally speaking, the more lock designed into the front geometry, the smaller the turning circle but the more the tires scrub as the wheels approach the extreme lock positions. The number of turns lock-to-lock refers to the number of rotations of the steering wheel required to go from one extreme lock position to the other. It can also refer to the locking mechanism usually contained in the steering column that locks the steering wheel when the ignition switch is turned off or the key is removed from the switch.

Steering overall ratio - Ratio of the degrees you must turn the steering wheel to turn the road wheels one degree from their straight ahead position.

Steering post or column - The shaft connecting the steering gear unit with the steering wheel.

Steering ratio - The ratio of the gearing within a steering system such as the rack to the pinion, or the worm gear to the recirculating nut.

Steering spindle - A journal or shaft upon which the steerable wheels of a rear-wheel-drive vehicle are mounted.

Steering system - The mechanism that allows the driver to guide the vehicle down the road and turn the wheels as he or she desires. The system includes the steering wheel, steering column, steering gear, linkages and the front wheel supports.

Stress - The force or strain to which a material is subjected.

Strut - See *MacPherson strut*.

Stud - A metal rod with threads on both ends. A stud usually screws into the cylinder block, the head, etc. on one end and has a nut placed on other end.

Stud puller - A tool used to remove or install studs.

Subframe - A partial frame that is sometimes bolted to the chassis of unit-body vehicles. It can be used to support the engine, transmission and/or suspension instead of having these components bolted directly to the main body structure. This more expensive design generally results in better road isolation and less harshness.

Sun gear - The central gear around which the other gears revolve in a planetary gear system. See *planetary gears*.

Supplemental Restraint System (SRS) - See *Airbag*.

Suspension - Refers to the various springs, shock absorbers and linkages used to suspend a vehicle's frame, body, engine and drivetrain above the wheels.

Sway bar - See *stabilizer bar*.

Swing axle - A type of independent rear suspension using half-shafts that have universal joints only at their inner ends on either side of the differential. The swing radius is small and thus there is great camber change with up-and-down wheel movements and the wheels are nowhere near vertical under many conditions. When cornering hard, vehicles with swing axles are prone to wheel jacking, which can lead to large positive camber at the outside rear wheel and can induce sudden oversteer. Earlier (pre-1969) VW Beetles used swing axles.

Synchromesh - A device used in transmission gearing to facilitate the meshing (engagement) of two gears by synchronizing the speed of both gears.

Synchronizer - A cone or sleeve that slides to and fro on the transmission mainshaft and makes the gears rotate at the same speed to prevent clash when the gears are about to mesh. When the vehicle is rolling, the transmission mainshaft is turning and the clutch gear is spinning. Even though the clutch is disengaged, the clutch gear continues to spin until friction slows it down or stops it. Thus when the driver shifts into another gear, he is trying to mesh gears that may be moving at different speeds. By using synchronizers, the possibility of broken or damaged teeth is reduced and shifting effort is reduced. A transmission using synchronizers is called a *synchromesh transmission*.

System - A group of interacting mechanical or electrical components serving a common purpose.

T

Tang - A lip on the end of a plain bearing used to align the bearing during assembly.

Tap - To cut threads in a hole. Also refers to the fluted tool used to cut threads.

Tap and die set - Set of taps and dies for internal and external threading - usually covers a range of the most popular sizes.

Tapered roller bearing - A bearing utilizing a series of tapered, hardened steel rollers operating between an outer and inner hardened steel race.

Temper - To change the physical characteristics of metal by applying heat.

Threaded insert - A threaded coil that's used to restore the original thread size to a hole with damaged threads; the hole is drilled oversize and tapped, and the insert is threaded into the tapped hole.

Throw-out bearing - The bearing in the clutch assembly that is moved in to the release levers by clutch-pedal action to disengage the clutch. Also referred to as a *release bearing*.

Thrust washer - A bronze or hardened steel washer placed between two moving parts. The washer prevents longitudinal movement and provides a bearing surface for thrust surfaces of parts.

Tie-rod - A balljoint connecting the steering linkage (on recirculating ball type steering gearboxes) or the rack (on rack-and-pinion type steering gearboxes) to the steering arm or steering knuckle. The tie-rod end (the balljoint part) is threaded so that it can be moved in or out in relation to the inner tie-rod to allow toe adjustments.

Tire - A tubular corded carcass covered with rubber or synthetic rubber, mounted on a wheel inflated to provide traction for moving and stopping the vehicle.

Tire slip - The difference between the speed of the vehicle and the speed between the tire and the ground, expressed in a percentage.

Toe - The highest point on the cam lobe; the part of the lobe that raises the lifter to its highest point. Also called the *nose*. The term also refers to the inside or smaller half of a gear tooth. It also refers to the end of the brake shoe against the anchor.

Toe-in - The amount the front wheels are closer together in front than at the rear when viewed from the front of the vehicle. A slight amount of toe-in is usually specified to keep the front wheels running parallel on the road by offsetting other forces that tend to spread the wheels apart.

Toe-out on turns - The related angles assumed by the front wheels of a vehicle when turning.

Tolerance - The amount of variation permitted from an exact size of measurement. Actual amount from smallest acceptable dimension to largest acceptable dimension.

Torque - A turning or twisting force, such as the force imparted on a fastener by a torque wrench. Usually expressed in foot-pounds (ft-lbs).

Torque converter - A fluid coupling that transmits power from a driving to a driven member

Torque tube drive - A driveline with the driveshaft enclosed inside a stiff tube anchored to the axle housing. Only one universal joint is used in the driveline, at the front end of the driveshaft. The torque tube prevents the axle housing from twisting when engine power or braking torque is applied. This generally results in an improvement in ride because the springs don't have to absorb any driving or braking torque; they only have to cushion the ride.

Torsional vibration - The rotary motion that causes a twist-untwist action on a vibrating shaft, so that a part of the shaft repeatedly moves ahead of, or lags behind, the remainder of the shaft; for example, the action of a crankshaft responding to the cylinder firing impulses.

Torsion bar - A long straight bar fastened to the frame at one end and to a suspension part at the other. In effect, a torsion bar is merely an uncoiled coil spring and spring action is produced when the bar is twisted. the main advantage of the torsion bar over the coil spring in the front suspension is the ease of adjusting the front suspension height.

Torque wrench - A special wrench which can accurately measure and indicate the tightening force applied to a fastener.

Track - The distance from the center of one front (or rear) tire (or wheel) to the other front (or rear) tire (or wheel) with the vehicle set to its normal ride height and wheel alignment specifications.

Track rod - See *panhard rod*.

Traction - The amount of adhesion between the tire and ground.

Trailing arm - A type of independent rear suspension in which the pivot axis is exactly across the vehicle, or perpendicular to the longitudinal axis. This means the wheels are always upright relative to the body, so they're leaning with the body in a corner. This design is widely used at the rear of front-wheel-drive vehicles.

Tramp - An oscillating motion and/or heavy vibration that occurs when the wheels are turning. See *axle wind-up*.

Transaxle - A transmission and differential combined into one integral unit. Transaxles are used on all front-wheel-drive vehicles, rear-engined vehicles and even on some front-engine, rear-wheel-drive sports cars.

Transfer case - An auxiliary device in a four-wheel-drive vehicle that allows power to be delivered to both axles. Normally, the transfer case incorporates a shifting device so that the front drive can be disconnected for running on pavement.

Transmission - A device which selectively increases or decreases the ratio of relative rotation between its input and output shafts in order to trade speed for power, or vice versa, through gearing or torque conversion. It includes various devices and combinations for changing ratio between engine revolutions and driving wheel revolutions.

Transverse arm - A suspension arm not split into two separate sections like an A-arm. Transverse arms are often used as the bottom locating link with MacPherson strut suspension systems. Also known as a *lateral arm*.

Tread - The portion of the tire that comes in contact with the road. Also, the distance between the center of the tires at the points where they contact the road surface.

Troubleshooting - A process of diagnosing possible sources of trouble by observation and testing.

Turning circle or turning radius - The diameter of a circle within which a vehicle can be turned around.

Twist drill - A metal cutting drill with spiral flutes (grooves) to permit exit of chips while cutting.

Two-wheel ABS - An anti-lock brake system that only operates on the rear wheels.

U

Understeer - A handling characteristic where the slip angles are larger at the front than at the rear. During hard cornering, the front tires slip while the rear tires maintain traction. The vehicle will tend to travel in a straight line, requiring more steering input to negotiate the turn. When a vehicle exhibits understeer it is said to "push."

Unit body - A type of body/frame construction in which the body of the vehicle, its floor pan and the chassis form a single structure. Such a design is generally lighter and more rigid than a vehicle with a separate body and frame.

Universal joint or U-joint - A double-pivoted connection for transmitting power from a driving to a driven shaft through an angle. A U-joint consists of two Y-shaped yokes and a cross-shaped member called the *spider*. The four arms of the spider are assembled into bearings in the ends of the two yokes. With the normal cross-and-two-yoke U-joint, there is some change in speed when the driveshaft and driven shaft are at an angle to each other. The change in speed occurs because the driven yoke and driven shaft speed up and slow down twice with every revolution of the driveline. The greater the angle between the drive and driven shafts, the greater the speed variation. To eliminate this speed variation, which results in increased wear of the affected parts, constant velocity joints are used on many vehicles.

Unsprung weight - The parts of the vehicle - wheels, axles, etc. - that isn't supported by the springs.

Upper and lower A-arms - A suspension system utilizing a pair of A-arms. See *double-wishbone suspension*.

V

Vacuum gauge - An instrument used to measure the amount of intake vacuum.

Vacuum-operated power booster - A power booster that uses engine manifold vacuum to assist the driver in the application of the brakes.

Viscosity - The thickness of a liquid or its resistance to flow.

W

Wandering - A condition in which the steering wheels of a vehicle tend to turn slowly first in one direction, then the other, interfering with directional control or stability.

Web - A supporting structure across a cavity.

Weight transfer effect - Because the center of gravity of a vehicle is located above the centers of wheel rotation, a sudden stoppage of the vehicle tends to cause the center of gravity to move forward, thus throwing more weight onto the front wheels.

Wheel - The circular steel or alloy assembly onto which the tire is mounted, and which is itself bolted to the axle or spindle.

Wheel alignment - See *alignment*.

Haynes suspension, steering and driveline manual

Wheelbase - The longitudinal distance between the centerlines of the front and rear axles.

Wheel cylinder - A small cylinder with one or two pistons that's fitted inside each brake drum. When brake fluid is forced from the master cylinder, it flows through the brake lines into the wheel cylinder, pushing the pistons apart. This pushes the brake shoes out, forcing the brake shoes tightly against the rotating brake drums. See *brake backing plate*.

Wheel hub and bearing assembly - The precision made assembly which houses the wheel bearings and to which the brake disc and/or wheel are bolted.

Wheel speed sensor - The component of an anti-lock brake system that picks up the impulses of the toothed signal rotor, sending these impulses to the ABS ECU.

Wheel tramp - See *tramp*.

Wing - An aerodynamic device, usually shaped like an upside-down aircraft wing, attached to a vehicle to induce downforce on either the front or rear end.

Wishbone - See *A-arm*.

Woodruff key - A key with a radiused backside (viewed from the side).

Worm gear - A shaft having an extremely coarse thread which is designed to operate in engagement with a toothed wheel, as a pair of gears.

Wringing-fit - A fit with less clearance than for a running or sliding fit. The shaft enters the hole by means of twisting and pushing by hand.

Index

A

Aerodynamics, 8-11
Air compressor, 2-21
Airbag system, general information, 7-22
Asbestos, 1-11
Automatic transmission/transaxle
    fluid
        and filter change, 4-16, 5-16
        level check, 4-8, 5-8
    troubleshooting, 3-6
Automotive chemicals, 1-10
Axle assembly (front), removal and installation, 5-48
Axle assembly (rear), removal and installation, 5-46
Axleshaft/driveaxle oil seal, replacement, 5-38

B

Balljoints, check and replacement, 6-43
Battery booster (jump) starting, 1-8
Battery, safety, 1-11
Beam type suspension with leaf spring, 6-62
Beam type suspension with strut or
    shock absorber/coil springs, 6-60
Beam type with coil spring and shocks, 6-57
Bench vise, 2-20
Brake and clutch fluid level checks, 4-6, 5-6
Brake-related tools, 2-26
Bushings, polyurethane, 8-7
Buying hand and general purpose tools, 2-4
Buying parts, 1-2
Buying tools, 2-3

C

Carrier and bearings, removal and installation, 5-52
Chassis lubrication, 4-12, 5-12
Cleaners, lubricants and sealants, 1-9
Clutch
    check, 5-12
    component disassembly, 1-5
    components, removal, inspection and installation, 5-12
    freeplay check and adjustment, 4-15, 5-15
    release bearing and lever, removal, inspection and installation, 5-14
    troubleshooting, 3-2

D

Dies, 2-18
Differential, check and adjustments, 5-49
Differential lubricant
    change, 4-17, 5-17
    level check, 4-12, 5-12
DOs and DON'Ts, 1-10
Drawbolt extractors, 2-19
Drills, 2-21
Driveaxle boot replacement and CV joint inspection
    and overhaul, 5-31

Driveaxles, removal and installation, 5-26
Driveline systems, 5-1 through 5-58
Driveline-related tools, 2-27
Driveshaft, 3-9
    center bearing, check and replacement, 5-25
    inspection, 5-18
    removal and installation, 5-18

E

Electricity and lights, 2-1
Environmental safety, 1-11
Equipment and tools, 2-3

F

Fastener sizes, 1-4
Fasteners, 1-3
Files, 2-16
Fire, 1-11
Front axleshaft and joint assembly (four-wheel drive models),
    removal, component replacement and installation, 5-47
Front driveaxles, 3-8
Front wheel bearing check, repack and adjustment, 4-20, 5-20

G

Gasket sealing surfaces, 1-6
General-purpose hand tools, 2-4
Glossary, GL-1
Grinder, bench-mounted, 2-21

H

Hacksaws, 2-15
Hammers, 2-13
Hand tools, 2-4
High-speed grinders, 2-24
Hose removal tips, 1-6
How to use this manual, 1-2
Hub and bearing assembly (front-wheel drive models), removal
    and installation, 7-19
Hub and bearing assembly, rear, removal and installation, 6-73
Hub/drum assembly and wheel bearings (full-floating axles),
    removal, installation and adjustment, 4-41, 5-41

I

Impact drivers, 2-9
Introduction, 1-1 through 1-12

J

Jacking and towing, 1-7

L

Limited-slip differential, 8-10
Lubricants, 1-9

M

MacPherson strut type suspension, 6-37
Maintenance
    schedule, 5-2
    techniques, 1-2
Maintenance-related tools, 2-25
Manual transmission/transaxle, 3-4
Manual transmission/transaxle
    lubricant
        change, 5-17
        level check, 5-11
Modifications, 8-1 through 8-18
Modified strut type suspension with separate coil springs, 6-33
Multi-link type suspension, 6-70
Maintenance schedule, 4-2
Manual transmission/transaxle lubricant change, 4-17
Manual transmission/transaxle lubricant level check, 4-11

P

Pilot bearing/bushing, inspection and replacement, 5-16
Pinion oil seal, replacement, 5-45
Place to work, 2-1
Pliers, 2-11
Power steering
    fluid level check, 4-9, 5-9
    pump, removal and installation, 7-6
    system, bleeding, 7-21
Power tools, 2-21
Pullers, 2-19, 2-20
Punches and chisels, 2-14

R

Ratchet and socket sets, 2-7
Rear axle, 3-10
    check, 5-36
Rear axleshaft
    bearing (semi-floating axles), replacement, 5-40
    full-floating axles, removal, installation and adjustment, 5-40
    semi-floating axles, removal and installation, 5-36
Ring and pinion, removal and installation, 5-54
Routine maintenance, 4-1 through 4-24

S

Safety first, 1-10
Safety items, 2-24
Screwdrivers, 2-12
Sealants, 1-9
Shock absorbers, 8-5
Solid axle suspension with control arms and coil springs, 6-50
Solid axle suspension with leaf spring, 6-54
Springs, 8-9
Stabilizer bars, 8-6

Steering systems, 7-1
    gearbox, removal and installation, 7-9
    knuckle, removal and installation, 7-15
    linkage, inspection, removal and installation, 7-10
    rack and pinion, removal and installation, 7-8
    steering wheel, removal and installation, 7-3
Steering tools, 2-33
Storage and shelves, 2-2
Support bearing assembly, check and replacement, 5-31
Suspension and steering systems, 3-11
Suspension kits, 8-10
Suspension modifications
    load capacity and trailer towing, 8-15
    off-road, 8-11
Suspension systems, 6-1 through 6-74
    steering and driveline check, 4-10, 5-10
    unequal length A-arm type, 6-13, 6-18, 6-27
    related tools, 2-32

T

Taps and dies, 2-17
Tightening sequences and procedures, 1-4
Tire and tire pressure checks, 4-7, 5-7
Tire
    maintenance, 8-4
    troubleshooting, 3-11
Tires, modification, 8-1
Tools and equipment, 2-1
Tools, storage and care, 2-3
Torque wrenches, 2-8
Towing the vehicle, 1-8
Trailing arm type suspension, 6-65
Transfer case, 3-7
    lubricant
        change, 4-17, 5-17
        level check, 4-12, 5-12
Troubleshooting, 3-1 through 3-14

U

Universal joints, replacement, 5-19
Using the manual, 1-2
Using wrenches and sockets, 2-9

V

Vehicle, how to improve handling, 8-1

W

Wheel alignment
    general information, 7-22
    specifications, WA-1
Wheel bearing check, repack and adjustment, 4-18, 5-18
Wheel stud, replacement, 6-49
Wheels, 8-5
Wheels and tires, general information, 7-21
Workbenches, 2-2
Working facilities, 1-6
Workshop, 2-1
Workshop requirements, 2-1
Wrenches and sockets, 2-4

Haynes Automotive Manuals

ACURA
*12020   Integra '86 thru '89 & Legend '86 thru '90

AMC
     Jeep CJ - see JEEP (50020)
14020   Mid-size models, Concord, Hornet, Gremlin & Spirit '70 thru '83
14025   (Renault) Alliance & Encore '83 thru '87

AUDI
15020   4000 all models '80 thru '87
15025   5000 all models '77 thru '83
15026   5000 all models '84 thru '88

AUSTIN-HEALEY
     Sprite - see MG Midget (66015)

BMW
*18020   3/5 Series not including diesel or all-wheel drive models '82 thru '92
*18021   3 Series except 325iX models '92 thru '97
18025   320i all 4 cyl models '75 thru '83
18035   528i & 530i all models '75 thru '80
18050   1500 thru 2002 except Turbo '59 thru '77

BUICK
     Century (front wheel drive) - see GM (829)
*19020   Buick, Oldsmobile & Pontiac Full-size (Front wheel drive) all models '85 thru '98
     Buick Electra, LeSabre and Park Avenue; Oldsmobile Delta 88 Royale, Ninety Eight and Regency; Pontiac Bonneville
19025   Buick Oldsmobile & Pontiac Full-size (Rear wheel drive)
     Buick Estate '70 thru '90, Electra '70 thru '84, LeSabre '70 thru '85, Limited '74 thru '79
     Oldsmobile Custom Cruiser '70 thru '90, Delta 88 '70 thru '85, Ninety-eight '70 thru '84
     Pontiac Bonneville '70 thru '81, Catalina '70 thru '81, Grandville '70 thru '75, Parisienne '83 thru '86
19030   Mid-size Regal & Century all rear-drive models with V6, V8 and Turbo '74 thru '87
     Regal - see GENERAL MOTORS (38010)
     Riviera - see GENERAL MOTORS (38030)
     Roadmaster - see CHEVROLET (24046)
     Skyhawk - see GENERAL MOTORS (38015)
     Skylark '80 thru '85 - see GM (38020)
     Skylark '86 on - see GM (38025)
     Somerset - see GENERAL MOTORS (38025)

CADILLAC
*21030   Cadillac Rear Wheel Drive all gasoline models '70 thru '93
     Cimarron - see GENERAL MOTORS (38015)
     Eldorado - see GENERAL MOTORS (38030)
     Seville '80 thru '85 - see GM (38030)

CHEVROLET
*24010   Astro & GMC Safari Mini-vans '85 thru '93
24015   Camaro V8 all models '70 thru '81
24016   Camaro all models '82 thru '92
     Cavalier - see GENERAL MOTORS (38015)
     Celebrity - see GENERAL MOTORS (38005)
24017   Camaro & Firebird '93 thru '97
24020   Chevelle, Malibu & El Camino '69 thru '87
24024   Chevette & Pontiac T1000 '76 thru '87
     Citation - see GENERAL MOTORS (38020)
*24032   Corsica/Beretta all models '87 thru '96
24040   Corvette all V8 models '68 thru '82
*24041   Corvette all models '84 thru '96
10305   Chevrolet Engine Overhaul Manual
24045   Full-size Sedans Caprice, Impala, Biscayne, Bel Air & Wagons '69 thru '90
24046   Impala SS & Caprice and Buick Roadmaster '91 thru '96
     Lumina - see GENERAL MOTORS (38010)

24048   Lumina & Monte Carlo '95 thru '98
     Lumina APV - see GM (38035)
24050   Luv Pick-up all 2WD & 4WD '72 thru '82
*24055   Monte Carlo all models '70 thru '88
     Monte Carlo '95 thru '98 - see LUMINA (24048)
24059   Nova all V8 models '69 thru '79
*24060   Nova and Geo Prizm '85 thru '92
24064   Pick-ups '67 thru '87 - Chevrolet & GMC, all V8 & in-line 6 cyl, 2WD & 4WD '67 thru '87; Suburbans, Blazers & Jimmys '67 thru '91
*24065   Pick-ups '88 thru '98 - Chevrolet & GMC, all full-size pick-ups, '88 thru '98; Blazer & Jimmy '92 thru '94; Suburban '92 thru '98; Tahoe & Yukon '98
24070   S-10 & S-15 Pick-ups '82 thru '93, Blazer & Jimmy '83 thru '94,
*24071   S-10 & S-15 Pick-ups '94 thru '96 Blazer & Jimmy '95 thru '96
*24075   Sprint & Geo Metro '85 thru '94
*24080   Vans - Chevrolet & GMC, V8 & in-line 6 cylinder models '68 thru '96

CHRYSLER
25015   Chrysler Cirrus, Dodge Stratus, Plymouth Breeze '95 thru '98
25025   Chrysler Concorde, New Yorker & LHS, Dodge Intrepid, Eagle Vision, '93 thru '97
10310   Chrysler Engine Overhaul Manual
*25020   Full-size Front-Wheel Drive '88 thru '93
     K-Cars - see DODGE Aries (30008)
     Laser - see DODGE Daytona (30030)
*25030   Chrysler & Plymouth Mid-size front wheel drive '82 thru '95
     Rear-wheel Drive - see Dodge (30050)

DATSUN
28005   200SX all models '80 thru '83
28007   B-210 all models '73 thru '78
28009   210 all models '79 thru '82
28012   240Z, 260Z & 280Z Coupe '70 thru '78
28014   280ZX Coupe & 2+2 '79 thru '83
     300ZX - see NISSAN (72010)
28016   310 all models '78 thru '82
28018   510 & PL521 Pick-up '68 thru '73
28020   510 all models '78 thru '81
28022   620 Series Pick-up all models '73 thru '79
     720 Series Pick-up - see NISSAN (72030)
28025   810/Maxima all gasoline models, '77 thru '84

DODGE
     400 & 600 - see CHRYSLER (25030)
*30008   Aries & Plymouth Reliant '81 thru '89
30010   Caravan & Plymouth Voyager Mini-Vans all models '84 thru '95
*30011   Caravan & Plymouth Voyager Mini-Vans all models '96 thru '98
30012   Challenger/Plymouth Saporro '78 thru '83
30016   Colt & Plymouth Champ (front wheel drive) all models '78 thru '87
*30020   Dakota Pick-ups all models '87 thru '96
30025   Dart, Demon, Plymouth Barracuda, Duster & Valiant 6 cyl models '67 thru '76
*30030   Daytona & Chrysler Laser '84 thru '89
     Intrepid - see CHRYSLER (25025)
*30034   Neon all models '95 thru '97
*30035   Omni & Plymouth Horizon '78 thru '90
*30040   Pick-ups all full-size models '74 thru '93
*30041   Pick-ups all full-size models '94 thru '96
*30045   Ram 50/D50 Pick-ups & Raider and Plymouth Arrow Pick-ups '79 thru '93
30050   Dodge/Plymouth/Chrysler rear wheel drive '71 thru '89
*30055   Shadow & Plymouth Sundance '87 thru '94
*30060   Spirit & Plymouth Acclaim '89 thru '95
*30065   Vans - Dodge & Plymouth '71 thru '96

EAGLE
     Talon - see Mitsubishi Eclipse (68030)
     Vision - see CHRYSLER (25025)

FIAT
34010   124 Sport Coupe & Spider '68 thru '78
34025   X1/9 all models '74 thru '80

FORD
10355   Ford Automatic Transmission Overhaul
*36004   Aerostar Mini-vans all models '86 thru '96
*36006   Contour & Mercury Mystique '95 thru '98
36008   Courier Pick-up all models '72 thru '82
36012   Crown Victoria & Mercury Grand Marquis '88 thru '96
10320   Ford Engine Overhaul Manual
36016   Escort/Mercury Lynx all models '81 thru '90
*36020   Escort/Mercury Tracer '91 thru '96
*36024   Explorer & Mazda Navajo '91 thru '95
36028   Fairmont & Mercury Zephyr '78 thru '83
36030   Festiva & Aspire '88 thru '97
36032   Fiesta all models '77 thru '80
36036   Ford & Mercury Full-size, Ford LTD & Mercury Marquis ('75 thru '82); Ford Custom 500, Country Squire, Crown Victoria & Mercury Colony Park ('75 thru '87); Ford LTD Crown Victoria & Mercury Gran Marquis ('83 thru '87)
36040   Granada & Mercury Monarch '75 thru '80
36044   Ford & Mercury Mid-size, Ford Thunderbird & Mercury Cougar ('75 thru '82); Ford LTD & Mercury Marquis ('83 thru '86); Ford Torino, Gran Torino, Elite, Ranchero pick-up, LTD II, Mercury Montego, Comet, XR-7 & Lincoln Versailles ('75 thru '86)
36048   Mustang V8 all models '64-1/2 thru '73
36049   Mustang II 4 cyl, V6 & V8 models '74 thru '78
36050   Mustang & Mercury Capri all models Mustang, '79 thru '93; Capri, '79 thru '86
*36051   Mustang all models '94 thru '97
36054   Pick-ups & Bronco '73 thru '79
36058   Pick-ups & Bronco '80 thru '96
36059   Pick-ups, Expedition & Mercury Navigator '97 thru '98
36062   Pinto & Mercury Bobcat '75 thru '80
36066   Probe all models '89 thru '92
36070   Ranger/Bronco II gasoline models '83 thru '92
*36071   Ranger '93 thru '97 & Mazda Pick-ups '94 thru '97
36074   Taurus & Mercury Sable '86 thru '95
*36075   Taurus & Mercury Sable '96 thru '98
*36078   Tempo & Mercury Topaz '84 thru '94
36082   Thunderbird/Mercury Cougar '83 thru '88
*36086   Thunderbird/Mercury Cougar '89 and '97
36090   Vans all V8 Econoline models '69 thru '91
*36094   Vans full size '92-'95
*36097   Windstar Mini-van '95-'98

GENERAL MOTORS
*10360   GM Automatic Transmission Overhaul
*38005   Buick Century, Chevrolet Celebrity, Oldsmobile Cutlass Ciera & Pontiac 6000 all models '82 thru '96
*38010   Buick Regal, Chevrolet Lumina, Oldsmobile Cutlass Supreme & Pontiac Grand Prix front-wheel drive models '88 thru '95
*38015   Buick Skyhawk, Cadillac Cimarron, Chevrolet Cavalier, Oldsmobile Firenza & Pontiac J-2000 & Sunbird '82 thru '94
*38016   Chevrolet Cavalier & Pontiac Sunfire '95 thru '98
38020   Buick Skylark, Chevrolet Citation, Olds Omega, Pontiac Phoenix '80 thru '85
38025   Buick Skylark & Somerset, Oldsmobile Achieva & Calais and Pontiac Grand Am all models '85 thru '95
38030   Cadillac Eldorado '71 thru '85, Seville '80 thru '85, Oldsmobile Toronado '71 thru '85 & Buick Riviera '79 thru '85
*38035   Chevrolet Lumina APV, Olds Silhouette & Pontiac Trans Sport all models '90 thru '95
     General Motors Full-size Rear-wheel Drive - see BUICK (19025)

(Continued on other side)

Haynes North America, Inc., 861 Lawrence Drive, Newbury Park, CA 91320-1514 • (805) 498-6703

Haynes Automotive Manuals (continued)

NOTE: New manuals are added to this list on a periodic basis. If you do not see a listing for your vehicle, consult your local Haynes dealer for the latest product information.

GEO

Metro - *see CHEVROLET Sprint (24075)*
Prizm - *'85 thru '92 see CHEVY (24060), '93 thru '96 see TOYOTA Corolla (92036)*
*40030 **Storm** all models '90 thru '93
Tracker - *see SUZUKI Samurai (90010)*

GMC

Safari - *see CHEVROLET ASTRO (24010)*
Vans & Pick-ups - *see CHEVROLET*

HONDA

42010 **Accord CVCC** all models '76 thru '83
42011 **Accord** all models '84 thru '89
42012 **Accord** all models '90 thru '93
42013 **Accord** all models '94 thru '95
42020 **Civic 1200** all models '73 thru '79
42021 **Civic 1300 & 1500 CVCC** '80 thru '83
42022 **Civic 1500 CVCC** all models '75 thru '79
42023 **Civic** all models '84 thru '91
*42024 **Civic & del Sol** '92 thru '95
*42040 **Prelude CVCC** all models '79 thru '89

HYUNDAI

*43015 **Excel** all models '86 thru '94

ISUZU

Hombre - *see CHEVROLET S-10 (24071)*
*47017 **Rodeo** '91 thru '97; **Amigo** '89 thru '94; **Honda Passport** '95 thru '97
*47020 **Trooper & Pick-up,** all gasoline models Pick-up, '81 thru '93; Trooper, '84 thru '91

JAGUAR

*49010 **XJ6** all 6 cyl models '68 thru '86
*49011 **XJ6** all models '88 thru '94
*49015 **XJ12 & XJS** all 12 cyl models '72 thru '85

JEEP

*50010 **Cherokee, Comanche & Wagoneer Limited** all models '84 thru '96
50020 **CJ** all models '49 thru '86
*50025 **Grand Cherokee** all models '93 thru '98
50029 **Grand Wagoneer & Pick-up** '72 thru '91 Grand Wagoneer '84 thru '91, Cherokee & Wagoneer '72 thru '83, Pick-up '72 thru '88
*50030 **Wrangler** all models '87 thru '95

LINCOLN

Navigator - *see FORD Pick-up (36059)*
59010 **Rear Wheel Drive** all models '70 thru '96

MAZDA

61010 **GLC Hatchback (rear wheel drive)** '77 thru '83
61011 **GLC (front wheel drive)** '81 thru '85
*61015 **323 & Protegé** '90 thru '97
*61016 **MX-5 Miata** '90 thru '97
*61020 **MPV** all models '89 thru '94
Navajo - *see Ford Explorer (36024)*
61030 **Pick-ups** '72 thru '93
Pick-ups '94 thru '96 - *see Ford Ranger (36071)*
61035 **RX-7** all models '79 thru '85
*61036 **RX-7** all models '86 thru '91
61040 **626** (rear wheel drive) all models '79 thru '82
*61041 **626/MX-6 (front wheel drive)** '83 thru '91

MERCEDES-BENZ

63012 **123 Series Diesel** '76 thru '85
*63015 **190 Series** four-cyl gas models, '84 thru '88
63020 **230/250/280** 6 cyl sohc models '68 thru '72
63025 **280 123 Series** gasoline models '77 thru '81
63030 **350 & 450** all models '71 thru '80

MERCURY

See FORD Listing.

MG

66010 **MGB** Roadster & GT Coupe '62 thru '80
66015 **MG Midget, Austin Healey Sprite** '58 thru '80

MITSUBISHI

*68020 **Cordia, Tredia, Galant, Precis & Mirage** '83 thru '93
*68030 **Eclipse, Eagle Talon & Ply. Laser** '90 thru '94
*68040 **Pick-up** '83 thru '96 & **Montero** '83 thru '93

NISSAN

72010 **300ZX** all models including Turbo '84 thru '89
*72015 **Altima** all models '93 thru '97
*72020 **Maxima** all models '85 thru '91
*72030 **Pick-ups** '80 thru '96 **Pathfinder** '87 thru '95
72040 **Pulsar** all models '83 thru '86
*72050 **Sentra** all models '82 thru '94
*72051 **Sentra & 200SX** all models '95 thru '98
*72060 **Stanza** all models '82 thru '90

OLDSMOBILE

*73015 **Cutlass** V6 & V8 gas models '74 thru '88

For other OLDSMOBILE titles, see BUICK, CHEVROLET or GENERAL MOTORS listing.

PLYMOUTH

For PLYMOUTH titles, see DODGE listing.

PONTIAC

79008 **Fiero** all models '84 thru '88
79018 **Firebird** V8 models except Turbo '70 thru '81
79019 **Firebird** all models '82 thru '92

For other PONTIAC titles, see BUICK, CHEVROLET or GENERAL MOTORS listing.

PORSCHE

*80020 **911** except Turbo & Carrera 4 '65 thru '89
80025 **914** all 4 cyl models '69 thru '76
80030 **924** all models including Turbo '76 thru '82
*80035 **944** all models including Turbo '83 thru '89

RENAULT

Alliance & Encore - *see AMC (14020)*

SAAB

*84010 **900** all models including Turbo '79 thru '88

SATURN

87010 **Saturn** all models '91 thru '96

SUBARU

89002 **1100, 1300, 1400 & 1600** '71 thru '79
*89003 **1600 & 1800** 2WD & 4WD '80 thru '94

SUZUKI

*90010 **Samurai/Sidekick & Geo Tracker** '86 thru '96

TOYOTA

92005 **Camry** all models '83 thru '91
92006 **Camry** all models '92 thru '96
92015 **Celica Rear Wheel Drive** '71 thru '85
*92020 **Celica Front Wheel Drive** '86 thru '93
92025 **Celica Supra** all models '79 thru '92
92030 **Corolla** all models '75 thru '79
92032 **Corolla** all rear wheel drive models '80 thru '87
92035 **Corolla** all front wheel drive models '84 thru '92
*92036 **Corolla & Geo Prizm** '93 thru '97
92040 **Corolla Tercel** all models '80 thru '82
92045 **Corona** all models '74 thru '82
92050 **Cressida** all models '78 thru '82
92055 **Land Cruiser** FJ40, 43, 45, 55 '68 thru '82
92056 **Land Cruiser** FJ60, 62, 80, FZJ80 '80 thru '96
*92065 **MR2** all models '85 thru '87
92070 **Pick-up** all models '69 thru '78
*92075 **Pick-up** all models '79 thru '95
*92076 **Tacoma** '95 thru '98, **4Runner** '96 thru '98, **& T100** '93 thru '98
*92080 **Previa** all models '91 thru '95
92085 **Tercel** all models '87 thru '94

TRIUMPH

94007 **Spitfire** all models '62 thru '81
94010 **TR7** all models '75 thru '81

VW

96008 **Beetle & Karmann Ghia** '54 thru '79
96012 **Dasher** all gasoline models '74 thru '81
*96016 **Rabbit, Jetta, Scirocco, & Pick-up** gas models '74 thru '91 & Convertible '80 thru '92
96017 **Golf & Jetta** all models '93 thru '97
96020 **Rabbit, Jetta & Pick-up** diesel '77 thru '84
96030 **Transporter 1600** all models '68 thru '79
96035 **Transporter 1700, 1800 & 2000** '72 thru '79
96040 **Type 3 1500 & 1600** all models '63 thru '73
96045 **Vanagon** all air-cooled models '80 thru '83

VOLVO

97010 **120, 130 Series & 1800 Sports** '61 thru '73
97015 **140 Series** all models '66 thru '74
*97020 **240 Series** all models '76 thru '93
97025 **260 Series** all models '75 thru '82
*97040 **740 & 760 Series** all models '82 thru '88

TECHBOOK MANUALS

10205 **Automotive Computer Codes**
10210 **Automotive Emissions Control Manual**
10215 **Fuel Injection Manual, 1978 thru 1985**
10220 **Fuel Injection Manual, 1986 thru 1996**
10225 **Holley Carburetor Manual**
10230 **Rochester Carburetor Manual**
10240 **Weber/Zenith/Stromberg/SU Carburetors**
10305 **Chevrolet Engine Overhaul Manual**
10310 **Chrysler Engine Overhaul Manual**
10320 **Ford Engine Overhaul Manual**
10330 **GM and Ford Diesel Engine Repair Manual**
10340 **Small Engine Repair Manual**
10345 **Suspension, Steering & Driveline Manual**
10355 **Ford Automatic Transmission Overhaul**
10360 **GM Automatic Transmission Overhaul**
10405 **Automotive Body Repair & Painting**
10410 **Automotive Brake Manual**
10415 **Automotive Detaiing Manual**
10420 **Automotive Eelectrical Manual**
10425 **Automotive Heating & Air Conditioning**
10430 **Automotive Reference Manual & Dictionary**
10435 **Automotive Tools Manual**
10440 **Used Car Buying Guide**
10445 **Welding Manual**
10450 **ATV Basics**

SPANISH MANUALS

98903 **Reparación de Carrocería & Pintura**
98905 **Códigos Automotrices de la Computadora**
98910 **Frenos Automotriz**
98915 **Inyección de Combustible 1986 al 1994**
99040 **Chevrolet & GMC Camionetas** '67 al '87 Incluye Suburban, Blazer & Jimmy '67 al '91
99041 **Chevrolet & GMC Camionetas** '88 al '95 Incluye Suburban '92 al '94, Blazer & Jimmy '92 al '94, Tahoe y Yukon '95
99042 **Chevrolet & GMC Camionetas Cerradas** '68 al '95
99055 **Dodge Caravan & Plymouth Voyager** '84 al '95
99075 **Ford Camionetas y Bronco** '80 al '94
99077 **Ford Camionetas Cerradas** '69 al '91
99083 **Ford Modelos de Tamaño Grande** '75 al '87
99088 **Ford Modelos de Tamaño Mediano** '75 al '86
99091 **Ford Taurus & Mercury Sable** '86 al '95
99095 **GM Modelos de Tamaño Grande** '70 al '90
99100 **GM Modelos de Tamaño Mediano** '70 al '88
99110 **Nissan Camionetas** '80 al '96, **Pathfinder** '87 al '95
99118 **Nissan Sentra** '82 al '94
99125 **Toyota Camionetas y 4Runner** '79 al '95

* Listings shown with an asterisk (*) indicate model coverage as of this printing. These titles will be periodically updated to include later model years - consult your Haynes dealer for more information.

Over 100 Haynes
motorcycle manuals
also available

5-98

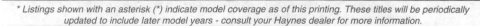

Haynes North America, Inc., 861 Lawrence Drive, Newbury Park, CA 91320-1514 • (805) 498-6703